W9-DEG-612

Intelligence Testing and Minority Students

Racial and Ethnic Minority Psychology Series

Series Editor: Frederick T. L. Leong, *The Ohio State University*

This series of scholarly books is designed to advance theories, research, and practice in racial and ethnic minority psychology. The volumes published in this new series focus on the major racial and ethnic minority groups in the United States, including African Americans, Hispanic Americans, Asian Americans, and Native American Indians. The series features original materials that address the full spectrum of methodological, substantive, and theoretical areas related to racial and ethnic minority psychology and that are scholarly and grounded in solid research. It comprises volumes on cognitive, developmental, industrial/organizational, health psychology, personality, and social psychology. While the series does not include books covering the treatment and prevention of mental health problems, it does publish volumes devoted to stress, psychological adjustment, and psychopathology among racial and ethnic minority groups. The state-of-the-art volumes in the series will be of interest to both professionals and researchers in psychology. Depending on their specific focus, the books may be of greater interest to either academics or practitioners.

Editorial Board

Please address all correspondence to the Series Editor:

Frederick T. L. Leong
The Ohio State University
Department of Psychology
142 Townshend Hall
1885 Neil Avenue
Columbus, Ohio 43210-1222
Phone: (614) 292-8219
Fax: (614) 292-4537
E-mail: leong.10@osu.edu

Intelligence Testing and Minority Students

Foundations, Performance Factors, and Assessment Issues

Richard R. Valencia
Lisa A. Suzuki

REMP
Racial & Ethnic
Minority Psychology

Sage Publications, Inc.
International Educational and Professional Publisher
Thousand Oaks ■ London ■ New Delhi

For information:

Sage Publications, Inc.
2455 Teller Road
Thousand Oaks, California 91320
E-mail: order@sagepub.com

Sage Publications Ltd.
6 Bonhill Street
London EC2A 4PU
United Kingdom

Sage Publications India Pvt. Ltd.
M-32 Market
Greater Kailash I
New Delhi 110 048 India

Printed in the United States of America

Library of Congress Cataloging-in-Publication Data

Valencia, Richard R.
    Intelligence testing and minority students: Foundations, performance factors, and assessment issues / by Richard R. Valencia and Lisa A. Suzuki.
        p.   cm. — (Racial and ethnic minority psychology)
    Includes bibliographical references and index.
    ISBN 0-7619-1230-4 (cloth: alk. paper)
    1. Minorities—Psychological testing—United States.
2. Intelligence levels—United States. 3. Intelligence tests—
United States. 4. Intellect—Genetic aspects. 5. Cognition.
I. Suzuki, Lisa A., 1961 – II. Title. III. Racial and ethnic
minority psychology series,
BF431.5.U6 V35 2000
153.9'3'086930973—dc21

                                                       00–008949

04 05 06 07 7 6 5 4 3 2

Acquiring Editor:	Jim Brace-Thompson
Editorial Assistant:	Anna Howland
Production Editor:	Diane S. Foster
Editorial Assistant:	Cindy Bear
Typesetter:	Tina Hill
Indexer:	Juniee Oneida
Cover Designer:	Michelle Lee

*To our wonderful families—Marta, Juan, and Carlos; John and Kaitlyn.
You're the greatest. With immense affection, Richard and Lisa.*

*To all children and youth of color in the United States,
we hope that this book encourages society
to view your intellectual capabilities as unlimited.*

*To all clinicians who are faced with the challenges
of assessing the intelligence of culturally/linguistically diverse children,
we hope that this book will prove valuable in informing your practice,
particularly in the very important task of
making nondiscriminatory assessments commonplace.*

Contents

Series Editor's Introduction

During the past two decades, Racial and Ethnic Minority Psychology has become an increasingly active and visible specialty area in psychology. Within the American Psychological Association, the Society for the Psychological Study of Racial and Ethnic Minority Issues was formed as Division 45. The Division has now acquired its own journal devoted to ethnic minority issues in psychology. In APA we also have seen the publication of five bibliographies devoted to racial and ethnic minority groups.

The first one was focused on Black males (Evans & Whitfield, 1988) and the companion volume was focused on Black females (Hall, Evans, & Selice, 1989). In 1990, the APA bibliography on Hispanics in the United States was published (Olmedo & Walker, 1990), followed by one on Asians in the United States (Leong & Whitfield, 1992). The fifth bibliography was focused on American Indians and published in 1995 (Trimble & Bagwell, 1995).

As another indication of the increasing importance of racial and ethnic minority psychology, a brief review of the studies published on racial and ethnic differences in journals cataloged by Psychlit revealed a significant pattern. Between 1974 and 1990 (16 years), Psychlit cataloged 2,445 articles related to racial and ethnic differences. Between 1991 and 1997 (6 years), the number of articles related to racial and ethnic differences was 2,584. As a convenient means of representing the increase in attention paid to racial and ethnic minority issues in psychology, one can easily divide the number of articles published during the time period by the number of years covered by that time period. Doing such a computation will reveal that between 1974 and 1990, an average of 153 articles were published each year on racial and ethnic differences. For 1990 to 1997, that number had jumped to an average of 430 articles published each year on racial and ethnic differences. All indications are that this pattern of growth will continue. There is converging evidence that racial and ethnic minority psychology is becoming an important and central theme in psychology in the United States. In recognition of this development, a new book series on **Racial and Ethnic Minority Psychology** (REMP) was launched at Sage Publications in 1995.

The REMP series of books is designed to advance our theories, research, and practice related to racial and ethnic minority psychology. It will focus on, but not be limited to, the major racial and ethnic minority groups in the United States (i.e., African Americans, Hispanic Americans, Asian Americans, and American Indians). For example, books concerning Asians and Asian Americans also will be considered. Books on racial and ethnic minorities in other countries will also be considered. The books in the series will contain original materials that address the full spectrum of methodological, substantive, and theoretical areas related to racial and ethnic minority psychology. With the exception of counseling and psychotherapeutic interventions, all aspects of psychology as it relates to racial and ethnic minority groups in the United States will be covered by the series. This would include topics in cognitive, developmental, industrial/organizational, personality, psychopathology, and social psychology. In covering psychology related issues on racial and ethnic minorities, the series will include both books that examine a single racial or ethnic group (e.g., Development of Ethnic Identity Among Asian Americans) as well as books that undertake a comparative approach (e.g., Occupational Stereotyping of Racial and Ethnic Minority Managers). As a series devoted to racial and ethnic minority groups in the United States and other countries, this series will not cover the usual cross-cultural issues and topics such as those covered by the **Sage Series on Cross-Cultural Research and Methodology Series** (Series Editors: Walter Lonner and John Berry).

Within the field of racial and ethnic minority psychology, one of the recurring controversial themes is concerned with the assessment of intelligence. From the earlier works of Arthur Jensen to that of *The Bell Curve* by Herrnstein and Murray, the question of the so-called "significant racial differences in intelligence scores" remains a vexing one for psychology. Despite the controversy, the continued use of intelligence tests has very real consequences for racial and ethnic minorities in this country. As scholars, it is important that we continue to use our scientific methodology to examine the meaning and use of intelligence tests so that biases can be corrected and a person's true abilities and talents can be identified and utilized. Given the importance of this topic in racial and ethnic minority psychology, it is not surprising that the first volume in the **Sage Series on Racial and Ethnic Minority Psychology** was devoted to a practical and constructive approach to the use of intellectual assessment tools with racial and ethnic minority children and adolescents. In the volume entitled *Assessing Intelligence: A biocultural approach,* Armour-Thomas and Gopaul-McNicol provide practitioners with an integrated, four-tier system for the assessment of intelligence with racial and ethnic minority children.

In keeping with this tradition of addressing important and significant issues in racial and ethnic minority psychology, regardless of the controversies, I am proud to introduce the third volume of the series and the second one devoted to the assessment of intelligence. In this volume entitled, *Intelligence Testing and Minority Students: Foundations, Performance Factors, and Assessment Issues,* Professor Richard Valencia of The University of Texas and Professor Lisa Suzuki of New York University bring to us a comprehensive and well-reasoned treatment of this important topic. This volume is a fitting companion to that of Armour-Thomas and Gopaul-McNicol. While Armour-Thomas and Gopaul-McNicol's volume serves as a valuable guide to practitioners by presenting an integrated biocultural approach to intelligence assessment with ethnic minority children, Valencia and Suzuki's volume provides a scholarly and scientifically grounded review of the topic. The authors do an admirable job of analyzing and dissecting the complex issues underlying this controversial topic. As leading scholars in this area, they have provided the field of psychology with a much-needed integrated and up-to-date review of the major issues in intelligence testing with ethnic minority students. I am confident that this volume will become "required reading" in graduate programs across the country that offer courses in the assessment of intelligence.

—Frederick T. L. Leong
Series Editor

References:

Evans, B. J., & Whitfield, J. R. (1988). *Black males in the United States: Abstracts of the psychological and behavioral literature, 1967-1987.* (Bibliographies in Psychology, No. 1). Washington, DC: American Psychological Association.

Hall, C. C. I., Evans, B. J., & Selice, S. (1989). *Black females in the United States: Abstracts of the psychological and behavioral literature, 1967-1987.* (Bibliographies in Psychology, No. 3). Washington, DC: American Psychological Association.

Leong, F. T. L. (1986). Counseling and psychotherapy with Asian-Americans: Review of the literature. *Journal of Counseling Psychology, 33,* 196-206.

Leong, F. T. L., & Whitfield, J. R. (1992). *Asians in the United States: Abstracts of the Psychological and Behavioral Literature, 1967-1991.* (Bibliographies in Psychology, No. 11). Washington, DC: American Psychological Association.

Olmedo, E. L., & Walker, V. R. (1990). *Hispanics in the United States: Abstracts of the psychological and behavioral literature, 1980-1989.* (Bibliographies in Psychology, No. 8). Washington, DC: American Psychological Association.

Sue, S., & Morishima, J. K. (1982). *The mental health of Asian Americans.* San Francisco: Jossey-Bass.

Trimble, J. E. & Bagwell, W. M. (1995). *North American Indians and Alaskan Natives: Abstracts of the psychological and behavioral literature, 1967-1994.* (Bibliographies in Psychology, No. 15). Washington, DC: American Psychological Association.

Preface

Intelligence Testing and Minority Students: Foundations, Performance Factors, and Assessment Issues is a psychometric tour de force. Its two authors, Richard R. Valencia and Lisa A. Suzuki, are individually well published in the field of psychometrics, especially in the area of testing of racial/ethnic minority populations, and have successfully collaborated on previous publications as well. In recent memory, their signal work in the *American Psychologist* (Suzuki & Valencia, 1997) was applauded as a scholarly recognition of the manifold nature of the problem by many of us in the business of improving assessments of minority students in the schools. In addition, it presaged the approach used in this comprehensive tome.

Perusing the first part of the book puts the reader in touch with American psychology's easy adaptation and widespread adoption of the IQ test so that the application of these new instruments to both immigrant and established populations of nondominant ethnic groups can be understood. The potential application of the IQ test for gatekeeping was not lost on politicians, segregationists, and ideologues. The wholesale testing that went on in schools and at Ellis Island supported a selectively exclusionary immigration policy against "undesirable" groups such as Jews, Catholics, and Southern and Eastern Europeans.

The second part of the book reviews the heredity-environment debate and looks at specific performance factors that can bias tests and their use and interpretation. As Valencia and Suzuki point out, even the phrase in common parlance, "racial/ethnic group differences in measured intelligence," itself contributes to misconceptions and stereotypes among the societal elites in our country given that most educated people do not have a good grasp of analysis of variance and the ratio of between-group to within-group variabilities. Another negative contribution of measurement psychology to common misunderstanding is the implicit belief that any IQ that can be specified in two—not three—places indicates a lack of ability that will curtail a person's productivity throughout life and, therefore, should exclude that person from certain educational or professional opportunities.

Testing also set into motion the streaming practices in American schools that have persisted into this new millennium, for example, the tracking that ensured the ascendancy not of true merit or ability but rather of privilege or even caste. What is more, all this was done under the guise of objective scientism, the bias that emphasized the "hidden curriculum" and values of the upper middle class. The same items that predict that one will succeed in school are the ones that value the consequences of wealth such as foreign travel, written expression, vocabulary developed from a ready access to books, the manipulation of printed symbols and numbers, and logical analysis.

Meanwhile, other important cognitive abilities go unmeasured in IQ tests. Social intuition, face-to-face communication, tonal or emphatic vocabulary (in speaking or listening), guessing and adaptive strategizing, foresight (e.g., weaving without a sketch, constructing without a blueprint), creativity, leadership, the accurate recounting of events, and estimating risk and successfully beating the odds in real situations all remain, to a great extent, unrecognized in the popular and psychological definitions of intelligence, even though they underlie important life skills and have much to do with vocational choice. It is, then, not surprising that IQ tests administered in school have little predictive validity for success during adulthood. They best predict school-based learning. One is left to ruminate about the possibility that the long-term use of tests of maximal performance for selection and placement over several generations has made even this aspect of their validity a self-fulfilling prophecy as institutions and programs adapt to the types of students they receive.

This book will fascinate psychologists who like honest, if alternative, interpretations of the history of psychological conundrums such as the controversies that attend the use of intelligence tests on minority populations with a view to setting educational policy and affecting educational placement and praxis. The book also examines the notion of a biological substrate and its relevance to the fundamentally unchanging hereditarian explanations of mean ethnic differences in IQ by generations of research psychologists, and it helps to explain the nagging dissatisfaction that practitioners in clinical and school practice face daily in testing culturally and, ever more often, linguistically diverse clients. The third part of the book puts into perspective the explanations that have competed for the attention of psychometricians, who design tests of cognitive ability, and other researchers and psychometrists, who use these tests in their work. Particular tests and assessment techniques are reviewed as well.

Although the book is divided into three parts, the authors manage to never let the reader forget the complex and sometimes beguiling nexus of scientific psychology, political ideology, and societal agendas. Indeed, the degree of cross-referencing in the book seems precisely geared to keep the reader from perceiving only certain parts of the picture without looking at the broader conceptual and consequential contexts in which psychological testing occurs.

Valencia and Suzuki's *Intelligence Testing and Minority Students* will give the reader a truly fresh opportunity to relearn and reconsider the nature of IQ and its strengths and limitations. Better yet, reading this book will give the psychometrician and the practitioner a good sense of what to do in the short run, and the book argues persuasively about what needs to be done on a priority basis if testing and education are ever to serve the needs of all Americans, responding fairly and validly to the growing diversity of this nation's citizens.

—Ernesto M. Bernal
Professor, Department of
Educational Psychology,
University of Texas–Pan American

Acknowledgments

Our book represents the contributions of a number of individuals. We gratefully thank Fred Leong, editor of the Sage Publications Racial and Ethnic Minority Psychology book series, who gave us support and encouragement throughout this project. Deep appreciation goes to Jim Brace-Thompson, editor at Sage, who provided us with excellent guidance and generous support during the long task of preparing the manuscript. Thanks also go to Anna Howland, editorial assistant, D. J. Peck, copy editor, and all the other fine professionals at Sage who helped to bring this book to fruition. To Ernesto Bernal, who took time out of his busy schedule to write the Preface, *mil gracias.* Sincere appreciation goes to our superb research assistants—Bruno Villarreal (The University of Texas at Austin) and Ellen Short and Lyndon Aguiar (both of New York University). Very special thanks are extended to Bruno, whose work was outstanding. His bibliographic searches, preparation of tables, and typing of most of the manuscript text were of the highest quality. Your work is deeply appreciated. Thanks also go to Christie Robbins (The University of Texas at Austin, Library and Information Sciences), who helped us track down some hard-to-obtain sources.

I, Richard Valencia, extend my affection and gratitude to my wonderful wife, Marta. Thanks, dear, for your solid and loving support while I worked on the book. To my pride and joy—my twin boys, Juan and Carlos—thanks, fellows, for being so patient while Daddy did his writing. *Mijos,* you're the best sons a father could have. I wish to thank the Center for Mexican American Studies at the University of Texas at Austin, which provided me with a 1998 research grant to undertake work on this book. Finally, I want to thank Manuel J. Justiz, Dean of the College of Education at The University of Texas at

Austin, who awarded me a Dean's Fellowship for the fall semester of 1999. Being free from teaching obligations provided me with valuable time to put this book to rest. *Muchísimas gracias*, Manny.

I, Lisa Suzuki, would like to express my deepest love and appreciation to my husband, John, for his continual support and understanding during the various phases of the preparation of this book. His editing and challenging of ideas presented in the book were invaluable. To my daughter, Kaitlyn, who was born shortly after the completion of the first draft of this book, you have added immensely to the richness of my life and give added meaning to my work. To the student leaders of my research team—Tamiko Kubo, Alex Pieterse, and Ellen Kim—your growing devotion to this area has inspired my thinking in this area over the years. Finally, I would like to express my appreciation to New York University and the School of Education for awarding me a Goddard Junior Faculty Fellowship. This supported me during the Fall 1999 semester so I too was free from teaching obligations to finish this book.

—Richard R. Valencia
Professor, Department of Educational Psychology, The University of Texas at Austin

—Lisa A. Suzuki
Assistant Professor, Department of Applied Psychology, New York University

Introduction

Few topics in the behavioral and social sciences have garnered so much interest and generated so much debate in the United States as has intelligence testing and racial/ethnic minority students.[1] Research on this topic has spanned nearly 90 years, beginning with the first reported study of "Negro"-White differences in intelligence by Strong (1913),[2] who administered the Binet scales to the participants (see Chapter 1 of the present book). Since then, a voluminous amount of literature has accumulated on racial/ethnic differences in "measured intelligence."[3] Although the focal racial/ethnic minority population of these research endeavors primarily has been African Americans, other racial/ethnic minority groups (hereafter referred to as "minorities") have been participants in these studies—American Indians, Latinos (especially Mexican Americans), and Asian/Pacific Islanders (A/PIs).

Numerous research investigations inform us that racial/ethnic mean differences in measured intelligence are among the most thoroughly documented findings in psychology (see, e.g., Brody, 1992; Garth, 1925; Loehlin, Lindzey, & Spuhler, 1975; Shuey, 1958, 1966; Valencia, 1985a; Vraniak, 1993). This body of research has shown that the mean difference between African Americans and Whites (as groups) on intelligence tests averages about 1 standard deviation (15 points).[4] Latinos and American Indians (as groups) tend to have mean scores approximately 0.67 of a standard deviation below the White mean of about 100, and Asian Americans score, on average, slightly above the White mean. One focus of this book is to explore those factors that may best explain these racial/ethnic group differences in average intelligence test scores, with particular emphasis on minority groups that perform below the White mean. The factors on which we focus are socioeconomic status (SES), home environment, cultural bias in tests, and heredity.[5] At this introductory stage in the book, it is important to clear up a common misconception demonstrated by laypeople and scholars alike when referring to "racial/ ethnic group differences" in measured intelligence.[6] This phrase is misleading and tends to perpetuate stereotypes. What the group differences actually refer to are *average* intellectual differences in performance of the *individuals* who are linked, usually through self-identification, to racial/ethnic group membership (e.g., Whites, African Americans, Mexican Americans). Thus, an important question to ask is, "How large are the average differences between groups relative to the range of variation that occurs within groups?" Although "different observers vary considerably in the emphasis they place on average differences between groups compared with variation within groups" (Loehlin et al., 1975, p. 15), the major finding is that "the majority of the variation in . . . levels of [intellectual] ability lies within U.S. racial/ethnic and socioeconomic groups, not between them" (p. 235). That is, for measured intelligence, average differences in performance between groups tend to be quite modest relative to the range of differences within groups. In any event, the indiscriminant use of the term *group differences* in referring to patterns in intelligence scores among racial/ethnic populations ignores the reality of overlap of individual scores between groups and perpetuates the myth that nearly everybody of one racial/ethnic group (e.g., Whites) performs higher than practically everybody of another group (e.g., African Americans).[7]

Although this book focuses on racial/ethnic *group differences* in intelligence (and with the understanding that within-group differences far exceed between-group differences), it is important to underscore that the field of psychology is primarily interested in *individual differences* in intelligence. This statement is true of *applied* psychology but not necessarily for all researchers working in the area of racial/ethnic group differences in intelligence. Many researchers continue to focus on group differences in the absence of a multicultural context. This is a major problem of the literature. In any event, research in the area of individual differences has indicated that "variations in scores on tests of intelligence are attributable primarily to characteristics of persons that are independent of their racial and socioeconomic background" (Brody, 1992, p. 281).[8] Furthermore, as Brody (1992) reminded us, one's performance on a measure of intelligence "is never defined or predictable from his or her group identity. Research on intelligence, properly considered, should invariably serve as an antidote to stereotypes" (p. 281). This point of individual differences is central to a second major focus of our book. In Part III, which covers a host of assessment issues (e.g., underrepresentation of gifted minority students, prereferral process), we offer insights as to why such concerns exist as well as suggestions on how assessments can be improved to accommodate *individual* minority students.

Earlier, we noted that because individual minority students are members of racial/ ethnic groups who, on average, perform below the White mean on intelligence tests, there is a likelihood that such students (e.g., Mexican Americans) will be stereotyped as low performers on measures of intelligence. These false perceptions of minority students, particularly those of low SES, often are grounded in "deficit thinking," which is a mind-set molded by a fusion of ideology and science. Deficit thinking is tantamount to "blaming the victim," a dynamic form of social thought positing that the poor school performance of some minority students is rooted, in part, in students' *cognitive* deficits (for a comprehensive explication of the deficit thinking paradigm, see Valencia, 1997a). In the present book, particularly in Chapter 1, we proffer that deficit thinking has had a major influence on how some people perceive and interpret the low intellectual performance of some minority students (see also Chapter 6). This book offers, throughout, a number of rival interpretations as to why minority students, on average, tend to perform lower than White students on intelligence tests (e.g., variations in SES, test bias, discriminatory assessments).

Purpose and Intended Audience

The existing literature on intelligence testing and minority students is scattered, lacks coherence, and is substantively shallow. This poor state of the art motivated us to prepare a comprehensive, up-to-date book. We also provide, in a number of instances, diverse opinions on controversial issues. We prepared this book for the reader who would want a book that is comprehensive in scope and with updated discussions, data, and references. We believe that this goal has been met. There are more than 1,000 references and more than 300 footnotes. When combined as a whole in our integrative analyses, all these provide a solid state-of-the-art book on intelligence testing and minority students. In our writing of *Intelligence Testing and Minority Students*, we were guided by the principle that to understand the many testing issues germane to minority students, one has to understand the "big picture." That is, minority students do not exist in a vacuum. Testing concerns that affect them are part of a wider context of testing issues in the United States. For example, in our chapter on gifted minority students (Chapter 8), the many issues that minority students face are imbedded in the general context of the gifted movement (e.g., how best to identify gifted students, equity vs. excellence). As such, a number of chapters in this book provide a historical, evolving context that sets the framework for understanding the many concerns of intelligence testing vis-à-vis minority students. In taking this general to specific course, it was necessary to be comprehensive in our analyses and syntheses. In sum, this book should meet the needs of scholars, researchers, students in graduate courses, and testing personnel who desire considerable breadth and depth on the subject of intelligence testing and minority students.

Organization

This book consists of 10 chapters, organized into three parts. Part I, "Foundations," provides a necessary substratum for understanding and illuminating the contemporary period regarding the sociopolitical aspects of intelligence testing, theories of intelligence, scholarly search for explaining mean differences in measured intelligence between minority and White students, and assessment issues germane to minorities. Two chapters make up Part I. In Chapter 1, "Historical Issues," we trace the importation of the first intelligence scales from Europe to the United States. Once culturally and psychometrically appropriated by U.S. psychologists, individually administered (and later group-administered) measures of intelligence became tools of numerous "race psychology" investigations during the 1920s in which minority and White children's intellectual performances were compared. The observed lower mean performances of American Indian, African American, and Mexican American students frequently were interpreted to have genetic bases. This chapter also covers the use of intelligence test scores in shaping curriculum differentiation, and hence inequities, in the schooling of minority and White students. We also examine the role of heterodoxy, that is, the early challenges by a number of minority and White scholars to the orthodoxy of hereditarianism of the 1920s. The chapter closes with a discussion of a number of contemporary testing issues that began to emerge during the 1920s and unfolded through the 1990s. Our thesis of Chapter 1 is as follows: To understand the contemporary issues in regard to intelligence testing and minority students, one must have a good sense of the historical context that helped to shape such concerns.

Chapter 2, "Multicultural Perspectives of Intelligence: Theory and Measurement Issues," highlights the complexities of defining and measuring intelligence. Theoretical frameworks that incorporate cultural aspects of intelligence are noted along with a brief discussion of emotional, practical, and creative facets of this construct. We also provide discussions of assessment-based models of intelligence and specific information regarding racial/ethnic group differences in intelligence. The complexities of understanding intelligence from a theoretical and measurement perspective, within a cultural context, are especially emphasized.

Part II, "Performance Factors," consists of four chapters. Here, we offer a comprehensive analysis and discussion of four major areas—SES, home environment, test bias, and heredity—that scholars have advanced as possible factors that might help to explain the persistent difference in mean intellectual performance between minority student groups (i.e., African Americans, Latinos, American Indians) and White students. In Chapter 3, "Socioeconomic Status," we examine the long-standing finding that SES of parents is positively correlated with their children's measured intelligence. This ubiquitous connection between SES and intelligence frequently has been reported with White participants. We examine two central questions here. First, what does the literature inform us about this relation *across* racial/ethnic groups? The next question has to do with a methodological concern. When White and minority participant groups are "matched" on SES, or SES is controlled, the mean difference in measured intelligence frequently is reduced, but a gap often remains. In these studies, do such findings suggest that attempts to control for SES in White and minority intellectual comparisons cannot truly be made? We proffer the hypothesis that such attempts are difficult, if not futile, because researchers fail to make a distinction between "class" and "caste." Conventional SES measures might be able to distinguish SES groups, but how these instruments are able to discern various groups based on caste-like societal arrangements is not clear. As such, the unmeasured effects of racism need to be considered in such research given that uncontrolled caste factors are likely to be operative.

In Chapter 4, "Home Environment," we examine the home intellectual environmental literature as to how it provides insights about minority children's intellectual performance. Although SES is an important (albeit weak) statistical predictor of children's measured intelligence, it is limited as a construct because of its distal nature. That is, it is a summarizing variable. It does not reveal the complexities and comprehensiveness of what transpires in the home environment regarding the ways in which parents create learning situations that foster children's cognitive development (e.g., parent reading to a young child). As such, researchers in developmental psychology have examined how the home intellectual environment shaped by parents serves as a fairly strong predictor of children's measured intelligence. Most of this research has been conducted using White families. In this chapter, we discuss a representative corpus of home environment research studies in which African American and Mexican American families served as participants. Our review of findings from this research base informs us that what parents *do* in the home environment is a stronger predictor of their children's measured intelligence than is what they *are* (i.e., SES). The major implication here has to do with intervention. Although home environment and SES are statistical predictors of children's intellectual performance, the home environment is much more easily modifiable than is SES (an ascriptive marker). That is, structuring the home environment to make it more conducive for stimulating intellectual growth is easier to do than is changing a family's status, for example, from low SES to middle SES. The home environment literature has much to offer in understanding between- and within-racial/ethnic group differences in measured intelligence. In Chapter 4, we also discuss some emerging research findings from behavioral genetics that challenge the major assertion by developmental psychologists that the family is the most important unit of cognitive socialization.

Chapter 5, "Test Bias," explores one of the most long-standing and contentious claims that has been advanced to explain mean differences in measured intelligence between minority and White groups. That is, intelligence tests, critics assert, are culturally biased against minority children and youths. After introducing this debate, we unpack the notions of "cultural loading" and "cultural bias," two concepts that are central to an understanding of the test bias controversy. The core of this chapter is a comprehensive review of 62 empirical studies of possible cultural bias regarding four psychometric properties of individually administered intelligence tests: reliability, construct validity, content validity, and predictive validity. Based on our review, we offer several critiques of this literature (e.g., most of what we know about cultural bias is confined to research on the Wechsler scales; many investigations have failed to control SES). Our major conclusion of the 62 test bias studies suggests that the majority of studies weigh in on the "nonbiased" side. We contend, however, that because about 30% of the investigations resulted in findings of "mixed" or "biased" conclusions, the subject of cultural bias in intelligence tests is an open research issue.

In Chapter 6, "Heredity," we provide comprehensive coverage of a highly controversial hypothesis, that is, that genetic factors are strongly implicated in explaining the mean differences in measured intelligence between minorities (e.g., African Americans) and Whites. In our attempt to illuminate this complex and disputatious subject, we discuss central concepts (and misconceptions) germane to this debate (e.g., heritability). Also, we offer an analysis of the contemporary discourse among the behavioral genetics, environmental, and interactionist perspectives on the nature-nurture debate. The core purpose of this chapter is to explore the conjecture that genetics plays a significant role in explaining mean differences in intelligence between African American and White groups. We do so by examining (a) the writing of scholars who have proffered genetic bases to explain such group differences (e.g., Jensen, the team of Herrnstein & Murray) and (b) findings from empirical research investigations (e.g., racial admixture studies, transracial adoption investigations). Our analysis of the literature informs us that the hypothesis of genetic differences between African American and White populations that has been advanced to explain differences in measured intelligence cannot be confirmed.

Part III, "Assessment Issues," consists of four chapters. The focus is on issues that are pertinent to the intellectual assessment of minority students. Here, we provide discussions of concerns germane to special education, the identification of gifted minority students, multicultural aspects of cognitive ability tests, and best-case practices regarding nondiscriminatory assessment. Chapter 7, "Race/Ethnicity, Intelligence, and Special Education," begins with a review of the numerous historical factors that have played roles in the development of testing practices in the schools. Our discussion includes Public Law 94-142, the Individuals with Disabilities Education Act amendments of 1997, and court cases focusing on testing of minority students. Current definitions of the special education categories most influenced by intelligence tests are provided, as is a review of the literature surrounding these classifications (i.e., educable mentally retarded [EMR], learning disabled [LD]). A review of the referral and testing process is provided, with specific attention given to the role of intelligence tests. Based on our calculations of data from the most recent survey of the Office for Civil Rights (OCR), disproportionate representation rates of minority students in the EMR and LD classifications are presented along with recommendations for special education programs in the future.

In Chapter 8, "Gifted Minority Students," we focus on a neglected population of minority students—the gifted. Unfortunately, minority students (i.e., African Americans, Latinos, American Indians) often are stereotyped as performing *below* the norm on standardized intelligence tests. This chapter is about those minority students who defy this stereotype, that is, minority students who perform substantially *above* the norm. We begin this topic by providing a historical sketch of giftedness in the United States, with a focus on how minority students have fit in. This is followed by a discussion of a number of issues regarding giftedness that affect minority students (e.g., narrow conceptions of giftedness, charges of elitism). In this discussion, a particularly significant topic is the severe underrepresentation of minority students in gifted programs. To explore this, we calculate national and state underrepresentation rates, using the most recent data of the OCR survey (U.S. Department of Education, OCR, 1997). The results of our analysis are not encouraging. For example, Latino and African American K-12 public school students are *underrepresented* in gifted programs in 48 of the 50 states. By sharp contrast, White students are *overrepresented* in 48 of the 50 states. Following this coverage, Chapter 8 contains discussions of select literature on gifted African American, Latino, A/PI, and American Indian students. We pay particular attention to those empirical studies investigating innovative conceptions and practices that show promise in increasing the number of gifted minority students.

In Chapter 9, "A Multicultural Review of Cognitive Ability Tests," we present information regarding many of the instruments currently used to assess intelligence and cognitive abilities more generally. The focus of this review is more broadly defined given that many test developers and writers have moved away from using the term "intelligence" or "IQ" to define their measures. Thus, the phrase "cognitive ability tests" often is used in reference to the instruments reviewed. Specific attention is paid to the areas assessed by cognitive measures, test development strategies currently used (e.g., norming, standardization), racial/ethnic profiles of abilities reported on popular measures, and current procedural modifications in the administration and interpretation of traditional cognitive tests (e.g., biocultural model of assessment, Gf-Gc Cross-Battery Assessment model).

Chapter 10, "Future Directions and Best-Case Practices: Toward Nondiscriminatory Assessment," highlights future directions based on our understanding of the field of intelligence measures and multicultural issues. Despite concerns from the field regarding issues of test bias and test unfairness, the use of standardized instruments continues to increase at accelerated rates given school reform measures and an emphasis on accountability. Our assessment practices have not been able to keep up with the diversity of our growing minority populations. As such, evaluators are in dire need of instruments designed to assess the abilities of language-minority students as well as individuals from cultures different from the U.S. majority. We assert that testing practices must be located within a multicultural context where theoretical knowledge, performance factors, and psychometric testing practices are integrated and merged. Training areas are outlined in view of necessary assessment practices.

PART I

Foundations

1

Historical Issues

The present issues surrounding intelligence testing and minority students can be best understood by providing a historical perspective. Although the knot between the past and the present cannot be snugly tied, there are some common themes and issues that connect these temporal periods. Current issues regarding cultural bias in tests, differential reliability and validity, racial/ethnic representation in norm samples, underrepresentation of minority students in gifted and talented programs, test uses in decision making, and the role of genetics in intellectual performance all have historical roots (Ayres, 1911; Bagley, 1922; Wallin, 1912; see also Chapman, 1988; Valencia, 1997b).

For our discussion of historical issues, we examine the following: (a) the emergence of the intelligence testing movement in Europe, (b) the importation and cultural appropriation of the Binet-Simon Intelligence Scale by U.S. psychologists, (c) the ideology of the intelligence testing movement, (d) "race psychology" studies of intellectual performance, (e) actual use of IQ tests in curriculum differentiation during the 1920s, (f) heterodoxy, (g) group intelligence testing taking on a life of its own, and (h) the emergence of contemporary testing issues.

AUTHORS' NOTE: Portions of this chapter are excerpted, with minor modifications, from Valencia (1997b, pp. 42-44, 51-53, 55-56, 58-59, 61-71, 73-78, 80, 86-87, 89-90, 92-94). For a comprehensive discussion of the early era, see Valencia (1997b).

The Emergence of the
Intelligence Testing Movement in Europe

English biologist Sir Francis Galton, half cousin of Charles Darwin, frequently is given credit for launching the psychological testing movement (Anastasi, 1988; Cohen, Swerdlik, & Smith, 1992).[1] In his late-19th-century book, *Inquiries Into Human Faculty and Its Development,* Galton (1883) described a number of sensorimotor techniques to measure human perception (e.g., visual, auditory, and weight discrimination; reaction time). In addition to his pioneering work in the field of psychological testing, Galton was a strong believer in hereditarianism, that is, the theory that individual and group differences in behavior (e.g., intelligence) can primarily be accounted for on the basis of genetics (Valencia, 1997b).

Most scholars would agree that the modern-day nature-nurture controversy can be traced to Galton's (1870) work, *Hereditary Genius: An Inquiry Into Its Laws and Consequences* (Chorover, 1979). The belief that heredity plays a monumental role in determining human behavior and outcomes was Galton's thesis. Using detailed genealogical data (e.g., the judges of England who served the courts between 1660 and 1865), Galton (1870) presented a case "so overpoweringly strong" (p. 3) that few scholars of the times would challenge his thesis. According to Galton, men of eminence (e.g., statesmen, military commanders, scientists, poets) were in such positions due to their "natural gifts" that were genetically inherited via their parents and ancestors who also held positions of eminence. Galton failed to consider, however, the rival explanation that in a highly socially stratified England, eminent people were in such fortunate positions because privilege, status, and wealth were *socially* inherited.

Galton (1870) was not remiss at presenting his views of racial differences in *Hereditary Genius.* His racist perspectives were forecast in the title of his chapter, "The Comparative Worth of Different Races." According to his ranking system, the ancient Greeks were at the top, Anglo-Saxons held an intermediate position, and the African Negro and "Australian type" (we assume that he was referring to the indigenous people of Australia) were at the bottom rungs. More specifically, Galton proffered, "The average [intellectual] ability of the Athenian race is, on the lowest possible estimate, very nearly two grades higher than our own—that is, about as much as our race is above that of the African negro" (p. 342). Commenting further on the intelligence of Negroes, Galton noted,

> The number among the negroes of those whom we should call half-witted men is very large. Every book alluding to negro servants in America is full of instances. I was myself much impressed by this fact during my travels in Africa. The mistakes the negroes make in their own matters were so childish, stupid, and simpleton-like as frequently to make me ashamed of my own species. (p. 339)

Galton's views on hereditarianism and his racist pronouncements about individual differences between Whites and people of color subsequently would enter the American discourse of the seminal intelligence testing movement. Such views had some impact on (a) how American psychologists approached the practice of testing minority children on intelligence tests (largely a practice of indifference to cultural and environmental differences) and (b) how behavioral scientists and applied psychologists attempted to explain White-minority group differences in intellectual performance (frequent genetic interpretations).

Although the psychological testing movement was launched in Europe by Galton, French psychologists Alfred Binet and Théodore Simon are attributed for supplying the initial fuel that set the testing movement forward on its maiden voyage. Binet and Henri (1895, cited in Anastasi, 1988) presented criticisms that existing psychological tests were too sensory in nature, unduly simple, and highly specialized. This led Binet and Simon to develop the first cognitively based intelligence measures (first in 1905, with revisions in 1908 and 1911). Some of the behaviors sampled in the Binet-Simon intelligence scales were memory, comprehension, attention, judgment, and reasoning (Anastasi, 1988). These functions were included, in part, in the 1916 Stanford-Binet intelligence test as well as in the development of subsequent intelligence tests.

The Importation and Cultural Appropriation of the Binet-Simon Intelligence Scale by U.S. Psychologists

From about 1908 to 1911, Henry H. Goddard, who had studied with Stanley Hall (a Galtonian hereditarianist), translated the Binet-Simon scale to English and concluded that it measured innate intelligence along a unilinear scale (Blum, 1978; Gould, 1981). Although Goddard was highly influential in launching the Binet-Simon scale in America, it was Lewis Terman (also a student of Hall) who is considered the chief architect of its popularity (Gould, 1981).[2] Terman's revisions (from about 1911 to 1915) of the Binet-Simon scale were substantial and included (a) nearly doubling the number of items; (b) extending the upward part of the scale to assess "superior adults"; (c) popularizing the ratio "IQ" (calculated as one's mental age divided by one's chronological age and multiplying by 100);[3] (d) establishing a standard deviation of 15 to 16 points at each age level; (e) scaling the test to approximate a Gaussian (normal, bell-shaped) curve; and (f) standardizing the scale near the Stanford, California, area (hence, the 1916 individually administered "Stanford-Binet" intelligence test). As such, historical analysis informs us that by 1916 the Binet-Simon scale had been imported, translated, culturally appropriated, psychometrically modified, and normed by American psychologists.

Notwithstanding the historical importance of the development of the Stanford-Binet intelligence test, two significant issues arose regarding minority children. First, there was the concern of the makeup of the standardization sample, which numbered 1,000 children. It appears that the norm group was almost exclusively White (Western European descent) and middle class. In his description of the Stanford-Binet standardization sample, Terman (1917) noted that the selected schools

> were such [that] almost anyone would classify as middle-class. Few children attending them were either from very wealthy or very poor homes. . . . Care was taken to avoid racial differences due to lack of familiarity with the language. None of the children was

foreign-born, and only a few were of other than western European descent. (p. 29, cited in Chapman, 1988, pp. 29-30)[4]

The deliberate exclusion of Mexican American, African American, and other children of color from the original Stanford-Binet standardization sample would be a criticism for decades to follow. The 1916 Stanford-Binet intelligence test was revised in 1937 (Terman & Merrill, 1937) and in 1960 (Terman & Merrill, 1960). In both revisions, only White children were included in the norm group. It was not until the 1972 revision (Terman & Merrill, 1973), *nearly six decades* after the original Stanford-Binet intelligence test was developed, that minority children were included in the standardization sample.[5] Thus, for well over half a century, minority children were routinely administered the Stanford-Binet scale, a test on which they were not represented in the norm groups.[6] The conscious omission of minority children from the standardization samples of prominent intelligence tests, particularly during the seminal era of the intelligence testing movement, informs us about the attitudes some of the test developers held toward these children. One such disposition, it appears, was influenced by ethnocentrism, resulting in a cavalier posture toward accommodating culturally and linguistically diverse children in the assessment of intellectual performance.[7]

The second critical issue that arose regarding minority children during the time of the development of the original Stanford-Binet intelligence test had to do with perceptions about the *intellectual level* and *educability* of minority children. In his *Measurement of Intelligence*, a guide for the clinical use of the Stanford-Binet intelligence test, Terman's (1916) views on these two topics were undeniably registered. In his discussion of borderline cases of intelligence (typically between 70 and 80 IQ, according to the convention of the time), Terman described the cases of M.P. and C.P., two Portuguese brothers with measured IQs of 77 and 78, respectively.[8] Terman's prognosis for M.P. and C.P. was disheartening, predicting that each brother would "doubtless become a fairly reliable laborer at unskilled work . . . and will probably never develop beyond the 11- or 12-year level [of intelligence] or be able to do satisfactory school work beyond the fifth or sixth grade" (p. 90). Regarding Terman's views toward "Indians, Mexicans, and negroes," his sentiments about these children were unequivocally racist:

What shall we say of cases like the last two [M.P. and C.P.] which test at high-grade moronity or at borderline . . .? Hardly anyone would think of them as institutional cases. Among laboring men and servant girls, there are thousands like them. They are the world's "hewers of wood and drawers of water." And yet, as far as intelligence is concerned, the tests have told the truth. These boys are uneducable beyond the merest rudiments of training. No amount of school instruction will ever make them intelligent voters or capable citizens in the true sense of the word. Judged psychologically, they cannot be considered normal.

It is interesting to note that M.P. and C.P. represent the level of intelligence that is very, very common among Spanish-Indian and Mexican families of the Southwest and also among negroes. Their dullness seems to be racial, or at least inherent in the family stocks from which they come. The fact that one meets this type with such extraordinary frequency among Indians, Mexicans, and negroes suggests quite forcibly that the whole question of racial differences in mental traits will have to be taken up anew and by experimental methods. The writer predicts that when this is done, there will be discovered enormously significant racial differences in general intelligence, differences that cannot be wiped out by any scheme of mental culture.

Children of this group should be segregated in special classes and be given instruction that is concrete and practical. They cannot master abstractions, but they often can be made efficient workers, able to look out for themselves. There is no possibility at present of convincing society that they should not be allowed to reproduce, although from a eugenic point of view they constitute a grave problem because of their unusually prolific breeding. (pp. 91-92)

Ideology of the Intelligence Testing Movement

As we have discussed, the intelligence testing movement, particularly during the 1920s, rested on a measurement foundation. Ideology played an equally influential function in shaping the movement (see, e.g., Blum, 1978; Chorover, 1979; Cravens, 1978; Degler, 1991; Fass, 1980; Gould, 1981; Guthrie, 1976; Marks, 1981; Valencia, 1997b).

To understand the conjecture that intelligence testing was, in part, driven by ideological or extrascientific factors, it is informative to examine the notion of intelligence testing itself. Drawing from Fass (1980), we use the term *intelligence testing* to denote collectively (a) the procedure of measuring intelligence, (b) that which is perceived to be the measure of intelligence (the IQ), and (c) the organizational and social mechanism referred to as the *intelligence testing movement.* In that the prime movers of intelligence testing were, for the most part, hereditarians, eugenicists, and even racists, it makes sense to conclude that intelligence testing—consisting of procedural, measurable, and social movement aspects—was a value-laden idea with significant implications for the stratification of schooling practices and outcomes.

With the preceding context in mind, the following assertions can be made about the role of ideology in the intelligence testing movement.

1. The confluence of measurement and ideology becomes critical in that this joined stream has influence over both the perceived arrangement and control of social relations. As Chorover (1979) commented,

After all, the power to measure is merely an extension of the power to define. The point is worth pondering because throughout its history, the measurement of human diversity has been linked to claims of human superiority and inferiority and has thereby been used to justify prevailing patterns of behavior control. (pp. 33-34)

2. The development of the construct of "individual differences," coined by J. McKeen Cattell in 1916, was a major fundamental accomplishment in the creation of psychology as a behavioral science.[9] The "new psychology" had no room in its perspec-

tive for ideas such as free will, the mind, and consciousness—interests of the "old psychology" (Marks, 1981). The construct of individual differences had as its focus behavior. Drawing from the frameworks of the biological and physical sciences, psychologists viewed heredity as biological and behavioral. The logical extension was that the construct of individual differences, when used to understand social behavior, went well beyond scientific credibility. "Psychologists viewed the behavior of people—including women, blacks, and immigrants—as inherently indicating their biological capacities rather than their socio-economic, cultural-biological conditions" (Marks, p. 9).

3. Terman's work as a scholar, researcher, and test developer is well known. However, we must not forget Terman as the applied psychologist, ideologue, and social reformer. As Marks (1981) wrote, Terman approached his work with "a burning faith in human improvability. . . . He maintained an unshakeable commitment to equality of opportunity and social amelioration. . . . Terman was involved in much more than creating and selling tests; he was involved in the construction of a meritocratic social reality" (pp. 167-168). At the surface level, Terman's conception of "equality of opportunity" appears democratic and inclusive. At a deeper level, however, his idea about this important notion is antithetical to the basic principles of democracy such as fair play, equality, and freedom. To Terman, to have equality of opportunity was for the individual to have education and training, commensurate with his or her *innate* ability, so as to attain his or her occupational status. Guided by deficit thinking, this perversion of equality of opportunity meant that people of color and the poor would be virtually locked out of the upward flow of educational and social mobility.

In sum, a strong case can be advanced that both measurement and ideological foundations were influential in molding the intelligence testing movement. Chapman (1988) commented that this union can be reduced to several essential elements:

> The ideology of the emerging test movement, as expressed by Terman's views, can be reduced to its most essential components. Intelligence could be measured by tests and expressed in a single numerical ratio. This ability was largely constant and determined by heredity. Class and racial inequality could be explained in large part by differences in intelligence. Used in schools, intelligence tests could be used to identify ability, prescribe curricula, and determine students' futures. (p. 92)

"Race Psychology" Studies of Intellectual Performance

According to Guthrie (1976), the first reported study of White-Negro differences in intelligence, as measured by the Binet-Simon scale, was by Strong (1913), who conducted his investigation in Columbia, South Carolina. The participants included 225 White and 125 Negro children. Strong, a graduate student, concluded that the "colored children are mentally younger than the White [children]" (quoted in Guthrie, 1976, p. 54). Strong even evoked the familiar "mulatto hypothesis," concluding that Negro children of lighter complexions outperformed their darker-skin peers. In a study a year later, Morse (1914), the mentor of Strong, commented on Strong's findings and wrote that although his student's investigation showed that "the colored children did excel in rote memory . . . . , [they] are inferior in esthetic judgment, observation, reasoning, motor control, logical memory, use of words, resistance to suggestion, and . . . orientation or adjustment to the institutions and complexities of civilized society" (quoted in Guthrie, 1976, p. 54). The tone of this sweeping deficit-thinking indictment of the alleged limited mental abilities of Negro children would reverberate for decades to come.

One of the first cross-racial research studies in which schooling prescriptions were actually suggested was the investigation by Phillips (1914). After testing 137 White and 86 Negro children with the Binet-Simon scales and finding the latter group to be inferior in intellectual performance, Phillips raised the question of providing separate education:

> If the Binet tests are at all a gauge of mentality, it must follow that there is a difference in mentality between the colored and the white children, and this raises the question: Should the two groups be instructed under the same curriculum? (quoted in Guthrie, 1976, p. 55)

With the development of the Stanford-Binet intelligence test in 1916, this instrument replaced the Binet-Simon scale as the instrument of choice in investigations of race differences in intelligence. Yet, given that the Stanford-Binet was an individually administered test, research on race differences was slow in developing due to the lengthy administration time. By 1916, group-administered intelligence tests—instruments that could be given to a group of participants in a single setting—were not available for use in the schools or for research. With such tests, the amount of data that could be gathered would increase greatly, compared to the amount of data collected from individual test administrations.

This problem would be addressed, however, with the arrival of the National Intelligence Tests (NIT) in 1920 (Valencia, 1997b). The NIT's development stemmed directly from the development of the Army Alpha and Beta mental tests designed to assess the intellectual ability of recruits during World War I.[10] The NIT developers, who asserted that the "scale [was a] measure of native ability" (quoted in Chapman, 1988, p. 77), sought to use the test for "classifying children in grades 3 to 8 with respect to intellectual ability" (Gould, 1981, p. 178). Regarding the norming of the NIT, the process was just as exclusive as that of the Stanford-Binet intelligence test—Whites only. It appears that *no* African Americans, Mexican Americans, Asian Americans, or American Indians were part of the standardization sample.[11]

With the popularity of the Stanford-Binet intelligence test soaring, the development of the NIT, the proliferation of other group-administered intelligence tests,[12] and the rise of the intelligence testing movement, the study of racial differences in mental ability escalated. The parent field of the study of cross-cultural research (the term commonly used today) was referred to during the 1920s as "race psychology" (apparently coined by scholars such as Thomas R. Garth). One major area of interest that preoccupied race psychologists was the investigation of the intellectual performance of White children/youths, compared to that of their peers of color.[13] Garth's (1925, 1930) literature reviews, titled "A Review of Race Psychology," are particularly informative.

Garth, a psychology professor at The University of Texas and later at the University of Denver, was a leading scholar in race psychology, a hereditarian, and a believer in the mental superiority of the White race (something he later would recant). In his first review (Garth, 1925), the period of research publication spanned from 1916 to 1924. Garth identified 45 studies in which 19 "racial groups" were studied.[14] In the 45 studies, 73 separate investigations were conducted. Of the 73 investigations, 55 (75%) focused on intelligence. The remaining investigations covered other "mental processing" aspects such as learning, mental fatigue, and memory. Regarding the 19 racial groups who served as participants in the 73 investigations, in descending order, the groups studied most frequently (with their respective numbers of investigations) were American Negro (25), American Indian (8), Chinese (6), Italian (5), Portuguese (4), Mexican (4),[15] and Japanese (3). All other groups (predominantly White ethnic such as Slav, Hungarian, English Scottish, German, Swedish, and French) served as participants only once or twice. With respect to a conclusion about race differences, Garth (1925) noted, "These studies taken all together seem to indicate the *mental superiority of the white race.* There may be some question, however, about the indicated inferiority of the yellow races" (p. 359, italics added).

In his second review of race psychology literature, Garth (1930) covered the period from 1924 to 1929. A major trend was the dramatic increase in the number of investigations and groups studied. Of the total 36,882 participants studied,[16] there were 7,158 Negroes (19%) in 25 of 132 studies. Next in frequency were American Indians (5,795 participants [16%] in 14 of the 132 investigations), followed by Mexicans (4,140 participants [11%] in 7 of the 132 investigations). Thus, we can see that African Americans, American Indians, and Mexican Americans—the three prime targets of deficit thinking—were studied the most frequently in race psychology studies from 1925 to 1929, accounting for nearly half of all participants tested (17,093 participants [46%]).

As was found in his first review, Garth (1930) observed that the measure of racial differences in intelligence was the most frequent mental processing aspect studied from 1925 to 1929. Regarding a major conclusion about racial differences in intelligence, Garth observed that a belief in the "hypothesis of racial inequality" was losing ground to the "hypothesis of racial equality":

> What then shall we say, after surveying the literature of the last five years, is the status of the racial difference hypothesis? It would appear that it is no nearer being established than it was five years ago. In fact, many psychologists seem practically ready for another, the hypothesis of racial equality. (p. 348)

Garth's (1930) conclusion about the race psychology literature regarding racial differences in intelligence was used as the basis for Cravens's (1978) point that although most race psychology studies by 1925 articulated a hereditarian explanation for differences, by 1930 no such studies did so. "The change was that rapid and total" (p. 239). Notwithstanding the decline of the White superiority contention by 1930, it still is informative to review some of the race psychology research of the 1920s that investigated group differences in intelligence. Next, we summarize representative studies in which Mexican American, African American, and American Indian children and youths served as participants. Our focus is on the 1920s, a time period that contains some of the principal examples of cross-cultural research on intellectual performance.

Mexican Americans

Regarding Mexican American children, Garth's (1925, 1930) race psychology reviews showed that these children were frequent participants in intelligence testing research. Based on Valencia's (1997b) analysis, and drawing from Sánchez (1932), there are eight such studies published between 1920 and 1929 in which Mexican American children were participants (Garretson, 1928; Garth, 1923, 1928; Goodenough, 1926b; Koch & Simmons, 1926; Paschal & Sullivan, 1925; Sheldon, 1924; Young, 1922). The point of most interest is that in all eight studies, the authors concluded that the lower intelligence test performance of the Mexican American children, compared directly to their White peers or White normative data, was due to heredity. In some cases, the hereditarian conclusions of inferior genetic constitution of Mexican American children was made explicit (Garretson, 1928; Young, 1922) or was suggestive (Goodenough, 1926b). Let us examine one of these eight investigations.

Garretson (1928) sought to explain the causes of "retardation" (i.e., being overage for one's grade level) among Mexican American children attending school in a small public school system in Arizona. Garretson compared the intellectual performance of 197 "American" and 117 "Mexican" pupils. The children, who were enrolled in Grades 1 to 8, were administered the NIT, the Pintner-Cunningham Primary Mental Test (a verbal scale), and the Myers Pantomime Group Intelligence Test (a nonverbal test).

The results of Garretson's (1928) study showed that the median IQs for the three tests were higher (no significance testing reported) for the White children at nearly all grade levels. After analyzing irregularity of school attendance and transiency as possible explanations for retardation of the Mexican American children, Garretson dismissed them as factors. Drawing from an investigation by Young (1922), Garretson (1928) proffered a genetic interpretation for the group differences in intelligence: "This [finding] apparently agrees with the conclusion of Young that the native capacity, if we assume that the intelligence quotients are indicative of native capacity, of the Mexican pupil is less than that of the American child" (p. 38). As to why the Mexican American children were overage for their grade levels and not making normal progress, Garretson suggested the underlying cause to be innate: "Probably the principal factor governing retardation of the Mexican child is his mental ability as measured by the group test" (p. 40).

Common to race psychology studies of this period, Garretson's (1928) investigation failed to control for key variables, nor did it discuss that group differences actually are in reference to *average* intellectual differences—meaning that overlap exists among individuals of each racial/ethnic group. To wit, (a) Garretson failed to control for socioeconomic status (SES), as the White children were overwhelmingly of higher SES[17]; (b) he did not fully consider the possibility of language confoundment (i.e., the depressed intellectual performance of the Mexican American children was due, in part, to limited English proficiency)[18]; and (c) he failed to address the finding of overlap (i.e., in Grade 7, the Mexican American children outperformed their White peers on the NIT [97.0 IQ vs. 96.3 IQ, respectively], and in Grade 3, the Mexican American children also out-performed their White peers on the Pantomime test [74.0 IQ vs. 72.0 IQ, respectively]).[19]

African Americans

Regarding race psychology investigations focusing on the intelligence of African American children and youths, they were by far the prime participants of study during the period from 1916 to 1930 (Garth, 1925, 1930; Price, 1934). As was the general status of intelligence testing research, such studies with African American participants prolifer-ated after the commercialization of group intelligence tests pursuant to the develop-ment of the Army Alpha and Beta tests. Explanations for the observed low average intel-lectual performance of African Americans, compared to that of their White peers or White normative data, frequently were hereditarian based. Let us examine one such study from the 1920s.

Florence Goodenough, the developer of the Goodenough Draw-a-Man Test, pub-lished a race psychology study in 1926 that deserves our attention. Goodenough (1926b) began by stating that there are two theories that had been advanced to account for racial differences in intelligence:

> The first ascribes the inferior showing made by the South Europeans and the negroes to such post-natal factors as inferior environment, poor physical condition, and linguistic handicaps. The second point of view, while it recognizes that the factors named may to some degree affect the test results, nevertheless holds that it is impossible to account for all the facts which have been observed upon any other hypothesis than that of *innate dif-ferences* among the groups under consideration. (pp. 389-390, italics added)

To shed light on the possible explanatory base of group differences in intelligence, Goodenough (1926b) administered her drawing test to 2,457 schoolchildren (Grades 1 to 4) representing 22 "racial stocks." Of the total participants, there were 682 Negroes (69 attending schools in California, 613 attending schools in Tennessee and Louisiana). The weighted mean IQ (computed by Valencia, 1997b) for the Negro group was 79.4, about 20 points lower than the mean IQ of 100.3 for the 500 "American" participants (i.e., White, primarily "North European stock").

Goodenough's (1926b) conclusion as how best to explain the observed Negro-White differences in IQ leaned in the direction of a genetic interpretation. She came to this con-clusion, it appears, by the following reasoning. First, the issue of the Negro child having any form of a linguistic handicap was irrelevant because the Draw-a-Man Test "is en-tirely independent of language" (p. 393). Second, the home environment and SES of the

Negro child is *both* a cause and an effect of inferior mental ability. In an interesting and convoluted form of deficit thinking, Goodenough commented,

> It seems probable, upon the whole, that inferior environment is an effect as least as much as it is a cause of inferior ability, as the latter is indicated by intelligence tests. The person of low intelligence tends to gravitate to those neighborhoods where the economic requirement is minimal; and, once there, he reacts toward his surroundings along the lines of least resistance. His children inherit his mental characteristics. (p. 391)

It seems that Goodenough's (1926b) conclusion, suggesting that Negro children were innately inferior in intelligence, was partially shaped by the hereditarian Zeitgeist of the times. The tone, language, and data presented in the study point to this. It also appears that her study and conclusions were influenced by self-interests and research bias, that is, Goodenough's need to validate and legitimize her newly developed drawing test, an instrument that she purported was completely free of language. Might it be that Goodenough viewed her test as a "perfect" or "culture-free" test? If so, then she was very likely wrong. For example, Klineberg (1935b), commenting about a decade later about the cultural boundedness of verbal intelligence tests, had this to say about performance tests:

> The non-linguistic, or performance, tests present similar difficulties. The Goodenough test of "drawing a man" is based upon the concept of a fully clothed man as seen in our society. When Porteus gave this test to the Australians, he found that they would almost invariably draw the man naked and so lose points given for correct drawing of the clothes. This test also assumes that a man is the figure most frequently drawn by children; the writer [Klineberg] found that among the Dakotas (Sioux), the horse was much more popular (pp. 158-159).[20]

More recently, Anastasi (1988) noted, "Such [cross-cultural, cross-ethnic] investigations have indicated performance on this test [Goodenough's Draw-a-Man] is more dependent on differences in cultural background than was originally assumed" (p. 306). It also appears that Goodenough (writing with Harris in 1950) a quarter century later was of the opinion that "the search for a culture-free test, whether of intelligence, artistic ability, personal-social characteristics, or any other measurable trait is illusory" (Goodenough & Harris, 1950, p. 399, quoted in Anastasi, 1988, p. 306).

American Indians

As discussed earlier, American Indian children and youths were studied with great frequency (second only to African Americans) during the 1920s race psychology era (for reviews, see Garth, 1925, 1930). Likewise, as with the case of African Americans, race psychologists spent considerable time in investigating White-American Indian hybrids. Next, we examine one representative study.

Garth (1923) administered the NIT to 941 boys and girls ages 12 to 18 years. The groups consisted of three full-blood American Indian groups, one mixed-blood Indian group, and a "Mexican" (Mexican American) group. For analyses, NIT IQs were based on aggregate scores of all participants by combining both sexes and all age levels for each group. The results showed that, with respect to highest to lowest median scores, the ranks were (1) mixed-blood, (2) Mexican, (3) Plains and Southeastern full-blood, (4) Pueblo full-blood, and (5) Navajo/Apache full-blood. Garth compared the median IQ score of each blood group to the median score of the Plains and Southeastern Indians, the group that scored intermediately. The author suggested a genetic hypothesis to explain the results: "The mixed breeds excel [i.e., exceed] the pure breeds in intelligence scores. . . . If these groups may be taken as representative of their racial stocks, the results indicate differences between their racial stocks in intelligence as here measured" (p. 401). Garth measured but did not control for educational attainment of the children in the five groups (an issue he admits). In fact, the school attainment by the five groups almost exactly corresponded to the ranking of their scores on the NIT. Valencia (1997b) calculated a Spearman rank order coefficient in which the five groups' ranks in IQ and schooling attainment were correlated. The observed ρ (rho = .99) was significant at the .01 level of confidence. The strong association between schooling attainment and IQ had been established by the time of Garth's study, yet he failed to control for the former. Thus, his suggestion of a racial admixture hypothesis denoting intellectual advantages of mixed-bloods having White ancestry was unwarranted. To be sure, 1920s race psychology research of racial/ethnic differences in intelligence was a prominent activity. As we have seen, the conclusions drawn about intellectual differences between White children and some children of color (i.e., African Americans, Mexican Americans, American Indians) were predominantly hereditarian based. Children of color, it was alleged, performed lower than their White peers largely because of inferiority in native intelligence. This research claim, in and of itself, is indeed a significant social statement of the nature of race relations during the height of the hereditarian era. The allegation of the intellectual inferiority of these racial/ethnic minority groups speaks to the descriptive and explanatory aspects of deficit thinking during this period. A scrutiny of the investigations by race psychologists also informs us that many of these researchers who studied intellectual performance across racial/ethnic lines made little effort to take social, cultural, linguistic, and educational factors into consideration in their conceptual frameworks, research designs, and discussions of findings.

Actual Use of IQ Tests in
Curriculum Differentiation During the 1920s

A major theme of this book is that intelligence tests do not exist in a vacuum. Historically, they frequently had practical functions and significant social consequences. One such important function, "sorting," was embedded in the "management of instruction" rubric (Resnick, 1979).[21] Here, IQ tests were administered prior to instruction, and students were assigned to relatively homogeneous groups in regular education and, in some cases, special education. This general practice is referred to as "curriculum differ-entiation," which Valencia (1997b) defined as "the sorting of students into instructional groups based on perceived and/or measured educability" (p. 71).[22]

As introduced earlier, Terman was a staunch advocate of the use of intelligence tests for curricular assignments via homogeneous grouping. In brief, let us examine some of his views. Terman's views on the prescriptive role of tests in school reform are most clearly articulated in his book, *Intelligence Tests and School Reorganization* (Terman, 1922). His plan, a universal proposal, contained the following features (Chapman, 1988). First, all children in Grade 1 should be administered an individual intelligence test. Second, beginning in Grade 3, children should be tested at least every other year with the use of group intelligence tests. Third, based on the test results, "homogeneous class groups" should be adopted.[23] Although in practice most school districts that used tests for curric-ulum differentiation employed a three-track tier, Terman believed that the ideal plan should use a "multiple track plan" with five groups. The five homogeneous groups (with their respective suggested percentages of students of the total student body) are "gifted" (2.5%), "bright" (15.0%), "average" (65.0%), "slow" (15.0%), and "special" (2.5%).[24] Terman's proposed system of using intelligence tests for classification, and the explicit need to eliminate heterogeneity in ability, led him to lay down this major princi-ple in *Intelligence Tests and School Reorganization:* "A reasonable homogeneity in the men-tal ability of pupils who are instructed together . . . is a sine qua non of school efficiency" (quoted in Chapman, 1988, pp. 86-87).

Thus, Terman contended that his universal plan could make it possible for intelli-gence tests to be used in some fashion so that both teachers and students would have a more clearly "differentiated course of study, as regards both content and method" (quoted in Chapman, 1988, p. 89). To be sure, Terman (1920) had great confidence in the power of intelligence tests to assess and predict a child's educability and course of study:

> The grade of school work which a child is able to do depends chiefly upon the level of mental development he has attained. . . . *The limits of a child's educability can be fairly accu-rately predicted in the first school year.* By repeated tests, these limits can be determined ac-curately enough for all practical purposes by the end of the child's fifth or sixth school year. (p. 21, italics added)

Terman not only was a strong supporter of the use of intelligence tests for educa-tional tracking via homogeneous grouping, he also argued for such tests to be used in an applied manner for workforce concerns and interests, that is, vocational guidance. In 1916, the year of the development of the Stanford-Binet intelligence test, Terman (1916) predicted, "The time is probably not far distant when intelligence tests will become a recognized and widely used instrument for determining vocational fitness" (p. 17). Ad-mitting that current intelligence tests were not infallible in revealing the occupation an individual is best fitted to follow, he still lauded the robustness of such present and future tests in determining the best match:

When thousands of children who have been tested by the Binet scale have been followed out into the industrial world and their success in various occupations [has been] noted, *we shall know fairly definitely the vocational significance of any given degree of mental inferiority or superiority.* Researchers of this kind will ultimately determine the minimum "intelligence quotient" necessary for success in each leading occupation. (p. 17, italics added)

Terman was very successful in his call for the use of intelligence testing in school reform. One of the reasons why Terman became so influential was through his professorial duties in teaching, supervising, and producing students with graduate degrees who carried on his mission.[25] For example, Virgil Dickson (M.A., 1917, and Ph.D., 1919, Stanford University) supervised the testing of 20,000 students in 1919-1920 in the Oakland School District in California. The accomplishments of Dickson are very noteworthy in the history of test use in schools, as it appears that Oakland was the first city in the nation to implement a district-wide program of mass mental testing and curriculum differentiation. Chapman (1988) observed, "By the middle of the decade, the use of intelligence tests for sorting students had been institutionalized in Oakland" (p. 108), and subsequently, "school districts around the country would be following Oakland's lead" (p. 64). Dickson's star status in intelligence testing and curriculum differentiation also was enhanced by the publication of his book, *Mental Tests and the Classroom Teacher* (Dickson, 1923), published under the auspices of the Measurement and Adjustment Series edited by Terman.[26]

Another case in point regarding the actual use of intelligence testing for curriculum differentiation involving minority students (mostly Mexican Americans) was the Los Angeles public schools (1920s). Educational historian Gilbert González's research has found that the institutionalization of mass intelligence testing, homogeneous groupings, curriculum differentiation, and counseling programs in Los Angeles schools were used in ways that effectively stratified students along racial/ethnic and SES lines (González, 1974a, 1974b, 1990). During the 1920s, mass intelligence testing in Los Angeles was indeed a big enterprise. By decade's end (1928-1929), a total of 328,000 tests were administered, for example, at the elementary level alone (González, 1974b). Primarily based on IQ test results, students were placed in one of four types of elementary classes: *normal* classes, *opportunity* rooms, *adjustment* rooms, or *development* rooms. Opportunity rooms were designed for both the mentally superior children (Opportunity A) and slow learners (Opportunity B [children whose IQs were above 70 but below the norm]). Adjustment rooms, on the other hand, were structured for normal children (i.e., average interval IQ) who had specific skill problems (e.g., remediation in reading, limited English proficiency) or were "educationally maladjusted" (most likely, this was what we currently refer to as emotional and behavior disorders). Development rooms (sometimes referred to as centers) were designed for children whose IQs were below 70; these children typically were referred to as "mentally retarded" or "mentally deficient." González (1974b) noted that the median IQ for Mexican American elementary school-age children in Los Angeles during the late 1920s was about 91.2.[27] He suggested, given the observed median IQ of 91.2, that "there was a very high probability that nearly one-half of the Mexican children would find themselves placed in slow-learner rooms and development centers" (p. 150). González's suggestion that about 50% of Mexican American elementary students were placed in classes for the mentally subaverage (Opportunity B classes) and for the mentally retarded (development centers) translates (based on our analyses) into an astounding overrepresentational disparity of 285% in that it has been estimated that Mexican American students constituted only 13% of the student population in the city and county of Los Angeles in 1927 (Taylor, 1929, cited in González, 1974b).[28]

In sum, historical analysis of IQ testing and curriculum differentiation during the 1920s informs us that many minority students' educational and subsequent occupational opportunities were severely limited. It appears that numerous Mexican Americans, for example, faced one of two equally unattractive educational pathways: (a) dead-end special education for alleged slow learners or (b) nonacademic vocational education that emphasized low-level skills. For the most part, either trail led to manual occupations, typically requiring minimal or no skills.

Heterodoxy

Thus far, our historical analysis of intelligence testing and minority students has covered the orthodoxy, that is, the established or conventional views. It must be underscored, however, that the status quo did not prevail without contestation during the first three decades (particularly the 1920s) of the 20th century. Valencia (1997b) examined this heterodoxy and provided rich examples of this discord. Drawing from his discussion, we present a brief overview of these heterodoxic perspectives and research efforts.

Interestingly, there was no color line among those scholars who challenged hereditarianism and who brought to light the insensitivity of race psychologists who did not consider cultural, linguistic, and environmental factors in the intellectual assessment of minority children and youths (Valencia, 1997b). Among White scholars, there were, for example, Franz Boas, Otto Klineberg, Ada Arlitt, and William C. Bagley. African American scholars (e.g., Horace Mann Bond, Herman C. Canady, Martha MacLear) also were an integral part of the heterodoxic camp. George I. Sánchez, a Mexican American educator, also was highly visible as an early critic of intelligence testing.

One significant scholar who made his mark, particularly challenging the belief in White intellectual superiority, was Klineberg, a Columbia University psychology professor (see Degler, 1991; Valencia, 1997b). Klineberg's industrious research program was quite varied in scope. For example, he traveled to villages in Europe to test the Nordic superiority hypothesis advanced by Brigham's (1923) analysis of the Army Alpha and Beta mental testing data. Klineberg (1932) also examined the school records of 500 Negro children who attended schools in Tennessee and subsequently migrated to northern cities. His intention here was to examine the "selective migration" hypothesis, the conjecture that the higher IQ scores seen among northern Negroes was due to the more intelligent Negroes who migrated from the South to the North (Klineberg, 1935a).

Much of Klineberg's original research publications and his criticisms of deficit thinking and hereditarian orthodoxy can be seen in his now classic book, *Race Differences*

(Klineberg, 1935b). In all, he was quite successful in debunking biologically based re-search regarding racial/ethnic group comparisons. In his book, which is written mainly for students and the "intelligent layman," it is clear that Klineberg was greatly influenced by Boas (to whom the book is dedicated). Methodically, in the introductory section of the book, he outlines and critiques racial superiority theories, rejects polygenetic theory, speaks to the unjustification of assuming that *race* and *nationality* are synonymous, and argues for the importance of culture in understanding group differences. Also, like Boas, Klineberg was ideologically committed in his scholarly work to notions of equality, inclusivity, and the application of research to social problems.

For our purposes, the two chapters on intelligence testing in Klineberg's (1935b) *Race Differences* are especially informative. In part, Klineberg presented cogent discussions of methodological confoundment in intelligence testing research (e.g., failure to control for SES, failure to consider the role of English-language skills). As a case in point, let us study the language issue. Klineberg noted that, beginning very early in the development of intelligence tests, it was known that an examinee's familiarity with the English language was important to consider. (Today, we refer to this concern as one of "cultural loading.") This issue was so critical that Robert M. Yerkes and his colleagues, in the development of the Army group intelligence tests, developed the Army Beta test, a non-verbal measure (pictorial) for illiterate recruits in which language does not appreciably play a role. Notwithstanding the caution about administering verbal tests to those who are limited or nonproficient in English, a substantial proportion of race psychology studies of group differences used verbal tests in the absence of first assessing the language status of examinees.[29] Children who were most likely to be penalized under these testing circumstances were those who were raised in environments where English was not the mother tongue and had limited second-language skills or were developing bilingually (e.g., American Indians, Italian Americans, Chinese and Japanese Americans, Mexican Americans).

To examine his conjecture that such children would be penalized on verbal intelligence tests, Klineberg (1935b), in a demonstration of his methodological adroitness, reviewed a number of studies in which IQ data were available for culturally/linguistically diverse groups (e.g., Italians) on both "linguistic" tests (i.e., verbally loaded tests such as the Stanford-Binet and NIT) and "performance" tests (i.e., nonverbal tests such as the Army Beta, actual performance tests such as the Goodenough Draw-a-Man Test and Porteus Maze Test). Klineberg then compared the grand mean IQ of the linguistic tests to the grand mean IQ of the performance tests. For all racial/ethnic groups studied, the mean IQ on the performance tests was higher than the mean IQ on the linguistic tests, thus supporting Klineberg's hypothesis that language was indeed a penalizing factor in the intellectual assessment of culturally/linguistically diverse groups. The groups examined (and the IQ gap between linguistic tests and performance tests) were Italian (8.0 points), Chinese/Japanese (12.4 points), American Indian (16.4 points), and Mexican (10.0 points).[30] Klineberg concluded, "If we compare the studies of various racial groups by these two types of tests, it becomes clear that the linguistic tests place many of them at a disadvantage" (p. 169). Implicit in Klineberg's conclusion about the language factor, as well as in his discussion on other control problems, is that any research-based statements about White superiority in innate intelligence are unwarranted in light of methodological flaws, particularly the failure to control key variables.

So far-reaching were Klineberg's heterodoxic goals and efforts that they led Degler (1991) to note categorically that Klineberg was to psychology what Boas was to anthropology. "He [Klineberg] made it his business to do for psychology what his friend and colleague [Boas] at Columbia had done for anthropology: to rid his discipline of racial explanations for human social differences" (p. 179).

Unbeknownst to many contemporary scholars of the intelligence testing movement of the 1920s, there was a cadre of African American researchers "who launched a concerted intellectual assault upon racist conclusions which white psychologists extrapolated from mental test data" (Thomas, 1982, p. 259). Perhaps the most comprehensive coverage of the role of African American scholars in the mental testing debate of the 1920s is William Thomas's (1982) article, "Black Intellectuals' Critique of Early Mental Testing: A Little Known Saga of the 1920s." Thomas's review discusses the contributions of nine African American scholars (all men) who published articles critical of intelligence testing, with particular focus on allegations by White scholars that African Americans were innately inferior in intelligence.[31] These nine scholars were Horace Mann Bond, Francis C. Sumner, E. Franklin Frazier, Charles H. Thompson, Charles S. Johnson, Howard H. Long, Ira De A. Reid, Joseph St. Clair Price, and Herman C. Canady.[32]

The majority of these 1920s scholars published their research in *Crisis* and *Opportunity*, periodicals of the National Association for the Advancement of Colored People and the Urban League, respectively. In that mainstream journals frequently were controlled by editorial boards and editors who were hereditarians (e.g., Terman's editorial control over the *Journal of Educational Psychology* and *Journal of Applied Psychology*), the new African American academic outlets, such as the *Crisis*, were welcome forums for the African American intellectual. This is not to say, however, that all African American scholars critical of differential race psychology research on intelligence testing during the 1920s were locked out of publishing their work in the standard journals. For example, Price (1929) and Sumner (1925) published in *School and Society*.

An example of one of these early investigations by African American scholars was that of Bond (1924). Bond, who earned his Ph.D. from the University of Chicago, sought to refute the assertion that African Americans were inherently inferior in intelligence by examining schooling effects on mental test performance (i.e., Army Alpha test). Using an ingenious research design, Bond turned the table on hereditarians. Thomas (1982) noted the following of Bond's study:

> By limiting his comparison to interracial differences among whites, he removed completely from his analysis the issue of differences due to race. This allowed him to question whether, as a result of differences in test performance, the exponents of intelligence testing as a discriminator of racial differences would assert that southern white draftees were inherently and racially inferior to whites in other regions of the country. (p. 272)

In an analysis of existing data, Bond (1924) compared the rank order of White draftees' intelligence test scores to the rank order of the "educational efficiency" (i.e., schooling achievement) of the respective states of the recruits. Thomas (1982), in a discussion of Bond's study, reported that the observed rank order correlation between test scores and state educational efficiency was a robust magnitude of .74. Regarding education in the South, Bond reported that nine southern states occupied the nine bottom ranks in educational efficiency. Likewise, the Army mental scores of the White draftees from these states also were among the lowest ranks. By contrast, states outside the South (e.g., California, Connecticut) were among the highest in educational efficiency, and recruits from these states were among the highest in intelligence test scores. Thomas's (1982) conclusions concerning Bond's study are well taken:

> Bond's astute observations put hereditarian proponents on the defensive. They had to admit either that the racial stock of these so-called racially pure [southern] states was distinctly inferior to whites in states having a higher percentage of southern and eastern European stock or that test performance was dependent on environmental conditions and cultural advancement reflected in schools. (pp. 272-273)

In conclusion, African American scholars of the 1920s were not silent on the allegations of deficit thinkers who asserted that African Americans were innately inferior in intelligence. As we have seen, using Thomas's (1982) account as a basis for discussion, African American intellectuals did not take lightly these frequent racial pronouncements of hereditarianism. Examining pointed hypotheses and using clever research designs and methodological rigor, a cadre of African American scholars of this period joined the rising heterodoxy. Although their research, rejoinders to the orthodoxy of hereditarian thought, and dissent have gone unrecognized by many scholars, the African American intellectual critique of early mental testing is a vital part of the history of challenges to deficit thinking in educational thought and practice.

Regarding Mexican American intellectuals of the 1920s who criticized mental testing research in which Mexican American students served as participants, there were no such scholars. This is ironic given that Mexican American students in the Southwest, as we have seen, were subject to frequent intelligence testing and resultant curriculum differentiation. For these children not to have someone of their own ethnic group to speak against contentions that they were innately inferior to White children in intelligence and to expose the inferior schooling they received via curriculum differentiation was, indeed, a sad state of affairs. This lamentable situation would change, however, in 1931 with the arrival of the indomitable George Isidore Sánchez on the academic scene.

One of Sánchez's most significant articles was his 1934 essay critical of mental measures and testing as they relate to Mexican American students, especially those who are limited in English competency. In this insightful analysis, Sánchez (1934) castigated myopic scholars and unsophisticated applied personnel who, without question, accepted intelligence tests as valid tools. The seven issues he raised were, in a sense, prognosticatory, as their relevance for Mexican Americans would hold for decades to come and would serve as beacons for reform in test development, individual assessment, and social concerns involving test use. Padilla and Aranda (1974), in their synopsis of Sánchez's essay, summarized these issues as follows:

1. Tests are not standardized on the Spanish-speaking population of this country.
2. Test items are not representative of the Spanish-speaking culture.
3. The entire nature of intelligence still is a controversial issue.
4. Test results from the Spanish-speaking continue to be accepted uncritically.
5. Revised or translated tests are not necessarily an improvement on test measures.
6. Attitudes and prejudices often determine the use of test results.
7. The influence of testing on the educational system is phenomenal. (p. 222)

Group Intelligence Testing
Taking On a Life of Its Own

By the mid-1920s, group-administered intelligence tests were used with great frequency in U.S. public schools, and bureaucracies arose to handle the mass testing and use of test results.[33] The U.S. Department of the Interior, Bureau of Education (1926), published a survey that sought to report the use of homogeneous grouping and the use of group intelligence tests in classifying students to ability groups (cited in Chapman, 1988). Based on data from 292 cities with populations ranging from 10,000 to more than 100,000, the percentages of cities reporting homogeneous ability grouping at the elementary, junior high, and high school levels were 85%, 70%, and 49%, respectively. In the same report, it was reported that 250 (86%) of the 292 cities surveyed used group intelligence tests in student classification. Thus, since the publication of the NIT in 1920 and by the mid-1920s, American public schools had become highly differentiated in curriculum, and the sorting mechanisms by far were group intelligence (and achievement) tests.[34] By 1932, 75% of 150 large cities made curricular assignments of pupils using the results of intelligence tests (Tyack, 1974).

Continuing into the 1940s and 1950s, group intelligence testing in most of the nation's public schools became a routine practice. Writing during the fall of 1949, Benjamin Fine, columnist for the *New York Times Magazine*, commented,

> Between now and June, 20,000,000 children will be subjected to tests to measure their intelligence. This figure indicates the position of influence to which IQ—Intelligence Quotient—tests have risen in little more than a generation in American school systems. In nearly all, they are used in greater or lesser degree to determine when a child should begin to read, whether another should go to college, and if a third is likely to grow up to be a dolt or an Einstein—that is, whether he is "worth worrying about" or "simply beyond help." (Fine, 1949, p. 7)

Although group intelligence testing became entrenched in the schools, research on intelligence testing dropped precipitously during this period. Haney (1981), in a review of frequency of periodical literature on intelligence testing, observed that such writings hit a peak during the 1920s, fell substantially during the 1930s, plummeted sharply during the 1940s, and rose only slightly during the 1950s.[35] This overall decline in research interest and inactivity in publishing on intelligence testing during the 1940s and 1950s might have been partially due to some scholars who were absent from the world of research as they served in the armed forces in World War II and the Korean War. Also, as Valencia and Aburto (1991) observed about the decline in intelligence testing research, part of this inattention was likely related to the development that group-based intelligence testing after the 1930s became widely implemented in the nation's schools and took on a life of its own—a life relatively free of controversy until the 1960s.

The Emergence of Contemporary Testing Issues

As we have seen, the first phase of criticisms against intelligence testing vis-à-vis minority students unfolded during the 1920s. Interestingly, the roots of the second phase of the testing controversy can be traced to one of the most monumental Supreme Court decisions in educational history—*Brown v. Board of Education of Topeka* (1954), which struck down the *Plessy v. Ferguson* (1896) "separate but equal" doctrine. Bersoff (1982) commented that for a dozen years after *Brown*, southern school systems attempted to forestall, or even circumvent, the Supreme Court's desegregation mandate through innovative strategies:

> Many of these tactics relied heavily on the use of intelligence and achievement tests. For example, in one major southern city, black children were not permitted to transfer to a "white" school unless their grade level score on an ability test was at least equal to the average of the class in the school to which the transfer was requested. Each of these dilatory mechanisms [was] challenged in the federal courts by minority plaintiffs and eventually struck down as unconstitutional. (p. 1046)

Bersoff (1982) noted that although minority plaintiffs were victorious in demonstrating how intelligence and achievement tests were used unconstitutionally against African Americans in the goal of desegregation, "in no case was the validity of the tests themselves attacked" (p. 1046). The reason for this inattention to validity issues, Bersoff asserted, is that the judiciary's only concern was whether standardized intelligence and achievement tests "were administered only to blacks or were used to make decisions solely on racial grounds" (p. 1046). Although such tests were routinely administered in the South, and elsewhere, to Whites and minority students during the decade after *Brown*, grouping practices (i.e., curriculum differentiation) and subsequent concerns about varying access to equal educational opportunity were free of judicial scrutiny.

In any event, the use of tests by southern schools to forestall the desegregation mandate was the impetus that set the second phase of the testing controversy in motion.[36] With the advent of the civil rights movement during the late 1950s and its peaking during the early 1960s, the rights of racial/ethnic minorities became focal points of national concern. Included in this debate was the role of group-administered IQ tests in the classification of minority students in the educational mainstream and the tributary of special education. Speaking of those years, Anastasi (1988) commented, "A common criticism of intelligence tests is that they encourage a rigid, inflexible, and permanent classification of pupils" (p. 67). So deep were these concerns that the use of IQ tests was discontinued, for example, in 1964 in New York City public schools (Gilbert, 1966).

In the remainder of this section, we introduce a number of contemporary intelligence testing issues germane to minority students. These concerns are (a) litigation, (b) neohereditarianism, (c) test bias research, (d) underrepresentation of minority students in gifted and talented programs, and (e) nondiscriminatory assessment.

Litigation

Hobson v. Hansen (1967) was the *first* case that spoke to the legality of using group intelligence tests in the curricular assignments of minority students. In *Hobson*, a federal district court in Washington, D.C., ruled in favor of the plaintiffs (African American students) that standardized group "aptitude" (i.e., intelligence) tests were used, in large part, to place many such students in the lower tracks. Such practices, the court ruled, created significant racial disproportionality in curricular assignments and subsequently led to diminished educational opportunity for the plaintiffs, compared to that for their White peers (Bersoff, 1982). Suffice it to say, *Hobson* would have a profound effect on the use of intelligence tests and in educational decision making to this day. On this, Sandoval and Irvin (1990) commented,

> The court decided that the standardized *group* aptitude tests were inappropriate (i.e., not valid) because the tests were standardized on white middle-class children and could not be generalized to black children. This notion of standardization became the first legal definition of test bias or lack of validity. In the decision, the court particularly disapproved of the inflexibility of the tracking system and its stigmatizing effect on black children. Because placement in a lower track was perceived to be harmful, the issue of equal educational opportunity was identified. In addition, the judge criticized the practice of using ability tests as the sole basis (or a major factor) for deciding on placement. (p. 89)

Several years after the landmark *Hobson* decision, there was a round of cases in which minority plaintiffs charged that intelligence tests were being used in a discriminatory fashion in the placement of students in educable mentally retarded (EMR) classes. These post-*Hobson* cases were particularly significant because of the targeted tests. Bersoff (1982) noted,

> Despite *Hobson*'s implicit approval of individual testing, these [post-*Hobson*] cases now began to attack the stately, revered, and venerated devices against which all other tests were measured—the individually administered intelligence scales such as the Stanford-Binet and WISC [Wechsler Intelligence Scale for Children]. (p. 1048)

The most significant of these post-*Hobson* cases are (a) *Diana v. State Board of Education* (1970), (b) *Covarrubias v. San Diego Unified School District* (1971), (c) *Guadalupe v. Tempe Elementary School District* (1972), and (d) *P. v. Riles* (popularly known as *Larry P. v.*

Riles, 1972, 1979). Collectively, these cases, which were brought forth by African American, Mexican American, and American Indian plaintiffs, were extremely influential in addressing the long-standing concern of overrepresentation of minority students in EMR classes.[37] Some outcomes of these post-*Hobson* cases involved, for example, the mandating of (a) assessment of dominant language *prior* to any form of psychological testing, (b) use of multiple data sources of assessment, and (c) due process.

Neohereditarianism

As discussed earlier, the dominant hereditarian thought of the 1920s lost its prominence around 1930. Foley (1997) advanced a number of explanations for its demise (e.g., due to the ascendancy of Hitler and Nazism, many U.S. hereditarians (especially eugenicists) did not want to align themselves with racist beliefs; the debunking of hereditarianism by Boas, Klineberg, and other critics).

Although genetic interpretations for racial/ethnic differences in intelligence were fairly silent during the 1930s, the 1940s, and most of the 1950s, such claims resurfaced during the late 1950s with Audrey Shuey's *The Testing of Negro Intelligence* (Shuey, 1958). Valencia and Solórzano (1997) traced the resurgence of hereditarianism, which they referred to as "neohereditarianism." Beginning with Shuey's (1958) work and ending with Richard Herrnstein and Charles Murray's *The Bell Curve* (Herrnstein & Murray, 1994), Valencia and Solórzano covered several decades of neohereditarianism and its implications for understanding minority intellectual performance. We include this coverage in our discussion of heredity in Chapter 6 of this book.

Test Bias Research

Another important contemporary issue, one that emerged during the late 1960s and peaked during the 1980s but still has currency, is whether intelligence tests are culturally biased against minority students. Although this concern was raised as far back as the 1920s, it has only been during the past 25 years that the measurement community has developed the statistical techniques to address the subject of bias in tests. Notwithstanding the current waning of research on cultural bias in individually administered intelligence tests, we contend that it still is important to press on with such investigations (Suzuki & Valencia, 1997). In Chapter 5 of this book, we provide a comprehensive review of 62 cultural bias studies spanning 25 years of research and draw several conclusions about the state of the art. Our major conclusion is that the subject of cultural bias in intelligence tests is an open issue.

Underrepresentation of Minority Students in Gifted and Talented Programs

Starting with the work of Terman (1925) about three fourths of a century ago, researchers and educators have expressed a strong interest in the identification and cultivation of intellectually gifted as well as talented students (e.g., Winner, 1996). Although there have been scattered demonstrations of such interest involving gifted and talented minority students (e.g., Jenkins's [1936] empirical study of Negro children of superior intelligence; Moreno's [1973] data on underrepresentation of Mexican American and African American students in California's public schools), such students have not been focal groups of attention as White students have been.

Such disinterest is evidenced by the pervasive persistent patterns of minority student underrepresentation in gifted and talented programs. For example, based on a report by the U.S. Department of Education's Office for Civil Rights (1997) on thousands of schools nationwide, African American and Hispanic (i.e., Mexican American, Puerto Rican, and other Latino) students were underrepresented by 50.6% and 50.8%, respectively, in these programs. In Chapter 8 of this book, we provide discussion on factors related to minority underrepresentation in gifted and talented programs. We also offer a number of suggested guidelines and strategies that, if judiciously applied, can increase the percentage of minority students in these programs.

Nondiscriminatory Assessment

Beginning with the *Hobson* case and the post-*Hobson* litigation, there has been an ongoing agitation for the identification, promotion, and implementation of ways in which to realize the nondiscriminatory assessment of minority students, particularly in the area of intellectual assessment. This concern has been so important that the judiciary, professional organizations, policymakers, researchers, and educators all have entered the discourse. Such discussions have centered on, for example, (a) debiasing intelligence tests, (b) reevaluating the use of current intelligence tests, (c) using alternative forms of intellectual assessment, and (d) modifying theories of intelligence (Suzuki & Valencia, 1997). Improving assessments for minority students is discussed in Chapter 10 of this book.

The present chapter reveals that researchers, as well as applied psychologists (testing personnel), continue to struggle with a number of issues raised decades ago about intelligence testing and minority students. For example, the nature-nurture debate, the role of culture and language, the role of SES, and test bias continue to be a part of current discussions. In the remaining chapters, our goal is to shed new light on these concerns.

2

Multicultural Perspectives of Intelligence: Theory and Measurement Issues

For centuries, attempts have been made to distinguish people who are intelligent from those who are less intelligent. Major efforts have been aimed at developing theories and measures that efficiently reflect and sample a variety of abilities believed to contribute to making a person smart and successful. Historically, concerns have been raised regarding the lack of sensitivity and test abuse that has affected the educational opportunities of racial/ethnic minorities. Tests that allegedly assess the potential of individuals have become long-standing gatekeepers of educational and employment opportunities. This is evidenced by the popularity of the Scholastic Aptitude Test, the Medical College Admission Test, and the Law School Admission Test that serve as integral parts of the admission criteria. As noted by Reschly (1990), aptitude tests such as these do not differ significantly from traditional intelligence measures. Currently, however, many contemporary theories and measures of cognitive abilities acknowledge the importance of cultural context.

The lack of an agreed-on definition of intelligence, and the historical separation of theory and psychometric test development, has proved to be problematic in the understanding of "intelligence" as a unified construct. In addition, the field has been plagued with accusations of cultural bias toward members of particular minority groups (e.g., African Americans, Mexican Americans) who tend to score lower on intelligence measures relative to White children and youths. We discuss the cultural bias research literature regarding individually administered intelligence tests in detail in Chapter 5 of this book.

Many dedicated researchers have attempted to understand the nature of racial/ethnic group average differences in measured intelligence. Socioeconomic status (SES), home environment, cultural bias in tests, and genetic explanations have been put forth, as noted in Part II of this book. The debate between environment and genetic perspectives continues today, as illustrated by the immediate popularity of texts focusing on racial/ethnic differences in intelligence. For example, Herrnstein and Murray (1994), authors of *The Bell Curve: Intelligence and Class Structure in American Life* (which we discuss in Chapters 3 and 6 of this book), suggested that differences in intelligence and social class may have genetic bases. Their writing spurred a number of publications to debate the authors' controversial conclusions (e.g., Fraser, 1995; Jacoby & Glauberman, 1995). It should be noted that, in spite of their focus on racial/ethnic group differences in one chapter of their book, Herrnstein and Murray (1994) did point out the following in their text: "The first thing to remember is that the differences among individuals are far greater than the differences between groups. If all the ethnic differences in intelligence evaporated overnight, most of the intellectual variation in America would endure" (p. 271). Thus, there exists a dilemma regarding this focus on racial/ethnic group differences. On the one hand, the discrepancies between *group* averages in measured intelligence are clear sources of concern to educators, researchers, and practitioners. On the other hand, application of these between-group differences to individual members of particular racial/ethnic groups is not warranted given considerable within-group differences. Debates regarding the meaning of racial/ethnic group differences in intelligence, particularly the nature-nurture controversy, continue to spark controversy. In Chapter 6 of this book, we examine this issue in detail.

Disagreements also have been raised regarding how the construct of intelligence is measured. For example, such debate is evidenced between Eysenck (1998) and Gould (1996). With regard to intellectual assessment, Gould wrote about his concerns regarding the interpretation of and credence given to test scores:

> The misuse of mental tests is not inherent in the idea of testing itself. It arises primarily from two fallacies, eagerly (so it seems) embraced by those who wish to use tests for the maintenance of social ranks and distinctions: reification and hereditarianism. (p. 185)[1]

Gould went on to present a number of arguments ranging from statistical concerns to measurement issues that illustrate problems regarding the interpretation of test scores.

Eysenck (1998), a supporter of intelligence testing, wrote, "I do not believe that anything that exists cannot be measured. As Thorndike said memorably, everything that exists, exists in some quantity and can therefore be measured" (p. 2). In addition, Eysenck emphasized that IQ correlates with performance in a variety of domains (e.g., achievement); thus, tests "measure something very important indeed" (p. 9).

In addition, a number of writers have interjected important information regarding the role of historical patterns of race relations into the intelligence debate. As noted in Chapter 3 of this book, Ogbu (1991) hypothesized that the racial/ethnic differences in intelligence and achievement found in America are based on whether a group experiences voluntary or involuntary minority status in this country. Immigrant or voluntary minorities are those who moved to their "present societies because they believed that the move would lead to more economic well-being, better overall opportunities, or greater political freedom" (p. 8). Involuntary minorities are those who were brought into "society through slavery, conquest, or colonization" (p. 9). Based on these distinctions, data on these two groups indicate that voluntary minorities attain higher levels of formal education and overall achievement (e.g., intelligence; Ogbu, 1987).

A major study conducted by Carroll (1993) attests to the large literature base that has accumulated in the area of cognition and intelligence. In his text titled *Human Cognitive Abilities: A Survey of Factor-Analytic Studies,* he examined approximately 1,500 data sets found in the literature that reported correlational or factor-analytic findings pertaining to cognitive abilities. Greater attention was given to those studies including "broader samplings of variables and individuals" with "lower priorities to data sets whose variables represented, for example, only tests from small, self-contained batteries like one of the series of Wechsler tests" (p. 79). More than 450 studies finally were selected for closer examination. In describing his final grouping of studies, Carroll stated, "Although it cannot be claimed that the file is truly exhaustive, it represents all or nearly all of the more important and classic factor-analytic investigations of the past 50 years or more as well as numerous others of potential interest" (p. 78). Carroll acknowledged that the data sets were not "sufficiently diverse with respect to ethnic or racial differences to support any definitive statement about differences in factor structures across these groups" (p. 685). His concluding statements, however, provide information regarding trends in the data examined. Based on his analysis, Carroll stated that there was little evidence that the factorial structures of cognitive abilities differ between cultural and racial groups. His findings instead indicate that "what differentiation occurs may be attributed to differences in schooling and other environmental experiences encountered by different sexes or different cultural and racial groups" (p. 687). In Chapter 5 of this book, our review of 23 construct validity studies that used White and minority participants reveals that 91% of the studies showed no cultural bias. This, in part, supports Carroll's finding of factorial equivalence across racial/ethnic groups. However, we do point out that, based on the *overall* picture regarding cultural bias, our findings in Chapter 5 lead us to conclude that cultural bias in intelligence tests is an open issue.

From the preceding paragraphs, it appears that understanding the meaning of racial/ethnic differences in cognitive abilities and intelligence is a complicated process based on many perspectives. These include, for example, disagreements about the roles of heredity and environment, measurement and meaning of group differences, immigrant status in the United States, and controversies surrounding assessment. The following sections of this chapter focus on illuminating issues related to these areas including (a) the search for a common definition of intelligence; (b) the split between theory and measurement of intelligence; (c) theoretical frameworks incorporating cultural aspects of intelligence; (d) emotional, practical, and creative aspects of intelligence; (e) assessment-based models of intelligence; (f) group differences in intelligence; (g) racial/ethnic profiles of intelligence; (h) cultural perspectives of intelligence; and (i) the complexities of multicultural perspectives of intelligence. Although this chapter is not exhaustive with respect to all of the literature in each of these sections, highlights of the work being conducted are provided.

The Search for a Common Definition of Intelligence

The construct of intelligence, and its associated abilities, has spawned a number of definitions. Researchers and theoreticians have debated as to what constitutes the core characteristics or factors of intelligence. The large number of models, theories, and definitions attest to the number of perspectives that currently exist. Vroon (1980) noted that intellectual abilities have been defined in various ways based on diverse perspectives ranging from abstract thinking, to critical thinking, to learning, to perception, to biological mechanisms (e.g., "changeability of the central nervous system" [p. 29]). "In short, we are confronted with words—in more or less jargon—which, in the form of sensory or non-sensory connected knowledge, have been written in philosophy for ages" (p. 29).

Realizations regarding the differences in definitions such as these have led some to state that intelligence might simply be what intelligence tests measure (Boring, 1923). Suffice it to say, this tautology has not been productive in unpacking the complexity of the intelligence construct.

A survey by Sternberg and Berg (1986) provided results regarding a comparison of two symposia held in 1921 and 1986 focusing on the definition and measurement of intelligence.[2] The study included a list of 27 attributes of intelligence noted in various definitions. The correlation between the 1921 and 1986 data was moderate (.50). Sternberg and Berg also reported that there exist a number of general trends in the area. First, researchers focused primarily on psychometric issues in 1921. In 1986, the concentration of effort seemed to be on "information processing, cultural context, and their interrelationships" (p. 162). In addition, researchers seemed to place greater emphasis on the importance of understanding behavior in comparison to issues of prediction. In general, Sternberg and Berg noted,

> On the one hand, few if any issues about the nature of intelligence have been truly resolved. On the other hand, investigations of intelligence seem to have come a rather long way toward understanding the cognitive and cultural bases for the test scores since 1921. (p. 162)

In addition, Sternberg and Kaufman (1998) noted that, over time,

> Learning and adaptive abilities retained their importance, and a new emphasis crept in: metacognition, or the ability to understand and control oneself. Of course, the name is new, but the idea is not, because Aristotle emphasized long before the importance for intelligence of knowing oneself. (p. 479)

The importance of knowing oneself reflects growing issues regarding social and emotional forms of intelligence that are discussed later in this chapter.

A more recent study by Snyderman and Rothman (1988) supports a general consensus regarding a definition of intelligence in terms of more cognitively based factors. They noted that most experts agreed that intelligence was related to abstract thinking as well as to reasoning, problem solving, and acquisition of knowledge.

The Split Between Theory and the Measurement of Intelligence

Early cognitive theorists viewed intelligence as a cognitive construct. Their attempts at theory construction, however, often resulted in failure to account for complex abilities, and so these theories "were either abandoned or attenuated" (Das, Kirby, & Jarman, 1979, p. 29). With the growth of the empirical testing movement, the assessment of intelligence evolved as a "technology divorced from psychological theory" (Das et al., 1979, p. 29). There existed a split between the process approach (the *how* and *why* of thinking) and the psychometric point of view (which emphasized description, selection, and prediction; Vroon, 1980). Tests of intelligence are comprised of items selected based on correlations with other measures, not based on "theoretical notions of what is being measured" (Das, Naglieri, & Kirby, 1994, p. 8). This assertion is, of course, controversial. A number of test developers have begun development with a theory of intelligence that has served as a guide (e.g., the Kaufman Assessment Battery for Children; Kaufman & Kaufman, 1983a, 1983b).

Eysenck (1982) noted that the definition and measurement of intelligence has been under the jurisdiction of psychometricians and concluded, "This is not and can never be enough if we are to gain a proper understanding of this important part of our mind" (p. 1). Thus, the need to merge the field of psychometric assessment and intellectual theory was recognized by researchers as reflected in the growth of research in this area. A recognition of the split between theory and measurement is important because it provides some degree of explanation for the current issues surrounding the understanding and measurement of intelligence.

Theoretical Frameworks Incorporating Cultural Aspects of Intelligence

The understanding of intelligence as a narrow set of abilities has been challenged by many who believe that the scope of measured intelligence is too limited.[3] Various theories attest to this notion in that they incorporate areas far beyond what is assessed traditionally. Most major theories include examination of multiple aspects, factors, or dimensions of intelligence as well as cultural components (McGrew & Flanagan, 1998). The critical importance of culture in understanding human abilities is highlighted by Armour-Thomas (1992):

Culture provides the content for attitudes, thought, and action; it allows for an idiosyncratic representation of knowledge among its people; it determines the kinds of cognitive strategies and learning modes that individuals use for solving complex problems within their society. (p. 558)

It is beyond the scope of this chapter to provide a comprehensive review of the voluminous literature on theories of intelligence. Therefore, this section focuses primarily on those that include specific references to cultural factors.

The g Factor. Spearman (1927, cited in Vroon, 1980), one of the developers of factor analysis, hypothesized that intelligence consisted of a general factor (*g*) and two specific factors (*s* verbal ability and fluency). Spearman's discovery of *g*, the first unrotated factor of an orthogonal factor analysis, fits well within his theory that intelligence is a general and hereditary ability. Tests with high *g* loadings were those that involved reasoning, comprehension, deductive operations, eduction of relations (determining the relationship between or among two or more ideas), eduction of correlates (finding a second idea associated with a previously stated one), and hypothesis-testing tasks. Tests with low *g* loadings focused on processes such as recognition, recall, speed, visual-motor, and motor abilities. With regard to racial/ethnic group differences, Spearman hypothesized that *g* differed in distribution between groups and that measures of *g* would reveal group differences. Jensen (1998), in his recent text titled *The g Factor: The Science of Mental Ability*, continued to support the *g* factor as accounting for individual and group differences in intelligence. Jensen addressed the question of whether the *g* factor is the same in Black and White groups. Based on his review of 17 studies, he concluded that the same *g* factor exists across various measures and for both Black and White samples.

Intelligences A, B, and C. Vernon (1950) used a factor-analytic technique in deriving and developing his theory of intelligence. Like others before him, he believed that general intelligence (*g*) needed to be considered in attempts to understand intelligence. He also cited the existence of factors in relation to environmental and social demands. His cross-cultural research led him to form a distinction among three types of intelligence (Vernon, 1969). The first two types (A and B) initially were identified by Hebb (1949). Intelligence A represents innate capacities of an individual that are primarily genetically determined. Intelligence B represents behavior that is societally recognized as being intelligent and might be more closely linked to culturally reinforced behaviors. Intelligence C represents those cognitive abilities that are measured by a traditional intelligence test. Thus, there exists a split between culturally reinforced abilities and those that are measured by mainstream intelligence instruments.

In his text *Intelligence and the Cultural Environment,* Vernon (1969) cited the importance of understanding the context of intellectual functioning. For example, Vernon noted that a measure of Intelligence C might have nothing to do with Intelligence A or B if it is not used in the appropriate cultural context. He stressed the importance of "a fresh interpretation of individual, social, class, and ethnic differences" (p. 6) that affect theories of intelligence.

Primary Mental Abilities. Thurstone (1941) was a major challenger of the *g* factor. Thurstone hypothesized that, instead of a general factor, there existed seven primary abilities or factors: verbal comprehension, word fluency, number facility, spatial visualization, reasoning (inductive and deductive), perceptual speed, and memory. He believed that intelligence could be divided into these seven factors, each with equal weighting. Although his statistical methods were challenged by others who attributed greater amounts of variance to a *g* factor (Eysenck, 1939), the notion of primary mental abilities has continued. Over time, other researchers have added to the notion of primary mental abilities. For example, Lynn and Hampson (1985-1986) noted in their review that some researchers expanded the number of abilities to more than 22.

A recent review of the theory of primary mental abilities conducted by Horn and Noll (1998) reported that "a system of more than 60 different abilities is needed to describe human cognitive abilities" (p. 63). In describing the Primary Mental Abilities (PMA) system, Horn and Noll described the following blending of various perspectives in the theoretical foundation: "Substantively, the theory incorporates many ideas and measures derived from Spearman's studies of *g,* Burt's hierarchical analysis, and Guilford's ideas about what ability tests measure" (p. 69).

Horn and Noll (1998) reported nine dimensions of ability that make up intelligence according to the PMA system. This finding was based on "patterns of intercorrelations among estimates of primary mental abilities" (p. 69). These abilities are fluid reasoning (Gf), acculturation knowledge (Gc), short-term apprehension-retention (SAR), fluency of retrieval from long-term storage (TSR), visual processing (Gv), auditory processing (Ga), processing speed (Gs), correct decision speed (CDS), and quantitative knowledge (Gq). Cultural components are incorporated under Gc that are measured in tasks indicating breadth and depth of knowledge of the dominant culture. Tasks that are associated with Gc in this model include general information, verbal comprehension, sensitivity to problems, syllogistic reasoning, behavioral relations, semantic relations, number facility estimation, mechanical knowledge, and verbal closure.

Crystallized and Fluid Abilities. Cattell (1963) suggested that intelligence (*g*) could be divided into two forms: crystallized and fluid abilities. He used Thurstone's tests of primary abilities and, through oblique rotation, determined the existence of fluid and crystallized abilities. Fluid abilities refer to the capacity required for problem solving (i.e., the capacity to perceive relationships and use logic). Specifically, fluid abilities may involve adaptive and new learning capacities involving mental operations and processes. Crystallized intelligence often involves established cognitive functions related to achievement. Crystallized abilities are those abilities influenced by formal and informal education. Jensen (1998) also discussed crystallized intelligence (Gc) as "consolidated knowledge" that was "most highly loaded in tests based on scholastic knowledge and cultural content where the relation-eduction demands are fairly simple" (p. 123). Fluid abilities are considered to be more nonverbal and culture reduced. Jensen also noted the following:

> Persons from very different cultural backgrounds, however, may differ markedly in the Gc appropriate to any one culture, even though they may be equal in Gf. But each person's Gc would closely parallel his or her Gf in the person's own culture. (p. 123)

With regard to assessment, it is crystallized intelligence that often is given greater weight, but not exclusively, on traditional intelligence tests. Traditional intelligence tests often are reflective of what an individual has learned within a particular environment. Thus, level of education is a major moderator of performance on intelligence measures. For example, Kaufman (1990) reported that college graduates obtain IQ scores 32.5 points higher than those of individuals with 7 years or less of schooling.

A related hierarchical model of intelligence developed by Jensen (1973b, 1974; Jensen & Reynolds, 1982) indicated that intellectual ability could be viewed as a combination of Level I and Level II abilities. Level I abilities are those involving simple associative learning. Level II abilities involve higher order conceptual learning similar to Cattell's concept of fluid intelligence. Jensen (1973b, 1974; Jensen & Reynolds, 1982) noted that although these two levels of intelligence are present in all populations, different social classes and ethnic groups demonstrate different patterns of abilities indicating varying strengths and weaknesses. According to his theory, Level I abilities are evenly distributed in all populations and are represented by activities involving rote learning and short-term memory. By contrast, Level II abilities are distributed differently in upper and lower social classes and varying ethnic groups. "The majority of children now called 'culturally disadvantaged' show little or no deficiency in Level I ability but are about one standard deviation below the general population mean on tests of Level II ability" (Jensen, 1970, p. 25). In Chapter 6 of this book, we return to further discussion of Jensen's Level I–Level II theory.

Multiple Intelligences. Gardner (1983) argued for the concept of multiple intelligences including interpersonal, intrapersonal, linguistic, logical-mathematical, bodily kinesthetic, musical, and spatial. Gardner (1987) noted that an important aspect of his multiple intelligences construct was the "ability to solve a problem or to fashion a product which is valued in one or more cultural settings" (p. 25).

Gardner's (1983) work also brought to the forefront personal intelligences—intrapersonal and interpersonal. These may be associated with socioemotional aspects of intelligence that are discussed in greater detail in the following section of this chapter. Specifically, Gardner noted,

> More so than any other realms, one encounters a tremendous variety of forms of interpersonal and intrapersonal intelligence. Indeed, just because each culture has its own symbol systems, its own means for interpreting experiences, the "raw materials" of the

sonal intelligences quickly get marshaled by systems of meaning that may be quite distinct from one another. (p. 240)

Gardner, Hatch, and Torff (1997) elaborated on the "symbol systems" approach that may integrate, and to some extent reconcile, biological and cultural perspectives of intelligence. They noted the importance of genes that directly "mediate" the development of the entire body including the brain. Although genes have control over "anatomical structures and physiological processes," there is no control or awareness of cultural factors (p. 249). The situation is reversed with respect to culture. "The culture exists in the practices of the group and in the minds that apprehend these practices. To survive, the culture must ensure that its thoughts, values, behaviors, and the like are perpetuated" (p. 249). Cultural understanding does not include awareness of the role of the genes or anatomical structures. Thus, biological and cultural perspectives both are critical in terms of understanding an individual's behavior, including intelligence.

By examining the use of symbols systems, "mechanisms that explain how messages contained in the genetic code and messages borne by the cultural languages may be identified" (p. 250). Thus, Gardner et al. (1997) noted that these two seemingly different perspectives (i.e., genetics vs. culture) may be reconciled and applied to the theory of multiple intelligences. For example, Gardner et al. provided a discussion of symbol system approaches to musical and spatial intelligences. In particular, they noted the neurobiology of musical abilities as well as the cultural impact of early musical development, exposure, and instruction.

The Triarchic Theory of Intelligence. Sternberg's (1997a) triarchic theory comprises three subtheories: (a) the *componential subtheory,* which focuses on the relationship between intelligence and the individual's internal world; (b) the *experiential subtheory,* which highlights the relationship between intelligence and the individual's immediate experience with tasks and situations; and (c) the *contextual subtheory,* which explains the relationship between intelligence and the individual's external world.

Related to these subtheories are components that form the foundation of intelligent thought and behavior. Sternberg (1997a) identified three types of information processing components: metacomponents, performance components, and knowledge acquisition components. Metacomponents are defined as "higher order executive processes" used to plan, monitor, and evaluate problem solving and behavior. Performance components are "lower order processes that execute the instructions of the metacomponents" (p. 94). Knowledge acquisition components are "used to learn how to do what the metacomponents and performance components eventually do" (p. 95). All of these components of intelligence are viewed as universal.

Thus, the components that contribute to intelligent performance in one culture do so in all other cultures as well. Moreover, the importance of dealing with novelty and automatization of information processing to intelligence are posited to be universal. But the manifestations of these components in experience [are] posited to be relative to cultural contexts. What constitutes adaptive thought or behavior in one culture is not necessarily adaptive in another culture. (p. 102)

Like other researchers cited in this section, Sternberg placed an emphasis on the importance of cultural relevance and context in understanding abilities.

Ceci's Bioecological Framework. Ceci's (1996) bioecological model acknowledges the joint role of biological (genetically determined cognitive potentials) and environmental influences on intelligence. The framework was noted to be "inherently developmental, biological, and contextual" (p. 93). Ceci noted, "One's cultural context is an integral part of cognition because the culture arranges the occurrence or nonoccurrence of events that are known to affect cognitive development" (p. 93).

In the preceding sections, all of these theorists have suggested, to some extent, the role of culture in understanding intelligence and cognitive abilities. Their theories have promoted diverse perspectives in this area and the importance of understanding the complexities of theoretical developments. These theoretical frameworks have begun to lead to the development of new measures and procedures in the assessment of intelligence that are discussed in the sections that follow.

Emotional, Practical, and Creative Aspects of Intelligence

Although the notion of social and emotional forms of intelligence have been noted in the literature for decades, it has only been recently that they have been touted in both scientific journals and the popular press as being important to successful achievement. One such construct made popular by Goleman (1995) is that of "emotional intelligence." The assessment of emotional aspects of intelligence has become in vogue as newer measures of intelligence often include subtests assessing the ability of participants to recognize and process emotional signals. In addition, measures of emotional intelligence are becoming more and more popular. As defined by Salovey and Sluyter (1997), emotional intelligence

> involves the ability to perceive accurately, appraise, and express emotion; the ability to access and/or generate feelings when they facilitate thought; the ability to understand emotion and emotional knowledge; and the ability to reflectively regulate emotions in ways that promote emotional and intellectual growth. (p. 23)

Salovey and Sluyter acknowledged the importance of culture and subculture in that "examining more complex manifestations of emotional intelligence (beyond that of the simple identification of emotion) often require[s] understanding the individual's own cultural framework" (p. 9).

Saarni (1997) pointed to the importance of emotional competence and the "demonstration of self-efficacy in emotion-eliciting social transactions" (p. 38). She commented that the importance of understanding the context of emotional responses is learned within a cultural context; therefore, the two are inseparable. Emotional maturity and competence are embedded in cultural context.

Cultural norms dictate appropriate emotional expression. Haviland-Jones, Gebelt, and Stapley (1997) pointed out that "cross-misuse" of emotional signals often results in an individual being labeled inappropriately by terms such as "hostile, rude, or infantile" (p. 241). Work by Izard (1980) suggested that culture plays a major role in emotional understanding. His findings suggested that although many characteristics of fundamental emotions are "innate and universal," there are cultural differences in "attitudes toward emotions and their expressions" (p. 216). Thus, given the growing emphasis being placed on emotional aspects of intelligence, it is important that these be understood within a cultural context. To date, there have been only a few studies that have been done in this area with regard to the assessment of emotional intelligence and the application to multicultural populations (Vraniak et al., 1998).

Practical and Creative Intelligence. Sternberg (1996) cited the importance of creative and practical areas of intelligence not commonly tapped in traditional measures. He emphasized that some aspects of intelligence are universal, whereas others differ between cultures. In particular, he reported that some members of racial/ethnic groups that do not perform as well on academic intelligence tests, on average, do compete successfully on measures of practical and creative intelligence. Berry and Irvine (1986) pointed out the importance of practical knowledge as the "highest possible relevant standard" (p. 271) with regard to the continuing survival of a group within an ecological "day-to-day" framework. In particular, they cited the importance of alertness, observation, vigilance, correct inference from environmental cues, honesty, humility, impartial use of authority, and respect for elders. "Without books, radio, or television, accumulated life experience is the foundation of all knowledge" (p. 277). African Americans and Latinos who generally have not done very well on traditional cognitive tests appear to do well on tests of creative and practical intelligence (Sternberg, 1996).

Moral Intelligence. Coles (1997) linked emotional intelligence to what he and his colleagues termed "moral intelligence." He noted, "We grow morally as a consequence of learning how to be with others, how to behave in this world, a learning prompted by taking to heart what we have seen and heard" (p. 5).

In Coles's text *The Moral Intelligence of Children,* he reported the importance of cultural influences in the family that affect the moral intelligence of children. He highlighted developmental issues regarding the "moral archaeology" of childhood and adolescence. He noted that some children's values are shaped early in life. Therefore, it is important to provide an environment that promotes the understanding of moral issues.

We end this discussion by raising a critical point. The constructs of emotional, practical, creative, and moral aspects of intelligence are viewed by many as contributions toward broadening the traditional construct of intelligence. Gardner (1999), however, recently challenged inclusion of so many intelligences. He noted that these concepts appear to indicate that "anything goes" in terms of a "New Age" concept of intelligence. Instead, "The challenge is to chart a concept of intelligence that reflects new insights and discoveries and yet can withstand rigorous scrutiny" (p. 73). This appears to be a parallel issue to what occurred earlier in the development of definitions of the term *intelligence.* To some, intelligence became a term with so many meanings that it "finally had none" (Spearman, 1927, cited in Vroon, 1980, p. 29).

Assessment-Based Models of Intelligence

A number of intelligence theories have been directly tied to the assessment process. These include (a) the Gf-Gc Cross-Battery Assessment model; (b) the Planning, Attention, Simultaneous, Successive (PASS) model; and (c) the biocultural model of intelligence. We discuss these models next.

Gf-Gc Cross-Battery Assessment Model. Although it is based on Cattell's (1963) original Gf-Gc theory, the Gf-Gc Cross-Battery Assessment model (McGrew & Flanagan, 1998) represents one of the newest models of the assessment of intelligence. According to McGrew and Flanagan (1998), the contemporary Gf-Gc theory promotes a multiple intelligences framework of human cognitive abilities. Based on this theory, an Intelligence Test Desk Reference (ITDR) was designed to evaluate current instruments according to a group of common theoretical criteria. Factor classifications of measures are provided with respect to the various ability areas highlighted in the model. The purpose of the ITDR is to assist clinicians in selecting the "most technically adequate and theoretically pure tests" (p. 63) for inclusion in the cross-battery assessment procedure. The overall model is made up of three strata. Stratum I includes narrow abilities including listening, general information, language development, lexical knowledge, general sequential reasoning, induction, quantitative reasoning, and speed of reasoning. These narrow abilities are divided into the Stratum II broad abilities of Gc (crystallized) and Gf (fluid) abilities. Stratum III is represented as the general or *g* level. Other variables affecting test performance are noted to be background/environmental variables, test directions, and the impact of the U.S. culture. McGrew and Flanagan further noted that the Gf-Gc structure of intelligence was found to be invariant across different cultural and racial groups based on Carroll's (1993) earlier review.

Jensen's (1998) work also supports the importance of consistency in the factors underlying *g* with regard to intelligence measures:

> The *g* is always influenced, more or less, by both the nature and the variety of the tests from which it is extracted. If the *g* extracted from different batteries of tests was not substantially consistent, however, *g* would have little theoretical or practical importance as a scientific construct. (p. 85)

As part of the ITDR, the cultural contents of various measures are evaluated based on the authors' "subjective evaluation of the degree to which U.S. cultural knowledge or experience is required to perform the task" (McGrew & Flanagan, 1998, p. 66). Measures were judged as high, medium, or low based on the degree to which test performance is influenced by exposure to American culture. In addition, cultural opportunities and experiences that impart knowledge of a culture were examined as part of the evaluation criteria incorporated in the ITDR. Additional information regarding the actual tests and the ITDR are provided in Chapter 9 of the present book.

The PASS Model. Das et al. (1994) posited that the PASS model, unlike other theories of intelligence, focuses on cognitive functions and processes. It is based on Luria's (1973) theory of neuropsychological functioning. Das et al. (1994) reported that these processes form a complex and interdependent system that aids the clinician in understanding the nature of individual differences, provides a conceptual framework for the assessment process, and leads directly to theory-based remediation.

The PASS model is made up of tasks assessing, first, Planning that provides the individual with the means to analyze cognitive activity, develop some method to solve a problem, evaluate the effectiveness of a solution, and modify the approach used as needed. Second, Attention enables the individual to focus on relevant stimuli and maintain efficiency on a particular task. Third, Simultaneous involves integration of stimuli into synchronous and primarily spatial groups. Fourth, Successive enables the integration of stimuli into temporally organized serial order. According to Das et al. (1994), all processes are influenced by knowledge obtained from the environment.

Simultaneous and sequential processing constructs have been studied cross-culturally. For example, Das (1973) and Das et al. (1979) hypothesized that one's culture and social class may create preferences for simultaneous and successive processing:

> The two parallel modes are available to an individual and are used according to the nature of the task as well as the bias of the individual toward one or the other method of information integration. Cultural or individual preference for the use of a specific mode may thus exist. But intelligence is not marked by a preference for one or the other mode. (Das, 1973, p. 108)

Das et al. (1994) related their theory to cultural settings, indicating that the Simultaneous and Successive factors have been found to exist in a range of culturally distinct settings. They cited a number of studies in which the theory has been applied to different populations including Chinese participants from Hong Kong, Canadians, Native Canadians, Australians, and Australian Aboriginal children. Das et al. concluded that these studies illustrate that simultaneous, successive, and planning tasks and processes have "functional similarity" across a spectrum of cultures, language, and socioeconomic classes.

Bicultural Model of Intelligence. Armour-Thomas and Gopaul-McNicol (1998) developed a bicultural model of intelligence. According to this model, intelligence is defined as "a culturally derived abstraction that members of any given society coin to make sense of observed differences in performance of individuals within and between groups" (p. 59). Cognitive potential is mediated by biological factors. The ways in which potential is developed and expressed, however, are based on cultural experiences. Thus, according to this model, intelligence is understood from a biocultural perspective. Four assumptions underlie this model. First, it is assumed that biology and culture are reciprocally interactive in a "dynamic weaving" (p. 60). Second, this perspective acknowledges the importance of knowledge and processing in cognitive development. Third, the development of cognitive potential is facilitated through instruction. Fourth, internal and external motivations are related to the development and expression of cognitive abilities. Understanding of these assumptions and the biocultural perspective has led to the development of an assessment system incorporating health, linguistic, prior experiences, and family issues. The system highlights the "cultural niche" of the child in terms of his or her home, school, church, and community. Information obtained from assessing these domains is integrated into the understanding of different forms of intelligence (e.g., musical intelligence, bodily kinesthetic intelligence) as well as traditional forms of cognitive ability. Armour-Thomas and Gopaul-McNicol also wrote a critical review of current assessment instruments in light of their biocultural model. In addition, they highlighted how an understanding of the model would affect actual interpretation of test scores and report writing.

The biocultural model appears to represent a more qualitative type of adjustment process in interpreting intelligence test scores based on information obtained regarding other forms of ability and testing practices. The model requires examiners to gather more information regarding the examinee's life history and current status to gain a more accurate estimate of overall abilities.

Group Differences in Intelligence

Prior to examining racial/ethnic group differences in intelligence, it is important to note that concerns regarding the use of intelligence measures with diverse populations have led to increased demands for cross-cultural validation of ability measures. In response, modifications have been made with various measures, and some new tests have been developed and advertised as being nonverbal and "culturally reduced." In sharp contrast to past decades, most current tests include proportional representation of diverse racial/ethnic minority groups in their standardization samples based on census data. Questions have arisen, however, as to the impact of such small numbers relative to the overall standardization sample. For example, the fourth edition of the Stanford-Binet Intelligence Scale included 3,691 Whites, 711 Blacks, 313 Hispanics, 107 Asians, and 82 Native Americans (Thorndike, Hagen, & Sattler, 1986a, 1986b; see also Chapter 9 of the present book). In this case, the small numbers of minority group members in the standardization sample will have little impact on the overall norming process given that the vast majority of individuals included are White. This position, however, leads in part to the argument surrounding racial/ethnic norming of tests. Nonetheless, some test developers (e.g., Wechsler, 1991) have included racial/ethnic oversampling to examine racial/ethnic group patterns.

Although the primary emphases of the present book are on factors related to minority intellectual performance and on racial/ethnic group differences in intelligence, it is important to note that within-group differences exceed between-group differences (see, e.g., Suzuki & Valencia, 1997). Thus, one must use caution in interpreting overall means or averages as reflective of individual members of these groups. We cannot emphasize this point enough.

Given these issues related to test development practice and within-group differences, concerns continue to be raised regarding group differences. One of the greatest controversies due to the growing emphasis on testing has been the consistent mean difference between particular racial/ethnic groups on intelligence measures. The most frequently used intelligence tests are standardized to have a mean of 100 and a standard deviation of 15. Overall IQ discrepancies between racial/ethnic groups were estimated by 52 professors identified as experts in intelligence and allied fields in a *Wall Street Journal* article titled "Mainstream Science on Intelligence" (1994). The findings included the following averages: Whites, 100; Blacks (African Americans), 85; Hispanics, midway between Whites and Blacks; and Asians and Jews, somewhere above 100. In addition, it has been estimated that American Indians score at approximately 90, on average (McShane, 1980). The discrepancies in scores have been attributed to a variety of factors including genetic bases, cultural and environmental differences, cultural bias in tests, SES, and some combination of these. In Part II of this book, we examine these factors in great detail.

There are several key concepts integral to understanding cross-cultural comparisons of children's measured intelligence: cultural bias, cultural loading, and cultural equivalence. Conceptions of cultural bias and cultural loading are discussed in Chapter 5 of this book. Here, our discussion focuses on cultural equivalence.

Cultural Equivalence

According to Poortinga (1983), "equivalence" or "comparability" is one of the most critical issues with respect to the methodology of cross-cultural or multicultural research. Equivalence refers to "the problem of whether, on the basis of measurements and observation, inferences in terms of some common psychological dimensions can be made in different groups of subjects" (p. 238). The issue of equivalence has proved to be noteworthy with respect to intelligence testing given racial/ethnic group differences in scores. Different forms of equivalence have been noted in the literature.

More recently, Helms (1992, 1997) summarized various forms of equivalence, incorporating the work of Butcher (1982), that affect cognitive ability testing. Helms (1992) cited several forms of equivalence indicating the following: (a) functional (i.e., test scores have a consistent meaning for different racial groups, and the construct measured occurs in equal frequency for each group); (b) conceptual (i.e., the information contained in the test items is equally familiar to different groups); (c) linguistic (i.e., the language presented in the test has similar meaning to different groups); (d) psychometric (i.e., the instrument measures "the same things at the same levels across cultural groups" [p. 1092]); (e) testing condition (i.e., testing procedures are familiar and "acceptable" to different groups); (f) contextual (i.e., environments "in which the person functions, the to-be-assessed cognitive ability is evaluated similarly" [p. 1092]); and (g) sampling (i.e., samples on which the test is developed and validated, and on which interpretations are based, are comparable for each group).

Another form of equivalence noted in an article published by the Laboratory on Comparative Human Cognition (1982, cited in Armour-Thomas, 1992) refers to the concept of functional stimulus equivalence. This form of equivalence "means that in order to make valid comparisons about the intellectual performance between two groups, the cross-cultural psychologist must ensure that the stimulus attributes of the task [are] the same for both groups" (Armour-Thomas, 1992, p. 556).

Examining equivalence issues in assessment is critical. For example, many researchers, educators, and clinicians rely on translated versions of tests, raising the issue of linguistic equivalence. Indeed, many test developers engage in state-of-the-art back-translation methods in which a test is translated into the target language and then back-translated into the original language. A comparison is then made between the original and back-translated versions. Although this procedure is widely accepted, it ensures only that the language is "equalized." There is no assurance that the translated measure maintains other forms of equivalence (for a discussion of issues related to test translations and the cognitive assessment of limited-English-proficient and bilingual children, see López, 1997). Continual examination of equivalence issues with regard to intelligence testing appears warranted despite increased sophistication in test development practices (see Chapter 9 of this book).

Racial/Ethnic Profiles of Intelligence

Examination of the overall IQ score provides only partial information regarding racial/ethnic group abilities. Many researchers have recognized that cultural differences between groups can influence the development of distinct patterns of mental abilities (Anastasi, 1958a; Eells, Davis, Havighurst, Herrick, & Tyler, 1962; Lesser, Fifer, & Clark, 1965; Vernon, Jackson, & Messick, 1988). Such distinct cross-racial/ethnic patterns can occur because each culture tends to focus on and encourage different skills and behaviors (Anastasi, 1958a). Kaufman (1990) noted that the "content of all tasks, whether verbal or nonverbal, is learned within a culture" (p. 25). This assumption again relates to the notion that all tests are culturally loaded. This is so because intelligence tests generally contain items relevant to the definition of "smart" behavior within a particular culture. Even some of the most popular tests indicate in their descriptions that caution should be used in interpreting the scores obtained by individuals who might have had "educational and cultural experience somewhat different from that of the traditional majority" (Educational Testing Service, 1994, p. 14).

The unique profiles of various racial/ethnic groups on traditional intelligence measures further attest to the potential importance of culturally reinforced abilities. For

example, researchers repeatedly have noted higher nonverbal reasoning abilities relative to verbal reasoning abilities, on average, for American Indians (McShane, 1980;
Sattler, 1988; Wilgosh, Mulcahy, & Watters, 1986) and for Latinos (McShane & Cook,
1985; Suzuki & Gutkin, 1993). Asian Americans' abilities also have reflected nonverbal
reasoning strengths along with high mathematical abilities. The profile for African
Americans has remained inconclusive because different samples yield different results.
There is some indication that African Americans tend to do better on verbal tasks than
on nonverbal reasoning measures (Vance, Huelsman, & Wherry, 1976). One needs to be
cautious, however, about drawing conclusions that minority populations can be characterized as being strong or weak in patterns of mental abilities. There probably are far
more similarities than differences in such patterns between racial/ethnic groups. Research repeatedly has indicated greater within-group variability than between-group
variability on test scores. Actual information regarding the profiles of abilities for various racial/ethnic groups with respect to particular instruments is provided in Chapter 9
of this book.

Cultural Perspectives of Intelligence

In addition to differences in overall intelligence test scores and profiles of abilities, understanding racial/ethnic group differences in this area has been complicated by findings indicating that various cultural groups might place different emphases on what aspects of intelligence are most valued. Many Western conceptions of intelligence focus on
verbal abilities as an important aspect of this construct. This does not necessarily hold
true, however, for other groups. Although our focus in this book is on domestic U.S. minority groups, some examples from cultural groups outside the United States can be
used to understand cultural perspectives of intelligence.

For example, in Chinese culture, the term *intelligence* is translated as *Chih li* (Chen &
Chen, 1988). Chen and Chen (1988) did a study of Chih li among 1st-year undergraduates at two Hong Kong universities that included students attending an English school
and those attending a Chinese school. They found that responses given by the students
indicated that intelligence comprised traditional component skills (i.e., verbal and nonverbal abilities) with regard to the domain of intelligence. There were other aspects of intelligence, however, that were weighted differently by the groups and also were deemed
important. The most highly valued skill was nonverbal reasoning; it was rated as significantly more relevant than verbal reasoning and social skills. In addition, the Chinese
school group rated verbal reasoning skills as less relevant than did the English school
group. This was explained in terms of school instruction. Chen and Chen noted that the
teaching style in traditional Chinese schools focused more on "silent mental activities,
whereas that of the English schools stressed more group discussions and verbal inquisitiveness" (p. 485).

These findings are supported by Irvine (1978), who noted that it is important to use
extreme caution in making inferences about an individual's cognitive processes based

on overt verbal behavior. Just because "people do not talk about something does not mean they are unable to think about it" (p. 308).

Azuma and Kashiwagi (1987) noted that "to be highly intelligent" translated into Japanese is *atama ga yoi.* Characteristics of an intelligent person included social competence, task efficiency, originality, reading, and writing. Gender differences also were noted, as females were ascribed greater intelligence based on social competence and writing. Thus, the authors pointed out that an intelligent person might be someone who is more adept in assuming his or her expected role.

In Gill and Keats's (1980) comparative research on Malay and Australian psychology students, differences in the definition of intelligence were found. In this study, students were asked to describe what intelligence meant descriptively and to provide specific facets in terms of particular behaviors, skills, and abilities. In general, findings indicated that Malays placed greater emphasis on "creativity and ability to adapt to different forms of a problem, while Australians stressed adaptability to new and changing matters" (p. 240). Other notable findings were the Malays' emphasis on social skills. Gill and Keats noted that the importance of social skills could have implications for testing in that Malays might focus on pleasing the examiner and provide answers that match their perceptions of what the examiner wants.

In Wober's (1974, cited in Gill & Keats, 1980) study on Ugandans, it was found that they linked intelligence with "slowness, gradualness, and taking one's time, whereas western-educated Ugandans and Indians in Uganda associated it with speed" (Gill & Keats, 1980, p. 234). In addition, links between attitude (i.e., obduracy, honorableness, and sociableness) and intellectual ability were noted. Wober also found that Muslim participants linked intelligence with being "friendly, honourable, happy, and public" (p. 241).

Similar findings regarding the importance of social aspects of intelligence were found in Dasen's (1984) study of the Baoule of the Ivory Coast:

> What the Baoule do value are social skills: being helpful, obedient, [and] respectful, but also being knowledgeable, taking responsibility, and showing initiative in tasks useful to the family and the community. More technological skills, such as sense of observation, quick learning, memory, [and] manual dexterity, are also valued, but only if they are put into the service of the social group. (p. 430)

Dasen noted that the relationship between cognitive and social attributes of intelligence has been observed in studies of participants in Nigeria, Uganda, and Zambia and might reflect an emic definition of intelligence that could be described as pan-African. This phenomenon may be found in the more traditional communities. Dasen reported, however, that the cognitive, competitive, and individualist values of educational systems may be "implanted in most African countries as an inheritance of colonial times" (p. 429).

The construct of intelligence translated into the Baoule language is *n'glouele.* In Dasen's (1984) analysis, he noted that when applying the term to children, the term is changed to *o yon' glouele foue* (literally "who will own intelligence"; p. 426). Baoule adults report that children are constantly engaged in a process of change; thus, they are reluctant to use a term indicating assessment in the present. Instead, they denote aspects of the children's behavior as it affects future intelligence. Other aspects of Baoule intelligence include terms indicating responsibility, initiative, obedience, honesty, politeness, verbal memory, speaking in a socially appropriate manner, maturity, wisdom, luck, observation, manual dexterity, and attention.

Dasen (1984) interpreted his findings with the Baoule in terms of Berry's (1976, cited in Dasen, 1984) ecocultural model, indicating that "people value and develop those skills and concepts that are useful in the daily activities required by each eco-cultural system" (Dasen, 1984, p. 411). For example, groups such as the Baoule, which live in agricultural communities requiring trade, tend to value quantitative concepts, whereas groups who are hunters and gatherers rely more on spatial skills given their more nomadic lives. For a more comprehensive review of work in this area, the reader is referred to Sternberg and Kaufman (1998).

The text earlier in this section highlighted only some of the literature in the area of cross-cultural definitions of intelligence. It appears that although different aspects of "intelligent behavior" were emphasized, similar cognitive skills also were noted (i.e., memory, nonverbal reasoning, reading, writing, intellectual ability, quick learning). Based on their recent review of this area, Sternberg and Kaufman (1998) concluded the following:

> Cultures designate as "intelligent" the cognitive, social, and behavioral attributes that they value as adaptive to the requirements of living in those cultures. To the extent that there is overlap in these attributes across cultures, there will be overlap in the cultures' conceptions of intelligence. Although conceptions of intelligence may vary across cultures, the underlying cognitive attributes probably do not. There may be some variation in social and behavioral attributes. As a result, there is probably a common core of cognitive skills that underlies intelligence in all cultures, with the cognitive skills having different manifestations across the cultures. (p. 497)

The Complexities of Multicultural Perspectives of Intelligence

From the preceding sections of this chapter, it appears that an understanding of the context of intellectual assessment from historical, cultural, and theoretical perspectives reveals the complexity of this construct that we know as intelligence. Although numerous theories and trends have been noted in the literature, a common core definition has not reached consensus among scholars. In addition, these complexities are exacerbated by the increasing variety of forms of intelligence being studied. The term now incorporates traditional cognitive abilities as well as social, emotional, moral, practical, and other aspects.

Although nearly all experts in the field of intelligence acknowledge the role played by cultural factors, the weighting placed on environmental, cultural, and genetic factors continues to be debated. Some continue to adhere to studies indicating that the overall construct of intelligence (*g*) remains constant across racial/ethnic groups. Others note that differences in cultural definitions of intelligence are important to understanding cognitive performance. As the number of definitions, theories, and measures increases, the complexities of multicultural perspectives will continue to raise challenges to our understanding of the construct of intelligence.

PART II

Performance Factors

3

Socioeconomic Status

> The question of the precise character of the relationship between certain socioeconomic variables and intelligence test score comprises one of the most persistent and perplexing problems in the entire field of mental testing. (Neff, 1938, p. 727)

Although this introductory passage from Walter Neff's article ("Socioeconomic Status and Intelligence: A Critical Survey") was written more than 60 years ago, it still has relevance today. Numerous studies over many decades, beginning with the seminal work of Galton's (1870) *Hereditary Genius: An Inquiry Into Its Laws and Consequences* and more recently with Herrnstein and Murray's (1994) *The Bell Curve: Intelligence and Class Structure in American Life,* have found the variable of socioeconomic status (SES) to correlate positively with intellectual performance. As one's SES increases, there is a tendency for one's intellectual performance to increase. This is not in dispute among scholars. How to explain average differences in measured intelligence between different SES *groups* (e.g., low vs. middle SES) is, however, a very different subject.

There are two general approaches that scholars typically use to investigate the relation between SES and measured intelligence (for a discussion, see Jensen, 1998). One method focuses on what is generally called "socioeconomic status of origin" (i.e., the current SES of parents who are raising a child). The other approach, "attained socioeconomic status," has to do with what an adult has attained (e.g., occupational status, years of schooling completed, some composite measure of socioeconomic indicators). Given that the focus in the present book is on children and youths, we confine ourselves to those studies that have examined the relation between SES of origin and intelligence. This is not to say that the relation between attained SES and intelligence is not important.

In fact, it is very significant. When schooling attainment (i.e., highest grade in school completed) is conceptualized as the predictor variable, and full-scale IQ is the criterion variable, the "sheer quantity of schooling . . . exerts a powerful influence on IQ" (Ceci, 1991, p. 703). Ceci (1991) noted that the correlations between schooling attainment and IQ frequently are quite large, often in excess of .80. Furthermore, even when likely confounding variables are controlled, correlations still are quite strong (between .60 and .80).

It is not always made explicit in the literature why parental SES is positively associated with children's intellectual performance (and school achievement). One major assumption is that, given their average economic earnings, parents of higher-SES backgrounds often are able to provide more learning resources (e.g., books, computers) than are their lower SES counterparts. Furthermore, given the higher educational attainment of socioeconomically advantaged parents, it is suggestive that the "skills and concepts that are implicit in school culture, and in the content of mental tests, may be passed on to children in proportion to the parents' own exposure to the culture of the schools" (Valencia, Henderson, & Rankin, 1981, p. 531). Nonetheless, it is important to underscore that SES is a summarizing variable that frequently "conceals the considerable range of variation among these characteristics [lifestyle variables] *within* a given socioeconomic status level (Henderson, 1981, p. 11).

The meaning of the SES and intelligence connection, however, is fraught with questions and, in some cases, controversy. What are these concerns? Once again, we turn to Neff (1938). The four questions he raised six decades ago remain germane today:

> How much of a role does this variable [SES] play in determining IQ? Which of the great variety of social data which accrue to a given individual in our present complex society correlate positively with intelligence? What bearing has this result upon the ubiquitous nature-nurture controversy? What implications has this generalization for education, for government, for the very mechanism for democracy itself? (p. 728)

On close examination, one can see that Neff's (1938) four questions, although couched in a general context, have considerable importance in our analysis of the role of measured SES as a predictor of measured intelligence among minority students. Specifically, Neff's concerns form a foundation for this chapter, and they are addressed when pertinent literature is available.

1. In general, what is the strength of the relation between SES and the dimension of intellectual performance?

2. Does SES correlate with measured intelligence across White and minority groups at the same magnitudes?

3. In studies that attempt to control for SES in White and minority intellectual comparisons, can SES be truly "matched?"

4. If SES is considered an "environmental" variable, then how does this ascribed social marker fit into the nature-nurture debate? Specifically, when SES is controlled in White

and minority studies, to what degree is the mean difference in measured intelligence reduced? If there remains a residual after mean reduction, then can one suggest that the residual has a genetic origin?

Before beginning our discussion of relevant literature, it is important to comment on why SES, as a measured predictor variable of intellectual performance, is a particularly significant area of scholarly inquiry with respect to minority students. First, SES and racial/ethnic status often covary, meaning that Whites tend to be overrepresented and underrepresented in the higher- and lower-SES levels, respectively, compared to minorities.[1] For example, regarding annual family income (an index of SES), a survey by the U.S. Department of Commerce, Bureau of the Census (1989, cited in Chapa & Valencia, 1993), found that 46% of African American families with young children (newborn to 4 years of age) had incomes of less than $10,000 and that 38% and 21% of Latino and Asian American families, respectively, had incomes of less than $10,000.[2] In sharp contrast to families of color, only 12% of White families had incomes of less than $10,000.[3] In sum, given the pervasive finding that low SES tends to correlate with low intellectual performance (and, of course, that high SES tends to correlate with high intellectual performance), the SES-intelligence connection carries special import for understanding variability of intellectual performance among minority students.[4]

Second, it is important to explore the SES-intelligence connection for minority students given the persistent average differences in intellectual performance, for example, between Whites and African Americans and between Whites and Mexican Americans. As we discussed in the Introduction of the book, these racial/ethnic differences have been thoroughly documented. It is critical to keep in mind, however, that many of these historical studies failed to control for SES (see Chapter 1 of this book). In addition, many contemporary studies of White and minority students have not controlled for SES (see Chapter 5 of this book). This has been the situation, for example, in studies of African American/White participants (e.g., Tate & Gibson, 1980) and in investigations of Mexican American/White participants (e.g., Valencia, 1985a). As a case in point, Valencia (1985a) examined 102 studies (spanning from 1923 to 1984) in which the intellectual performances of White and Mexican American children were compared. He found that nearly half (46%, *n* = 48) of the total investigations did *not* report the SES of the participants. As such, the question arises: In White/minority investigations of intellectual performance, to what degree can the reported average differences be due, in part, to artifacts based on SES differences?

A third reason why the SES-intelligence connection is an important area of investigation for minority students has to do with "deficit thinking," the construct discussed in the Introduction of this book. As noted in Chapter 1 of this book, deficit thinking, in part, is a mind-set molded by the fusion of ideology and science that locates problems (e.g., low performance on an intelligence test, school failure) within the student. As noted by Foley (1997), Pearl (1997), and Valencia and Solórzano (1997), low-SES minority students often are viewed by deficit thinkers as having limited intelligence and, hence, low educability. Suffice it to say that, given such deficit thinking views toward low-SES minority students as a group, it is vital to explore the SES-intelligence connection in the context of rigorous scientific standards of empirical confirmation. If this is not done, then it is likely that individual low-SES minority students may be perceived, a priori, to be low in intellectual functioning.

For our coverage of SES as a correlate of intellectual performance, we provide discussions in the following areas: (a) the meaning and measurement of SES; (b) SES and intelligence: historical perspectives; (c) research findings: African Americans; (d) research findings: Mexican Americans; and (e) SES and intelligence: contemporary perspectives. We initially intended to review SES-intelligence research findings on other minority groups (e.g., Puerto Ricans, Asian Americans), but our literature searches uncovered very little or no research on these groups.

The Meaning and Measurement of SES

SES (and its related term, social class) is a social construction describing human subgroups in a society. SES, an ascribed characteristic of groups, typically is viewed as a measure of prestige within a social group frequently based on schooling attainment, income, and occupation (Slavin, 1997). When SES is measured, one or some combination of these variables often is used. In addition to viewing SES as a measure of prestige, "along with social class goes a pervasive set of behaviors, expectations, and attitudes which intersect with and are affected by other cultural factors" (p. 114). It is beyond the scope of this chapter to explore this "way of life" thesis of social classes and the sociological debate over "class cultures" (for discussions, see Foley, 1997, and Valencia & Solórzano, 1997). Rather, as we introduced earlier in this chapter, our major interest is to examine SES as a performance factor in explaining mean differences in measured intelligence between White and minority students.

The ways in which SES has been measured in educational research comprise a litany of indicators. White (1982), using meta-analysis techniques that examined the relation between academic achievement and intelligence, reported a wide range of different variables that were used as SES indicators. In 143 studies, White found that more than 70 different variables were used (either singularly or in some combination). Given some overlap, White collapsed the variables into 43 broad categories, which in turn were collapsed into 4 specific categories. In descending order of frequency of use, they were (a) "traditional SES," (b) "home atmosphere," (c) "school resources," and (d) "miscellaneous." In the traditional SES category (which consisted of 9 variables), the number one variable in frequency of use was "occupation of parents" (used in 88 of 143 studies, 62%). This was followed by "education of parents" (40% of the studies) and "income of family" (33%). In the home atmosphere category (which consisted of 11 variables), the top three variables were "amount of cultural activities in which family participates" (11%), "reading materials in the home" (10%), and "aspirations of parents for child" (8%).[5] In the school resources category (which contained 16 variables), the top three variables were "instructional expense per pupil" (4%), "salary of teachers" (3%), and "percent[age] of teachers with [M.A. degrees]" (3%). Regarding this plethora of SES indicators, White voiced this concern:

In the absence of a widely accepted and precise definition of SES, this is not a surprising phenomenon. Nevertheless, the almost indiscriminate inclusion of whatever pleases a particular researcher as a measure of SES seriously weakens its validity as a research tool. (p. 473)

SES and Intelligence: Historical Perspectives

It appears that the earliest attempt to examine the relation between SES and mental ability was by Galton (1870), who in his *Hereditary Genius* traced the social origins of famous Englishmen (for further discussion, see Chapter 1 of the present book). It was not until several decades later, with the advent of the Binet-Simon scale, that research began in earnest on the association between SES and intelligence. Decroly and Degand (1910, cited in Bridges & Coler, 1917, and Neff, 1938) administered the Binet-Simon scale to 45 private school students in Brussels, Belgium, and found that none of the children performed below his or her age, 9 scored at their age, and the remainder scored 1 to 3 years above their age. Binet thought that the most logical explanation lay in social class. The Belgian children were from well-to-do families and attended a school that offered individual instruction in small classes. By contrast, the Parisian schoolchildren who formed the norm group for the Binet-Simon scale were from lower middle-class families. As such, Binet concluded that social class must be associated with mental ability (Neff, 1938).[6]

Available literature reviews of early research on the relation between SES and intelligence suggest tension between hereditarian and environmental camps (see, e.g., Ellis, 1932; Neff, 1938; Pintner, 1931). To illustrate these contrasting perspectives, we have examined a dozen studies published between 1912 and 1936. These investigations, along with their germane information, are listed in Table 3.1.

The 12 studies we chose for analysis were not randomly selected. They were chosen on the basis of providing the reader with an understanding of the nature-nurture controversy that was stimulated by this body of research.[7] As Neff (1938) noted in his review of the literature, researchers in the field of mental ability "are faced with major questions which are still subjects of controversy, with reference to which adherents of one systematic position or another have tended to take vociferous sides" (pp. 727-728). Also, we selected studies in which only White children served as participants (so far as we can ascertain, they were English-speaking children).[8] Our intention is to provide the reader with a sense of the wider body of research on the relation between SES and intelligence before discussing the representative research on minority students.

Before we discuss the findings of several representative studies listed in Table 3.1, a few observations regarding methodology are in order. First, the ways in which SES was measured varied considerably (e.g., school type [quality, location, SES background of students enrolled], father's occupation, parental tax assessment, several SES classification scales). Second, the measures of intelligence used varied considerably. In fact, in the 12 studies, 14 different intelligence tests were used. Third, of the 12 studies, only 4 (33%)

TABLE 3.1 Select Historical Studies of SES as a Predictor of Intellectual Performance

Study	N	Age or Grade	SES Measure	Intelligence Measure	r^a	Interpretation
Weintrob and Weintrob (1912)	210	8-11 years	School type	Binet-Simon[b]	nr	Equivocal
Bridges and Coler (1917)	301	6-12 years	School type; father's occupation	Yerkes-Bridges Point Scale	nr	Equivocal
Pressey and Ralston (1919)	548	10-14 years	Father's occupation	Cross-Out Tests	nr	Genetic
Pressey (1920)	520	6-8 years; 10-14 years	Father's occupation	Primer Scale	nr	Genetic
Dexter (1923)	2,782	Grades 1-8	Father's occupation	Dearborn tests; National Intelligence Tests	nr	Equivocal
Haggerty and Nash (1924)	8,121	Grades 3-8; high school	Taussig's classification	Haggerty Intelligence Exam; Delta 2	nr	Equivocal
Chapman and Wiggins (1925)	632	Grades 6-8	Chapman-Sims Scale	National Intelligence Tests	.31[c]	Environmental
Goodenough (1928)	380	1.5-4.5 years	Taussig's classification; Barr Scale	Kuhlman-Binet[d]	nr	Genetic
Stroud (1928)	1,057	5-18 years	Parental tax assessment	Pressey Classification Tests	.25	None
Chauncey (1929)	243	Grades 8-9	Sims Score Card	McCall Multi-Mental Scale	.19-.21	Environmental
Hildreth (1935)	1,161	Grades 1-8; high school	Taussig's classification	Stanford-Binet	nr	None
Byrns and Henmon (1936)	100,820	High school seniors	Taussig's classification	Several[e]	.18	None

NOTE: SES = socioeconomic status; nr = no r reported.

a. r refers to reported correlation between SES and intelligence.
b. Revisions by Henry Goddard.
c. Language factor eliminated.
d. Preschool version.
e. The measures of mental ability (with their respective years of administration) are the Ohio University Psychological Test (1929), the American Council Psychological Examination (1930, 1931), and the Henmon-Nelson Test of Mental Ability (1932, 1933).

reported correlation coefficients between SES and intelligence. Suffice it to say that, given such methodological differences, it is difficult to draw clear and general conclusions from this body of research as a whole.

Let us now examine a representative study from each of the four categories of data interpretation regarding the SES-intelligence connection, as seen in the final column of Table 3.1 (we coin these "genetic," "environmental," "equivocal," and "none"). Regarding a genetic interpretation in explaining the observed association between measured SES and intelligence, Goodenough (1928) is a good example. In her investigation, Goodenough administered the Kuhlman revision of the Binet-Simon scale to 380 preschoolers (190 boys, 190 girls; ages 2, 3, and 4 years).

The study was carried out at the University of Minnesota's Institute of Child Welfare. The SES measure was the Taussig's (1920) classification. In descending order, the groupings (based on father's occupation) were professional (Group I), semiprofessional (Group II), clerical and skilled trades (Group III), semiskilled and minor clerical (Group IV), slightly skilled (Group V), and unskilled labor (Group VI). The mean IQs of the children whose fathers belonged to the various SES categories were 125 (Group I), 120 (Group II), 113 (Group III), 108 (Group IV), 107 (Group V), and 96 (Group VI).[9] Although data are not presented, Goodenough commented, "The clearest separation of the occupational groups [in intelligence] was found among the two-year-olds" (p. 293). In her discussion of findings, Goodenough introduced an environmental explanation but dismissed it immediately:

> While one may plausibly advance the hypothesis that the home of low cultural standards does not afford a sufficient opportunity or an adequate stimulus for the acquisition of the more complex and precise language concepts or the fund of general information necessary to achieve a notable degree of success with the tests standardized at the upper age levels, it is less easy to understand the process by which these factors serve as a handicap to the two-year-old who is judged upon the basis of his response to such simple commands as, "Throw the ball to me"; his ability to name simple objects; or [his ability] to draw a rough circle with considerable help. Even the least cultured of modern city homes provide, it would seem, adequate opportunity for the acquisition of these early accomplishments of the young child. (p. 294)

Although her assumption is tenuous that lower-SES children, as compared to their higher-SES peers, have the same cultural experiences, have the same learning opportunities and materials in the home, and are equally prepared and motivated to perform correctly on the items on an intelligence test, Goodenough (1928) nevertheless drew this hereditarian conclusion:

> The fact that children of different social classes show as great differences in their performance of these extremely simple tasks as they afterward manifest in regard to the relatively complex problems of later life lends support to the theory that under ordinary conditions of modern life, variations of mental growth are more directly dependent upon innate factors than upon differences in post-natal opportunity or stimulation. (p. 294)

A good example of a study from Table 3.1 that has an environmental interpretation is Chauncey (1929). The author administered the McCall Multi-Mental Scale to 243 8th- and 9th-graders who attended a junior high school in Stillwater, Oklahoma. The Stanford Achievement Test also served as a criterion variable. SES was measured by the Sims Score Card for Socio-Economic Status (Sims, 1927); this instrument gathers data on factors such as father's occupation, parental schooling attainment, rooms per house, and number of books in home. Chauncey reported zero-order correlations of .21 (8th-graders) and .19 (9th-graders) between SES and intelligence as well as values of .30 (8th-graders) and .35 (9th-graders) between SES and achievement. The author noted, "The values are too low to be significant in both groups. We may conclude that home status, as measured by the Sims Score Card, plays a less important part in determining mental level than in determining school achievement" (p. 90). Notwithstanding Chauncey's incorrectness in implying causality from correlational analyses, her points on (a) the importance of the home environment and schooling in promoting academic success and (b) the overlap of SES groups in intellectual and academic performance are well taken:

> Two suggestions of considerable practical value emerge from a study of the above data and conclusions. First, progressive parents will find objective evidence justifying current efforts to create optimum conditions in the home, looking toward greater success of their children in the work of the school. Second, educators will find grounds for both encouragement and caution. The value of the association for home status and school achievement, while significant, is relatively low. It is apparent, first, that while a certain pupil has rated low in home status, his school achievement may have been satisfactory. Evidently, the school facilities have partially made up the deficiency. This is the educator's opportunity. It is further apparent that while a certain pupil has rated high in home status, his achievement may have been mediocre or low. In this case, the educator has failed, in a measure, to harness the possibilities for training which are implied in the good home situation. This suggests the educator's need of caution. (p. 90)

With respect to studies that led to what we refer to as equivocal interpretations of the observed positive relation between SES and intelligence, the Bridges and Coler (1917) investigation listed in Table 3.1 is representative of research that acknowledged the nature-nurture debate but did not lean in either direction in data interpretation. In Bridges and Coler's study, 165 children (6 to 12 years of age) attending a "favored school" and 136 children (also 6 to 12 years of age) attending an "unfavored school" in Columbus, Ohio, were administered the Yerkes-Bridges Point Scale, a measure of intelligence.[10] As expected, the authors found the ubiquitous positive association between SES and intelligence (no correlations, however, were reported). For example, based on general norms for the Yerkes-Bridges Point Scale, only 1.8% of the favored school group performed below the norm group (first quartile), whereas 32.4% of the unfavored school group scored below the norm group (first quartile).[11]

Table 3.1 also contains a fourth interpretative category that we label as "none." What is meant here is that the author(s), in the introduction or discussion sections to the article, did not even mention the nature-nurture issue. As such the author(s) merely reported the positive association between SES and intelligence (low correlations, by the way, in Hildreth, 1935, and Byrns & Henmon, 1936) and did not explicitly offer possible explanations about environmental or genetic factors (although in two of the three "none" studies, the author[s] appeared to hint at environmental aspects by discussing the finding of overlap). Hildreth (1935) commented, "A significant finding is the wide range of mental ability represented by the children of any one occupational group and, especially in the private school, the large amount of overlapping of ability in all occupational groups" (p. 157). In a similar vein, Byrns and Henmon (1936) concluded, "There is great overlapping of scholastic ability within the various parental occupational groups, and . . . the differences within every group are greater than the differences between the groups" (p. 291).

In sum, the question of the roles of heredity and environmental factors in explaining the character of the relation between SES variables and performance on intelligence tests occupied a central position in research endeavors during the first three decades of the 20th century. For those researchers who contended that the mental ability differences found among children of different social classes were largely due to innate factors, Pintner's (1931) conclusions in his book *Intelligence Testing: Methods and Results* succinctly capture their point of view. After reviewing pertinent literature on the SES-intelligence connection (including eight studies we reviewed and list in Table 3.1[12]), Pintner, a noted authority on mental testing, drew this sweeping conclusion:

> We may sum up these studies of the relationship of the intelligence of the child to the occupational or social status of the parent by saying *that they fit very well with the theory that intellectual potentiality is largely inherited.* As we have seen in a previous chapter, the IQ is not wholly impervious to environmental influence, yet on the whole it seems to measure roughly that potentiality for intellectual development which seems to be inherited. (p. 518, italics added)[13]

With respect to the environmental perspective on the positive relation between SES and intelligence, Neff's (1938) conclusions in his critical survey of the literature exemplified the field during the historical era. After discussing a number of environmental considerations (e.g., fluctuations in IQ on retesting, increases in children's IQ when going from "poor" to "good" environments, societal inequalities in social and economic opportunities), Neff posited,

> All these points taken together lead to a conclusion which we feel is forced and inescapable. Just as Klineberg [1935b] has shown that the standard intelligence tests are inadequate instruments for measuring the native ability of different races, *so do we find that these tests cannot be used for measuring the capacity of different social levels within our own society.* It should be emphasized that we are not raising here the question as to whether all of the abilities measured by the tests are determined exclusively by the environment, although it should be clear from the studies analyzed above that environmental variables are important. Rather, we are raising the question as to whether the generally obtained relationship between social status and intelligence test score requires any other explana-

tion than an environmental one. We feel that there is a strong, perhaps decisive, evidence for a negative answer. (p. 754)

Now that we have provided a historical backdrop for the SES-intelligence connection, we turn to discussions of how minority students have fared in this research base.

Regarding the sheer volume of research studies on the relation between SES and intellectual performance, African American students have been by far the most frequently investigated. Mexican Americans are a distant second, and there is a paucity of research on other minority students. The following discussions are not intended to be comprehensive reviews of the literature. Such endeavors are well beyond the scope and interest of this chapter. What we do provide are syntheses of findings from representative studies. In so doing, we use the following format: (a) an examination of *within-racial/ethnic group* comparisons, a cluster referring to studies in which a minority sample (e.g., African Americans) was the sole participant group and SES varied (i.e., SES served as the independent variable); and (b) an examination of *between-racial/ethnic group* comparisons, a cluster containing representative studies in which minority *and* White samples served as the participant groups and SES varied within racial/ethnic groups (in some of these between-racial/ethnic group studies, minority and White students were matched on SES). The representative investigations that we examine are arranged chronologically.

Research Findings: African Americans

Examples of *within*-racial/ethnic group studies include Beckham (1935); Robinson and Meenes (1947); Kennedy, Van De Riet, and White (1963); and Green and Rohwer (1971). Here, we briefly discuss two of these studies.[14]

Beckham (1935)

Beckham's (1935) study included 1,100 Black adolescents (for its time, the largest number of African American participants in a single study). These public school students, 12 to 16 years of age, were from the following cities: New York ($n = 100$); Washington, D.C. ($n = 753$); and Baltimore, Maryland ($n = 147$). In addition, there were 100 institutionalized delinquent boys (from the Blue Plains Industrial Home, Washington, D.C.). The SES measure was the Taussig's (1909) classification (based on occupation, education, and income of parents).[15] Based on the Stanford-Binet, the aggregate mean of IQ was 95.2, which was about one third of 1 standard deviation below the White normative mean;[16] most previous investigations had reported a 1 standard deviation difference (15 points) between Whites and African Americans.

The data presented in Table 3.2 show a positive relation between SES and intellectual performance. "It reveals an increase in the IQ through the groups until the fifth group is reached" (Beckham, 1935, p. 84). Beckham's point that combining the groupings makes for more convincing data is well taken:

TABLE 3.2 Mean IQs of Adolescents by SES Group

	I *(laborers)*	*II* *(unskilled)*	*III* *(skilled)*	*IV* *(clerical)*	*V* *(professional)*
Mean IQ[a]	93	95	97	101	98
n[b]	173	301	448	117	61
Percentage	15.7	27.4	40.7	10.6	5.4

SOURCE: From "A Study of the Intelligence of Colored Adolescents of Different Social-Economic Status in Typical Metropolitan Areas," by A. S. Beckham, 1935, *Journal of Social Psychology, 4,* p. 83. Copyright 1935 by Heldref Publications. Adapted with permission of the Helen Dwight Educational Foundation.
NOTE: SES = socioeconomic status.
a. Grand mean IQ = 95.2.
b. N = 1,100.

> When Groups I and II are combined, the mean IQ is 94, and if Groups IV and V are combined, the mean IQ is 99. The difference between Groups I and III is 4, which is significant. There is also an obvious difference between Group IV and Groups I, II, and III. (p. 84)[17]

Notwithstanding possible problems of selection, Beckham's study provides some evidence for the positive correlation between SES and African American students' intellectual performance.[18]

Kennedy, Van De Riet, and White (1963)

The study by Kennedy et al. (1963) was one of the most ambitious investigations of the intellectual performance of African American schoolchildren. The main purpose of the study was to gather normative intellectual data, based on the 1960 revision of the Stanford-Binet Intelligence Scale (Terman & Merrill, 1960), on a stratified random sample (N = 1,800) of Negro elementary school students (ages 5 to 16 years) in five southeastern states: Alabama, Florida, Georgia, South Carolina, and Tennessee. SES was measured by parental occupation (1950 Census data), and such data were then divided into seven classifications based on the McGuire-White Scale (where 1 is the highest classification and 7 is the lowest).

For the N of 1,800, Kennedy et al. (1963) reported a mean Stanford-Binet IQ of 80.7 (SD = 12.5). This mean was considerably lower than the mean IQ of 101.8 (SD = 16.4) reported for the 1960 Stanford-Binet normative sample (exclusively Whites). However, when Kennedy et al. disaggregated the Negro N of 1,800 by SES level, the familiar pattern of a positive association between SES and intelligence emerged (Table 3.3). The data show that the mean IQ difference between the highest SES group (Group 1) and the lowest SES group (Group 5) is very large—25.9 IQ points. It also should be noted that the mean IQ of 80.7 for this southeastern Negro sample was 14.5 points lower than what Beckham (1935) reported for his N of 1,100 colored adolescents (mean IQ of 95.2 based on

TABLE 3.3   Mean Stanford-Binet IQs of Southeastern Negro Students by SES Level

	1 (highest)	2	3	4	5 (lowest)[a]	Unknown
Mean IQ[b]	105.3	92.1	87.8	83.6	79.4	78.4
n[c]	6	28	112	304	918	432
Percentage	0.33	1.6	6.2	16.9	51.0	24.0

SOURCE: From "A Normative Sample of Intelligence and Achievement of Negro Elementary School Children in the Southeastern United States," by W. A. Kennedy, V. Van De Riet, & J. C. White, Jr., 1963, *Monographs of the Society for Research on Child Development, 28,* p. 80. Copyright 1963 by the Society for Research on Child Development. Adapted with permission.
NOTE: SES = socioeconomic status.
a. The authors did not report IQ data for SES Levels 6 and 7.
b. Grand mean IQ = 80.7.
c. *N* = 1,800.

the 1916 Stanford-Binet) in the cities of Washington, Baltimore, and New York.[19] This large difference suggests that the intellectual performance of Negro students during past decades was strongly related to differences in schooling conditions in the North and South (see, e.g., Klineberg, 1935a, 1935b).

Next, we turn to coverage of representative studies that sought to examine *between-racial/ethnic* comparisons of the relation between SES and intellectual performance of African American and White students. Examples of such investigations include Sunne (1917); Arlitt (1921);[20] Anastasi and D'Angelo (1952); Bird, Monachesi, and Burdick (1952); Higgins and Sivers (1958); Nichols and Anderson (1973); and Tate and Gibson (1980). Here, we briefly discuss three of these investigations.

Arlitt (1921)

So far as we can ascertain, the investigation by Arlitt (1921) was the first study published in which SES was systematically varied as a correlate of intellectual performance among White and African American children. The participants were "native-born" White ($n = 191$), Italian ($n = 87$), and Negro ($n = 71$) primary-grade students "attending the same grades in the same school" (p. 181).[21] To measure SES and intelligence, the study used the Taussig classification and Stanford-Binet intelligence test, respectively.[22] The five SES groups, in descending order, were (a) Professional, (b) Semiprofessional (and higher business), (c) Skilled, (d) Semiskilled, and (e) Unskilled. For reasons not too clear, Arlitt relabeled the occupational groupings using value-laden categories.[23] The first group became "Very Superior," the second group became "Superior," the third group became "Average," and because of small sample sizes, the fourth and fifth groups were combined and labeled "Inferior."

Based on aggregated IQ data analysis by race/ethnicity (i.e., SES was free to vary within each of the three racial/ethnic groups), the median IQs for the native-born White, Italian, and Negro samples were 106.5, 85.0, and 83.4, respectively.[24] Thus, the native-born White group's median IQ was 21.5 points higher than that of the Italian group and

23.1 points higher than that of the Negro sample. In that about 92% of the Italian and Negro children were of Inferior social status, Arlitt (1921) was unable to compare IQ scores by SES within these two groups. However, because of adequate sample sizes and SES variability, she was able to do so with the native-born White group. As seen in Table 3.4, the observed median IQ scores for the four occupational groups, in descending order, were Very Superior (125.9), Superior (118.7), Average (107.0), and Inferior (92.0). As noted in Table 3.4, the median IQ difference between the Very Superior and Inferior native-born White children was 33.9 points—a substantial gap. Arlitt then compared the median IQ (92.0) of the native-born White Inferior group with the median IQs of the Italian (85.0) and Negro (83.4) children of the same social class (Inferior). Table 3.4 shows that the median IQ of the native-born White group was 7.0 points higher than that of the Italian group and 8.6 points higher than that of the Negro group—strikingly less than what was seen between the Very Superior and Superior native-born White groups.

Arlitt's (1921) findings led her to conclude,

> The difference in median IQ which is due to race alone is in this case at most only 8.6 points, whereas the difference between children of the same race but of Inferior and Very Superior social status may amount to 33.9 points. It is apparent that such differences as we have between the negro and Italian children and between these and children of native-born white parents are not nearly so striking as the difference between children of the same race but of different social status. Of the two factors, social status seems to play the more important part. To such an extent is this true that it would seem to indicate that *there is more likeness between children of the same social status but different race than between children of the same race but different social status.* (pp. 182-183, italics added)

In sum, Arlitt's (1921) study was highly significant in the history of cross-racial/ethnic research on the relation between SES and intellectual performance. Her finding that SES transcended race/ethnicity as a correlate of intellectual performance would be consistently observed in later research, as we discuss shortly. Furthermore, Arlitt's finding was a blow to hereditarianism that prevailed during the 1920s (Valencia, 1997b), but its impact was slow in coming.[25]

Bird, Monachesi, and Burdick (1952)

In the investigation by Bird et al. (1952), the intellectual performance of Negro (*n* = 20) and White (*n* = 63) children (3rd- to 5th-graders) attending a school district in Minneapolis, Minnesota, was compared. Based on the Chapin Social Status Scale (Chapin, 1933), the White group was rated higher than the Negro group on SES, but the difference in mean scores was not significant; it appears that the groups were deemed middle class. Bird et al. (1952) reported mean IQs of 109.8 and 103.6 for the White and Negro samples, respectively.[26] The mean difference of 6.2 points, favoring Whites, was considerably less than the typically reported mean difference of 15 IQ points between White and Negro samples when SES was not controlled.[27]

TABLE 3.4 The Relations Between SES and Intelligence and Between Race/Ethnicity and Intelligence

Relation Between SES and Intelligence			Relation Between Race/Ethnicity and Intelligence		
Same Race/Different SES (White race)	*Median IQ*	*n*	*Different Race/Same SES (Inferior SES)*	*Median IQ*	*n*
Very Superior	125.9	24			
Superior	118.7	48			
Average	107.0	76			
Inferior	92.0	43	White	92.0	43
			Italian	85.0	81
			Negro	83.4	71
Difference between Very Superior and Inferior SES = 33.9 points			Difference between White and Italian groups = 7 points; difference between White and Negro groups = 8.6 points		

SOURCE: Adapted from Arlitt (1921).
NOTE: SES = socioeconomic status.

In addition to examining the relation between SES and intelligence among Negroes and Whites, a major focus of Bird et al. (1952) was to investigate "the acceptance or rejection of Negroes by white parents and their children" (p. 689) in an ethnically mixed area of Minneapolis. A summary of major racial attitudes is as follows:

> White parents . . . made little effort to inculcate racial tolerance in young children. . . . A small proportion of the white adults, approximately 20 percent, showed few if [any] antipathies toward Negroes in answering a questionnaire. A larger proportion were definitely antagonistic and resentful of what they considered infringements by Negroes upon the rights of white people. . . . They [Negroes] had achieved a level of formal education approximately equal to that of the white respondents, and . . . in one district, their level of social status, as judged by the arrangements and cultural objects within the homes, was not significantly different from the social status of white persons. It seems, however, that Negro men have not been given adequate opportunities to utilize their education in appropriate employment. Most of the Negro men were employed in positions requiring less skill than they seemed capable of manifesting. (p. 698)

The results of Bird et al. (1952) that Whites revealed considerable prejudice toward Negroes, and the authors' suggestion that White antipathy created barriers to Negro social mobility, are key points in understanding the fairly consistent finding that although African Americans and Whites might be equated on SES, IQ differences persist. A dominant theme in the literature is that any typical SES index "does not equate the samples from the two races on socioeconomic level as this concept is defined in conventional treatments of social classes" (Sperazzo & Williams, 1959, p. 273) given that "Negro life in a *caste society* is considerably more homogeneous than is life for the majority group. This makes it extremely difficult ever really to match racial groups meaningfully on class status as the context and history of social experiences are so different" (Deutsch & Brown, 1964, p. 27, italics added). In sum, due to racial prejudice and discrimination, being a member of a caste-like minority in the United States has restricting and confining consequences in social mobility. As such, an African American middle-class person, compared to a middle-class White person, might interpret his or her status from a very different phenomenological vantage point. More on the class and caste issue is discussed later.

Nichols and Anderson (1973)

In an attempt to examine how SES is related to the intellectual performance of White and Negro children, Nichols and Anderson (1973) selected large representative samples of women who registered for prenatal care at 12 university medical centers in Boston, Baltimore, and Philadelphia. The White and Negro women in the Boston sample were extremely similar in SES background (high), and the Baltimore and Philadelphia participants also were extremely similar in SES background (low).[28] In this longitudinal study, each child was administered the Stanford-Binet (short form, L-M) at 4 years of age and

TABLE 3.5   Mean IQs of Children, by Race and SES

Test	White				Negro			
	High SES		Low SES		High SES		Low SES	
	n	IQ	n	IQ	n	IQ	n	IQ
Stanford-Binet IQ (short form, L-M) (age 4 years)	6,475	108.0	937	96.8	797	102.0	7,471	92.6
Wechsler Intelligence Scale for Children IQ (Full Scale) (age 7 years)	4,721	104.2	535	95.3	492	100.0	4,121	91.2

SOURCE: From "Intellectual Performance, Race, and Socioeconomic Status," by P. L. Nichols & V. E. Anderson, 1973, *Social Biology, 20*, pp. 370-371. Copyright 1973 by *Social Biology*. Adapted with permission.
NOTE: SES = socioeconomic status.

the Wechsler Intelligence Scale for Children (WISC) at 7 years of age. Intellectual performance (as measured by IQ) between race and SES groups are shown in Table 3.5.

Several major points can be gleaned from the data presented in Table 3.5:

1. Previous reviews of the literature have reported a mean IQ difference of about 1 standard deviation (15 points), favoring Whites over African Americans, when studies are collapsed together into an undifferentiated aggregate (e.g., Shuey, 1966). Regarding the data shown in Table 3.5, we calculated the aggregated, grand mean Stanford-Binet IQ for Whites (i.e., combined low- and high-SES groups) as well as for Negroes. For the White and Negro children, the grand means were 106.6 and 93.5, respectively. This difference of 13.1 IQ points is very close to the 1 standard deviation difference commonly found in studies where SES is left free to vary.[29] When SES is controlled, however (as in Nichols & Anderson, 1973), the IQ gap between Whites and African Americans shrinks considerably. As seen in Table 3.5, the IQ difference on the Stanford-Binet for high-SES Whites and Negroes is only 6.0 points; for the low-SES groups, the Stanford-Binet IQ difference is just 4.2 points. Regarding WISC IQ differences, a very similar pattern is observable for the high- and low-SES White and Negro comparisons.

2. As can be discerned in Table 3.5, there is a clear pattern of descending order of intellectual performance when race interacts with SES. The highest performing children on the Stanford-Binet are high-SES Whites (mean IQ = 108.0), followed by high-SES Negroes (102.0), low-SES Whites (96.8), and low-SES Negroes (92.6). The identical descending order also is noted for WISC IQ performance. This pattern of high-SES White > high-SES minority > low-SES White > low-SES minority has been found in a small num-

ber of studies (e.g., Christiansen & Livermore, 1970; Oakland, 1978). Later in this chapter, we offer some possible explanations for this pattern.

3. Although we do not report such data in Table 3.5, Nichols and Anderson (1973) found consistent within-racial/ethnic group differences in IQ by varying SES. The authors divided scores on their SES index into three intervals for the high- and low-SES analyses by race and found a pervasive pattern. For example, on the Stanford-Binet, the mean IQs for the low-SES Negro group were 90.9, 93.4, and 96.2 for the lowest, intermediate, and highest SES intervals, respectively. The same pattern was observed for the high- and low-SES White groups.[30]

In conclusion, the investigation by Nichols and Anderson (1973) is particularly valuable given the research rigor involved, that is, its use of large representative White and African American samples, the attention paid to controlling SES, the SES distribution of African Americans and Whites being the same and the IQ distribution of the African American and White children also being the same, the use of individually administered intelligence tests, and the longitudinal nature of the study. On their findings, the authors offered their conclusion about the relation between SES and intelligence:

> In the present study, racial difference in intellectual performance in samples of young white and Negro children whose parents had similar socioeconomic scores, lived in the same city, and had gone to the same hospitals for prenatal care were much smaller than often reported. The fact that differences were minimized by comparing similar socioeconomic scores suggests that socioeconomic factors are largely responsible for the usually reported Negro-white differences in intellectual performance. (p. 374)

Research Findings: Mexican Americans

As we noted earlier, the number of studies on the relation between SES and intelligence among Mexican Americans is very small. Our search for pertinent research identified only seven studies. Examples of *within*-racial/ethnic group investigations include Ellis (1932), González (1932), and Valencia et al. (1981). Examples of *between*-racial/ethnic group studies include Christiansen and Livermore (1970); Oakland (1978); Valencia (1979); and Valencia, Rankin, and Livingston (1995). Next, we discuss several representative studies from this body of research.

González (1932)

González's (1932) investigation, a master's thesis completed at the University of Texas at Austin, used Mexican American participants in San Antonio and measured SES using the Sims Score Card for Socio-Economic Status (Sims, 1927). The 200 participants, all described as "Spanish-speaking," were from five elementary schools "representing different social levels" (González, 1932, p. 10).[31] For example, one school was described as being in a good residential section, having paved streets, and where most of the families owned their homes. Two other schools were located "in the worst district of San Antonio, [called] 'Little Mexico,' where housing conditions . . . are deplorable. . . . The houses are close together and mostly in 'vecinadas,' which means two-room apartments connected like a freight train. They usually have a community bathroom" (p. 11). The participants, all 3rd-graders, were administered two non-language intelligence tests: the Pantomime Group Intelligence Test (Myers, 1922) and the Non-Language Mental Test (Pintner, 1927). Tests were not administered in a counterbalanced design.

Based on the Sims Score Card, González (1932) reported a median SES rating for all 200 participants to be 2.5 on a continuum of 0 (*indeterminately low*) to 10 (*indeterminately high*). The median of 2.5 fell between *low* and *very low*. Regarding IQ data for the total sample, the participants' mean IQs on the Pantomime Test and Non-Language Mental Test were 93.0 and 101.4, respectively (normative means on both tests were 100).[32] The results of a correlational analysis between SES ratings and IQs showed a very low positive association (*r* = .15).[33] As concluded in a very similar study by Ellis (1932), González (1932) commented that the low correlation seen in her study was likely related to severe range restriction (i.e., piling up of scores in the lower end of the SES continuum). In fact, González reminded the reader that the very low SES ratings found in her study were likely connected to the Great Depression that was creating economic havoc for millions of people across the country: "Many families [in her study] were forced to give up their homes and to move into cheaper quarters on account of the lack of employment" (p. 29).[34]

In sum, González's (1932) thesis is particularly noteworthy in that her study is a reminder of the difficulty in studying the relation between SES and intelligence between minority populations during past decades when measured SES was so restricted in range. Furthermore, her study is informative in demonstrating the aspect of overlap: "As a final word, it should be emphasized that socioeconomic status [alone] is not a sufficient basis for predicting intelligence. Bright children often come from the most unpromising home conditions" (p. 42).

Christiansen and Livermore (1970)

Participants in the study by Christiansen and Livermore (1970) included 46 "Anglo-American" (i.e., White) and 46 "Spanish-American" (i.e., Mexican American) junior high school students, ages 13 to 14 years.[35] The sample was randomly selected from a larger pool of 200 students, first identified as being lower or middle class based on father's occupation.[36] The final sample consisted of 23 lower- and 23 middle-class White participants as well as 23 lower- and 23 middle-class Mexican American participants. The measure of intelligence was the WISC (Wechsler, 1949).

The results of a 2 (Race) × 2 (SES) analysis of variance performed by Christiansen and Livermore (1970) are shown in Table 3.6. The major points that can be drawn from the data shown in Table 3.6 are as follows:

TABLE 3.6 Mean IQs of White and Mexican American Students by SES Level

Mean Wechsler Intelligence Scale for Children IQ	White		Mexican American		F Values	
	Lower SES (n = 23)	Middle SES (n = 23)	Lower SES (n = 23)	Middle SES (n = 23)	Ethnicity	SES
Full Scale	99	116	91	111	12.93*	95.82*
Verbal Scale	95	120	89	111	12.66*	82.89*
Performance Scale	102	109	96	108	3.60 (n.s.)	28.86*

SOURCE: From "A Comparison of Anglo-American and Spanish-American Children on the WISC," by T. Christiansen & G. Livermore, 1970, *Journal of Social Psychology, 81,* p. 12. Copyright 1970 by Heldref Publications. Adapted with permission of the Helen Dwight Educational Foundation.
NOTE: SES = socioeconomic status.
*Significant at .01 level; n.s. = not significant.

1. For the race/ethnicity variable (White $n = 46$, Mexican American $n = 46$), in which SES was free to vary, Whites performed significantly higher than did Mexican Americans on Full Scale IQ and Verbal Scale IQ, but no significant difference on Performance Scale IQ was observed (as expected). Based on our calculations, the aggregated mean IQs, as well as mean IQ differences, are in the following table:

White	Mexican American	Difference
FS IQ = 107.5	FS IQ = 101.0	6.5
VS IQ = 107.5	VS IQ = 100.0	7.5
PS IQ = 105.5	PS IQ = 102.0	3.5

NOTE: FS = Full Scale; VS = Verbal Scale; PS = Performance Scale.

2. For the SES variable (lower-class $n = 46$, middle-class $n = 46$), in which race/ethnicity was free to vary, the higher-SES group scored significantly higher on all three WISC scales. We calculated the aggregated mean IQs and mean differences:

Middle Class	Lower Class	Difference
FS IQ = 113.5	FS IQ = 95.0	18.5
VS IQ = 115.5	VS IQ = 92.0	23.5
PS IQ = 108.5	PS IQ = 99.0	9.5

NOTE: FS = Full Scale; VS = Verbal Scale; PS = Performance Scale.

3. The calculations we performed in the preceding two paragraphs, plus the observed *F* values in Table 3.6, clearly support the assertion by Christiansen and

(1970) that "social class was a more important factor in differentiating Ss [i.e., participants] on WISC measures than [was] ethnic origin" (p. 12).[37] This conclusion is reminiscent of the study by Arlitt (1921), nearly half a century earlier, who also found that SES was a more powerful predictor of intellectual performance than was race/ethnicity.

4. A clear pattern in intellectual performance can be seen in Table 3.6 when SES × Race/Ethnicity is considered. In descending order, the mean IQs for the White middle-class, White lower-class, Mexican American middle-class, and Mexican American lower-class students are as follows:

FS IQ	VS IQ	PS IQ
WMC = 116	WMC = 120	WMC = 109
MAMC = 111	MAMC = 111	MAMC = 108
WLC = 99	WLC = 95	WLC = 102
MALC = 91	MALC = 89	MALC = 96

NOTE: FS = Full Scale; VS = Verbal Scale; PS = Performance Scale; WMC = White middle class; MAMC = Mexican American middle class; WLC = White lower class; MALC = Mexican American lower class.

This pattern of rankings is similar to what was observed in the previously discussed study conducted by Nichols and Anderson (1973), who found the following descending order of Stanford-Binet IQ rankings in their investigation: high-SES Whites, high-SES Negroes, low-SES Whites, and low-SES Negroes. We return to this important pattern later.

Oakland (1978)

Of particular interest in Oakland's (1978) study is the author's use of White, Mexican American, *and* Black participants. SES was determined on the basis of head of household's occupation by the use of the scale developed by Warner, Meeker, and Eells (1949). Six SES × Racial/Ethnic groups were chosen through a stratified sampling design: middle- and lower-class White, Mexican American, and Black 2nd-graders attending public schools in Austin, Texas.[38] The measure of intelligence was the Slosson Intelligence Test (Slosson, 1963), a group-administered test.[39]

The mean Slosson IQs reported by Oakland (1978) for the racial/ethnic groups, in descending order (with SES free to vary), were White (104.7), Black (96.1), and Mexican American (91.2). When scores were disaggregated by SES levels across the racial/ethnic groups, the mean IQs were middle-class White (114.9), middle-class Black (101.4), middle-class Mexican American (98.3), lower-class White (93.9), lower-class Black (90.9), and lower-class Mexican American (84.4). As can be seen, middle-class children in each racial/ethnic group performed higher on the Slosson than did their lower-class peers, demonstrating that SES is a stronger predictor of intelligence than is racial/ethnic background. Furthermore, the data show that even when SES is controlled, White children outperformed their minority peers across the same SES levels. These findings support

TABLE 3.7 Comparison of White and Mexican American Groups on K-ABC Intelligence Scales

K-ABC Intelligence Scale	White Sample (n = 100)		Mexican American Sample (n = 100)	
	M	SD	M	SD
Sequential Processing	96.9	11.4	94.3	10.1
Simultaneous Processing	102.3	11.5	101.0	11.4
Mental Processing Composite	100.0	10.9	97.9	10.7
Nonverbal	100.6	11.1	99.7	11.5

SOURCE: From "K-ABC and Content Bias: Comparisons Between Mexican American and White Children," by R. R. Valencia, R. J. Rankin, & R. Livingston, 1995, *Psychology in the Schools, 32*, p. 159. Copyright 1995 by John Wiley and Sons. Adapted with permission of first author.
NOTE: K-ABC = Kaufman Assessment Battery for Children. All mean differences between the White and Mexican American samples are nonsignificant.

similar results from studies that we have reviewed (Christiansen & Livermore, 1970; Nichols & Anderson, 1973).

Valencia, Rankin, and Livingston (1995)

In this study, Valencia et al. (1995) stratified White and Mexican American sample as closely as possible on grade level (5th), SES (low- to low-middle SES), and language (English).[40] Given that direct SES data (e.g., parental occupation) were not available, the authors used an alternative rough proxy of SES—children enrolled in their district's free or reduced-fee lunch program.[41] The final selected sample was made up of 100 White and 100 Mexican American boys and girls.[42] To measure intelligence, Valencia et al. administered the Kaufman Assessment Battery for Children (K-ABC; Kaufman & Kaufman, 1983a). The intelligence scales of the K-ABC consist of the Sequential Processing Scale and Simultaneous Processing Scale. When these two scales are combined, they form the Mental Processing Composite, a global estimate of intellectual functioning. Finally, there is a special Nonverbal Scale that contains overlapping subtests of the two processing scales. Each of these four K-ABC scales has a standardization mean of 100 and a standard deviation of 15.[43]

Table 3.7 shows the means and standard deviations for the K-ABC intelligence scales for the two groups. The results support a small body of research showing that when White and Mexican American populations are quite similar or equated on SES, mean differences in intelligence test performance are considerably reduced and in some cases extremely negligible, as seen in this investigation by Valencia et al. (1995).

With this selective review of research findings on the relation between SES and intelligence among minority students now behind us, we turn to a discussion of contemporary perspectives.

SES and Intelligence: Contemporary Perspectives

In this final section, we provide brief discussions of several contemporary perspectives on the relation between SES and measured intelligence. The areas covered are (a) modern hereditarian perspectives, (b) class versus caste, (c) status consistency of SES indexes and differential prediction of intelligence, and (d) the unit of analysis: aggregated versus confounded versus student.

Modern Hereditarian Perspectives

Recall that during the early period of research on the relation between SES and intelligence, particularly during the 1920s, some scholars (e.g., Pintner, 1931) concluded that intellectual differences found between different social classes were largely due to hereditary factors. To a small extent, this position has resurfaced during the past quarter century. As cases in point, we refer to the views of Jensen (1973a) and Herrnstein and Murray (1994).

Jensen's (1973a) book *Educability and Group Differences* contains two chapters ("Social Class Differences in Intelligence" and "Equating for Socioeconomic Variables") that are germane to our discussion. Regarding social class differences in intelligence, Jensen made several assertions that are tenuous. First, he stated (in the absence of any supporting references) that the average correlation between SES and phenotypic intelligence is .40 to .60 in various investigations. Later in this chapter, we present data (from White, 1982) that the average correlation is about half this magnitude when the appropriate unit of analysis is used.

Second, Jensen (1973a) contended that various social groupings "are breeding populations differing in gene frequencies, especially for genetic factors related to ability and very likely for the genetic component of those personality traits which favor the development, educability, and practical mobilization of the individual's intellectual potential" (p. 152).[44] The foundation of this assertion for a substantial genetic component for social class differences in intelligence is that such breeding populations occur in societies that have no inequalities in shaping social mobility. Jensen holds the following opinion:

> In any society which provides *more or less equal educational opportunities and a high degree of social mobility, and in which social stratification is based largely on education, occupation, and income*, the abler members of the society will tend to move upwards and the less able gravitate downwards in the SES hierarchy. In doing so, they of course take their genes for intelligence with them. (p. 152, italics added)

The problem with Jensen's (1973a) view is that it is not applicable to the United States, which we argue is a society that does not provide equal educational opportunity for all and does not allow for smooth and unfettered social mobility. There is a voluminous amount of literature to support this assertion regarding African American students (Feagin & Booher Feagin, 1999; Kozol, 1991; Lomotey, 1990), Mexican American students (Donato, 1997; Moreno, 1999; San Miguel, 1987; San Miguel & Valencia, 1998; Valencia, 1991), Puerto Rican students (Feagin & Booher Feagin, 1999), and American Indian students (Feagin & Booher Feagin, 1999). For Jensen to suggest that in the United States there are "more or less equal educational opportunities" for all its citizens is patently false. In the preceding quote, Jensen also implied that social stratification in the United States is largely formed by individual efforts and attainments (e.g., schooling attainment). This perspective, which is at the heart of historical and contemporary meritocratic frameworks of social mobility, is just *one* view of how social stratification arises in the United States. It is a model, we contend, that is simply inaccurate. It is beyond the bounds of this chapter to have a sustained critique of meritocracy. There is much literature that the reader might wish to review to examine the thesis that social stratification arises from structural inequalities in society, not variation in individual efforts (Bowles & Gintis, 1976; Fischer et al., 1996; Massey & Denton, 1993; Moore, 1989; Pearl, 1991; Takaki, 1990).

A third point that Jensen (1973a) raised is about social class differences in measured intelligence:

> The correlation between phenotypes (the measurable characteristic) and genotypes (the genetic basis of the phenotype) is the square root of the heritability [i.e., heritability coefficient, h^2], or h. An average estimate of h for intelligence in European and North American Caucasian populations is 0.90. (p. 155)

Here, Jensen has selected the highest value in estimating heritability of intelligence (i.e., .90). Scarr (1981) noted, "Heritability estimates range between .00 and .90 in reported research, but most values are in the middle range" (p. 259). Even if the heritability estimate of intelligence was of high and comparable magnitude across SES classes (e.g., low- and high-SES groups), could these within-group estimates explain between-group differences in measured intelligence? No. Why not? Because heritability estimates of intelligence *within* populations have nothing to say about the causes of mean differences of measured intelligence *between* populations. This is a major point that cannot be emphasized enough. In Chapter 6 of this book, we explicate the logic of this assertion.

In sum, Jensen's (1973a) position about social classes in the United States, through implications, is that (a) equal educational opportunities are comparable for all racial/ethnic and social classes in the United States, (b) social mobility and resultant social stratification are largely determined by individual efforts, and (c) mean intellectual differences between social classes are significantly genetic in origin. By sharp contrast, we argue that based on (a) historical and contemporary scholarship pertaining to denial of equal educational opportunities to students of color, (b) structural inequality paradigms of social class stratification, and (c) knowledge from behavioral genetics about the heritability of intelligence (a within-population trait), Jensen's theorizing about a genetically driven argument to explain social class differences in intelligence is specious.

Regarding Jensen's (1973a) views on equating for SES variables, we also have concerns about his arguments. First, Jensen draws heavily from Shuey (1966), who reported the findings of 42 studies that attempted to control SES in measuring the intelligence of Black and White participants. Shuey noted,

With two exceptions, the colored [groups] averaged below the white groups in mental test performance in all of the 42 investigations. Average IQs were reported in 33 of the studies including a total of 7,900 colored and 9,300 white Ss, and from these a mean difference of *11 points* favoring the whites was obtained[45] . . . in contrast [to] a mean difference of 15-16 IQ points when random or stratified samples have been used. (pp. 518-519)

Both Shuey and Jensen made the erroneous assumption that controlling SES between Blacks and Whites does indeed equate the two populations on SES. Neither author considered the distinction between class and caste and the implications that these conceptions carry in measuring the intellectual performance of Black and White students. As such, the unmeasured effects of being a member of a subordinate group (i.e., a caste) must be considered in any discussion of SES × Race/Ethnicity interactions. We return to this issue a bit later.

A second issue that we raise concerning Jensen's (1973a) discussion of equating for SES has to do with the Race × SES findings. Once again, Jensen drew from Shuey (1966), who found the following:

The combined mean difference in IQ differences between the 617 colored Ss of higher status and their 1,504 white counterparts is 20.3, in contrast [to] the combined mean difference of 12.2 between the 3,374 colored and 2,293 white children of low status. . . . The consistent and surprisingly large difference of 20.3 IQ points separating the high-status whites and the high-status colored[s] is accentuated by the finding that the mean of the latter group is *2.6 points below* that of the low-status whites. (pp. 519-520)

Jensen (1973a) commented that since the publication of Shuey's (1966) review, these findings have been repeated in three large sample investigations,[46] leading Jensen (1973a) to conclude that when Black and White children are classified by the same SES criteria "into from 3 to 5 categories according to parental SES, the *mean mental test scores of the lowest SES White group exceeds the mean IQ of the highest SES Negro group*" (p. 240). Although Jensen, of course, did not have access to later investigations, we previously discussed two studies (Nichols & Anderson, 1973; Oakland, 1978) in which Black higher SES students did, in fact, perform higher than the lower-SES White children. Therefore, the preceding assertion by Jensen (1973a) needs to be tempered with caution in light of more recent data.

Like Jensen (1973a) in *Educability and Group Differences,* Herrnstein and Murray (1994) in *The Bell Curve* made a dramatic shift, as seen in other hereditarian-bent writings, in the directionality of the SES and intelligence association.[47] In conventional research, SES typically is viewed as an independent variable, and measured intelligence is deemed the dependent variable, that is, SES ($\rightarrow$) intelligence. By sharp contrast, Herrnstein and Murray reversed directionality and asserted that measured intelligence is the independent variable and SES is conceptualized as the dependent variable, that is, intelligence (ρ) SES. As such, one can see that such a model carries enormous theoretical implications and practical applications in understanding societal stratification and class structure in the United States.

Herrnstein and Murray (1994) claimed that cognitive differentiation among Americans (within and between racial/ethnic groups) has resulted in a bifurcated society at the extreme levels of the IQ continuum—the emergence of the "cognitive elite" (top 5%, IQs of 120 or higher) and the "very dull" (bottom 5%, IQs of 75 or lower).[48] In some cases, Herrnstein and Murray included the "dull" (IQs from 75 to 89) as part of the "cognitive underclass." The authors contended that such "cognitive partitioning" is strongly linked to "socially desirable behaviors" (e.g., high educational attainment, prestigious occupational status, high income level) as well as "socially undesirable behaviors" (e.g., poverty, high school dropout, unemployment, divorce, illegitimate birth, welfare dependency, malparenting, crime, poor civility and citizenship). Thus, Herrnstein and Murray's cognitive partitioning thesis contends that (a) having a high IQ greatly improves one's life chances of social mobility and possessing desirable behaviors and (b) having a low IQ places one at substantial risk for possessing undesirable behaviors.

Among a number of points discussed by Herrnstein and Murray (1994) that formed their thesis was the contention that intelligence is endowed unequally within the White population and possibly between racial/ethnic groups (e.g., White-Black and White-Latino comparisons). The authors claimed a 40% to 80% heritability estimate of intelligence in the White population. Several points are noteworthy here. First, the discussion of intelligence being substantially inherited and linked with cognitive ability and social behaviors and outcomes (poverty, dropouts, etc., as seen in Chapters 5 through 12 of *The Bell Curve*) focused *exclusively on Whites* (p. 125, n. 11). Second, discussion on Black-White and Latino-White differences in cognitive ability is confined to Chapter 13 of *The Bell Curve*. Regarding the role of genetics in possibly explaining the higher mean IQ performance of Whites relative to those of Blacks and Latinos, Judis (1995) commented, "The authors [Herrnstein and Murray] . . . claim agnosticism on the question of whether genes or environment cause low IQ scores, but their analysis is heavily weighted toward genetic causes" (p. 128). Third, as we saw in the case of Jensen (1973a), Herrnstein and Murray's (1994) leap to suggest that mean differences in intelligence between minority and White populations are based on a within-group heritability estimate for the White population is unwarranted. Herrnstein and Murray's perspectives on the nature-nurture debate are explicated in Chapter 6 of the present book.

Our interests here are to examine Herrnstein and Murray's (1994) perspectives on SES and intelligence regarding minorities. On this, the authors raise two questions. First, they ask, "Are the differences in overall black and white test scores [IQ] attributable to differences in socioeconomic status?" (p. 286). Drawing from the data of the National Longitudinal Survey of Youth (NLSY; Ohio State University, Center for Human Resource Research, n.d.), Herrnstein and Murray (1994) reported that the Black-White difference in IQ was 1.21 standard deviations. When both race and parental SES were entered into a regression equation, the IQ gap shrunk to 0.76 standard deviation. The authors concluded, "Socioeconomic status explains 37% of the original B/W [Black-White] difference [in IQ]" (p. 286). Hedges and Nowell (1998), using mothers' and fathers' schooling attainment and their family income as SES indicators, reported that the Black-White gap in cognitive test scores was reduced by 0.30 standard deviation.

Notwithstanding the finding of mean Black-White IQ reduction, one that typically has been found in numerous studies, Herrnstein and Murray (1994) noted,

> But the remaining difference is not necessarily more real or authentic than the one we start with. This seems to be a hard point to grasp, judging from the pervasiveness of controlling for socioeconomic status in the sociological literature on ethnic differences. But suppose we were asking whether blacks and whites differed in sprinting speed and controlled for "varsity status" by examining only athletes on the track teams in Division I colleges. Blacks would probably still sprint faster than whites on the average, but it would be a smaller difference than in the population at large. Is there any sense in which this smaller difference would be a more accurate measure of the racial difference in sprinting ability than the larger difference in the general population? We pose that as an interesting theoretical issue. (p. 287)

We fail to see the logic of this example. Trying to control a variable (i.e., varsity status) to examine sprinting speed among elite Black and White runners, and then to draw comparisons to sprinting speed among the general Black and White populations, makes no sense regarding sampling design and external validity (i.e., generalizability).

The second question asked by Herrnstein and Murray (1994) is, "*As blacks move up the socioeconomic ladder, do the differences [i.e., in IQ] with whites of similar socioeconomic status diminish?*" (p. 287). Drawing again from the NLSY data, the authors hypothesized, "As blacks advance up the socioeconomic ladder, their children, less exposed to these environmental deficits, will do better and, by extension, close the gap with white children of their class" (p. 287). After presenting data on the initial hypothesis, Herrnstein and Murray concluded, "This expectation [closing the gap] is not borne out by the data" (p. 287). The authors presented data that they claimed showed the opposite effect: "It [the gap] gets larger as people move up from the very bottom of the socioeconomic ladder" (pp. 287-288). We ask the following: Might a plausible explanation for the absence of a total gap reduction (and the point that the gap increases as SES increases) be that Blacks and Whites never can be truly equated on SES because of the former group's caste-like status, regardless of their SES level? We now turn to this issue.

Class Versus Caste

We opened this chapter by noting that SES (and its related term, social class) is a social construction that is used to describe human subgroups in a given society. This social marker, a measure of prestige, frequently is based on schooling attainment, income, and occupation. As such, can SES (e.g., middle class), when used to describe a White family and an African American family in a particular community, be truly isomorphic across race/ethnicity? Some scholars think not (e.g., Bird et al., 1952; Blau, 1981; Deutsch & Brown, 1964; Dreger & Miller, 1960, 1968). For example, Dreger and Miller (1960), in their review of literature that has investigated the relation between SES and intellectual performance among Whites and Blacks, commented,

The various indices of socioeconomic status already devised or those at present conceivable on the same principles are intended to distinguish social *classes* from one another. How they can be employed to compare individuals in different *castes*, except very roughly, is difficult to see. (p. 366)

The point made by Dreger and Miller (1960, 1968) is well taken. In studies that have attempted to "match" Blacks and Whites on SES, true matches have not resulted because the social positions of Blacks, which have been tightly structured around patterns of racial discrimination imposed by Whites, have not been considered. Four decades ago, Dreger and Miller (1960) asserted that such caste arrangements had been documented both in the South (Dollard, 1949) and in the North (Brown, 1944; Long, 1957).[49] For example, Dreger and Miller (1960) recalled,

In the state of Florida where the writers reside, there are a number of Negroes whose social and economic statuses exceed those of most white persons. These Negroes, however, cannot yet sit in the same seats on public transportation (in most places) [or] go to the same hotel, restaurant, club, school, church, social events, or even restrooms. (p. 367)

In a study during the 1930s, Long (1934) measured the intelligence of 1,495 Negro schoolchildren (1st, 3rd, and 5th grades) born and raised in Washington, D.C. The Negro children attended segregated schools. Compared to the White schools, the Negro schools received comparable funding. Long made the point, however, that such matching of funding "must not lead us into believing that the cultural opportunities of the two races are equal" (p. 220). Commenting on the social and cultural separation of Negroes and Whites in Washington, he noted,

Culturally and socially, the colored people of Washington live to themselves except for those unavoidable circumstances inherent in spatial contiguity which neither racial group, even if it wished, could very well avoid. We must realize, however, that these unavoidable advantages are very significant. Negroes have separate churches, schools, hospitals, Christian associations, lodges, and recreation centers. A colored person may not enter and be served in the so-called "white" hotels and restaurants. Tickets to theatres in Washington are regarded as licenses and, therefore, may be taken from the purchaser if during the performance it becomes known that he has a small amount of Negro blood in his veins. In some theatres, he may buy seats in the second balcony or gallery. At public concerts, one is likely to observe conspicuous attempts at segregating Negroes into certain seating sections. In any event, it is a courageous soul that can brave, even for culture's sake, the patronizing glances or whispered comments of his white fellow citizens seated about him. (p. 221)

The preceding observation by Long (1934) led him to conclude, "The colored citizen is surrounded by a highly organized social and economic structure in which he is not allowed to compete on equal terms with his [White] fellows" (p. 222). Finally, Long noted that the colored students in his study had a mean IQ (Kuhlman-Anderson Test) of 95.4

(compared to a mean IQ of 100 for White schoolchildren in Washington). He concluded, "The wonder is not that the colored children of Washington fail to equal the whites in IQ score, but that their IQs are as high as they are" (p. 222).

More recent scholarship has provided support for the assertions that certain racial/ethnic minorities can be described in terms of caste. The leading proponent of this perspective is educational anthropologist John Ogbu of the University of California, Berkeley. Ogbu, in numerous writings (e.g., Ogbu, 1978, 1983, 1986, 1991, 1994), classified racial/ethnic minority groups in the United States as either "immigrant minorities" (e.g., some Latinos from Central America, Koreans, Japanese) or nonimmigrant or "involuntary minorities" whose current societal status is rooted in slavery (e.g., African Americans), conquest (e.g., American Indians), or conquest and colonization (e.g., Mexican Americans, Puerto Ricans). Sometimes referring to these involuntary minorities as "caste-like," Ogbu (1991) asserted that members of these groups "resent the loss of their former freedom, and they perceive the social, political, and economic barriers against them as part of their undeserved oppression" (p. 9). Ogbu (1991) also argued that involuntary minorities experience frequent discriminatory treatment with respect to being "confronted with social and political barriers, given inferior education, and derogated intellectually and culturally, and they may be excluded from true assimilation into the mainstream society" (p. 9).[50]

In sum, there is ample and rather convincing scholarship that some minority groups (i.e., involuntary minorities) are caste-like in characterization. Given this, the class versus caste issue does provide a means of understanding the long-standing finding that even when White and minority participants (e.g., African Americans) are "matched" on SES, there still remains a mean difference in intellectual performance favoring Whites. In the many investigations that have sought to control for SES in intellectual comparisons of Whites and involuntary minorities, it is not surprising that the mean differences still remain. A strong case can be made that a sampling bias exists in these investigations given that uncontrolled caste factors are operative. These unmeasured effects of racism need to be acknowledged in such research.

On a final note, we need to point out that there has been some criticism of the method for controlling for SES and then interpreting that the racial/ethnic gap reduction in mean IQ that inevitably occurs is due to environmental effects. Jensen (1998) opined,

> Adults' attained SES (and hence their SES as parents) itself has a large genetic component, so there is a genetic correlation between SES and IQ, and this [is] so within both the white and black populations. Consequently, if black and white groups are specially selected so as to be matched or statistically equated on SES, they are thereby also equated to some degree on the genetic component of IQ. Whatever IQ differences remain between the two SES-equated groups, therefore, [do] not represent a wholly environmental effect. (Because the contrary is so often declared by sociologists, it has been termed the sociologist's fallacy.) (p. 491)

It is widely acknowledged that *individuals* within SES groupings tend to marry people similar to themselves. This process, known as "assortative mating" (or nonrandom mating), involves a number of characteristics including phenotypic intelligence. Citing Jensen (1978), Plomin, DeFries, McClearn, and Rutter (1997) noted, "Assortative mating for *g* is substantial, with average spouse correlations of about .40. . . . In part, spouses select each other for *g* on the basis of education. Spouses correlate about .60 for education, which correlates about .60 with *g*" (p. 143). Thus, there appears to be some genetic influence on intelligence *within* social class. This is not an issue. Our quibble with Jensen (1973a, 1998) is his implication that such within-group heritability can be used to explain mean differences in intelligence *between* Whites and Blacks when SES is controlled. Jensen's (1998) claim that the residual in mean differences in intelligence when SES is equated between Blacks and Whites "does not represent a wholly environmental effect" is speculative. Much of Jensen's theorizing about a genetic hypothesis when it comes to SES and environmental effects rests on default and straw person arguments. (For further discussion on how Jensen's notion of the "sociologist's fallacy" can be refuted by a "hereditarian fallacy," see Mackenzie, 1984.)

Status Consistency of SES Indexes and Differential Prediction of Intelligence

Blau's (1981) book, *Black Children/White Children: Competence, Socialization, and Social Structure*, is one of the most comprehensive investigations undertaken that has identified the social processes that statistically predict intellectual performance and academic achievement among Black and White children. Here, we confine our discussion to Blau's findings on the relation between four constituent SES variables and IQ.

The participants in Blau's (1981) study were 579 Black and 523 White mothers and their 5th- and 6th-grade children in three communities in the Chicago metropolitan area. Data were gathered in 1968. Measured intelligence (IQ) was based on school records (mostly intelligence tests from school records, with names not reported). SES data were decomposed into four variables: (a) parents' educational attainment (mother's and father's combined schooling attainment), (b) occupational status of the parent with the higher Duncan's Socio-Economic Index (Reiss, Duncan, Hatt, & North, 1961), (c) social milieu (middle-class exposure score[51] plus the average educational attainment of mother's three closest friends), and (d) mother's and father's SES origin (i.e., having working- or middle-class origins).

The findings of Blau's (1981) investigation that are most germane to our discussion are as follows:

1. Regarding SES origin, 82% of the Black middle-class mothers had working-class origins; by sharp contrast, 32% of the White middle-class mothers had working-class origins. Similar findings for Black and White middle-class fathers were observed. Given these findings, Blau commented, "The origins of black middle-class parents closely

resemble those of white working-class parents, *not* those of white middle-class parents" (p. 21). This led Blau to conclude, "The black middle class [in 1968] is very largely a new middle class" (p. 193).

2. The three measures of family status (i.e., parental education, occupational status, and social milieu) demonstrated stronger status consistency among Whites than among Blacks. That is, the intercorrelations of these three variables were stronger among the White group than among the Black group. For example, the intercorrelation between parental education and occupational status was .71 for Whites and .57 for Blacks. This stronger status consistency of the measures of family status led Blau (1981) to conclude, "Therefore, the use of any single index of socioeconomic status is more likely to reflect the social environment of white children than of black children" (p. 24).[52]

3. Blau (1981) found that the three measures of family status (plus a demographic origin-variable) predicted Black children's and White children's IQs differently. Using regression analysis, it was found that for the White sample, the best predictor was parental education, followed by parents' demographic origins.[53] Occupational status entered the equation, but it added less than 1% of the explained variance. In total, these three variables explained 12% of the variance in White children's IQs.

For the Black sample, social milieu exerted the largest effect, followed by SES origin and then extent of mother's exposure to White friends and coworkers. Parental education entered the equation, but it did not significantly increase the amount of explained variance. In total, 14% of the variance in the Black children's IQs was accounted for by these four variables.

This investigation by Blau (1981) is extremely important because it demonstrated empirically that social status and social milieu, in particular, carry very different meanings as predictors of intellectual performance among White and Black children:

> The results of the analysis indicate that the use of either occupational status or parents' education (or even both of these variables) as [a control] does not achieve the intended purpose of equalizing the social environment of families in the two races. The social milieu of white families is more consistent with the status achievements of white upwardly mobile parents than of black ones. Consequently, social milieu supersedes any other component of social status as a predictor of black children's IQ scores. (p. 25)

The Unit of Analysis: Aggregated Versus Confounded Versus Student

White's (1982) article, the most comprehensive review of the literature to date on the relation between SES and measures of academic achievement and intelligence, identified a number of methodological problems in the reviewed studies. One of these issues

TABLE 3.8   Mean Correlations Between SES and Measures of Achievement and Intelligence

Category	Verbal		Math		Composite Achievement[a]		IQ	
	M	*N or n*[b]	*M*	*N or n*	*M*	*N or n*	*M*	*N or n*
All correlations	.31	225	.25	143	.37	66	.40	102
Aggregated	.68	35	.70	14	.64	15	.73	18
Confounded	.29	16	n.a.	n.a.	n.a.	n.a.	.34	10
Student	.23	174	.20	128	.27	46	.33	74

SOURCE: From "The Relation Between Socioeconomic Status and Academic Achievement," by K. R. White, 1982, *Psychological Bulletin, 91*, p. 469. Copyright 1982 by the American Psychological Association. Adapted with permission of the author.
NOTE: SES = socioeconomic status; n.a. = nonavailable data.
a. Composite Achievement measure refers to data in which total achievement was assessed.
b. *N* or *n* refers to the number of correlations.

centered on the "unit of analysis." White found that in the 101 studies he reviewed, researchers varied in their units of analysis. White coded these units as "aggregated," "confounded," or "student." He described them as follows:

> The *unit of analysis* used in computing the correlation coefficient was coded as aggregated, confounded, or student. When an aggregated unit (such as school or district) was used, both the SES measure and the achievement measure were averages for the aggregated unit, and the correlation was computed between the average scores. When the unit of analysis was confounded, SES was measured at an aggregated level, and achievement was measured at the student level or vice versa. For instance, all students in the same school might be given the same SES rating but could have individual achievement scores. The student was identified as the unit of analysis when both SES and achievement were measured separately for each student. (p. 465)

In his review, White (1982) found that the magnitude of the correlation between SES and academic achievement/intelligence was significantly related to a number of variables (e.g., year of study, grade level of participant, *N* participants of study, number of items in the SES measure). He reported that the most dramatic difference in correlation magnitude was related to the unit of analysis. That is, when correlations were computed from aggregated data, the resultant magnitudes were considerably higher than correlations computed using individual students as the unit of analysis. White noted that it is well known among statisticians that using aggregated units of analysis results in spuriously high correlations.[54] To illustrate, White presented data of mean correlations between SES for measures of academic achievement and intelligence. Table 3.8 presents these data.

It is exceedingly clear from the data in Table 3.8 that using the aggregated unit of analysis results in substantially higher correlations than does using the student unit of analysis, the more valid unit (Robinson, 1950). Of the various measures (Verbal, Math, Composite Achievement, and IQ), when the more accurate unit of analysis (i.e., student) is used, the r values range from a low of .20 (Math) to a high of .33 (IQ); these are low magnitudes. The observed r of .33 between SES and intelligence is indeed lower than the unsubstantiated claim made by Jensen (1973a) that the average r is between .40 and .60.

On a final point of interest, White (1982) collapsed his data across all coding variables and for all studies, and he found the mean correlation coefficient between SES and "achievement" (which included IQ measures)[55] to be .35. The median, which is the best estimate of central tendency in a skewed distribution, was only .25. When White analyzed the data with student as the unit of analysis, the mean and median were .25 and .22, respectively. By sharp contrast, when White analyzed the data using aggregated units of analysis, the mean and median jumped to .68 and .73, respectively.

In sum, White's (1982) comprehensive review of the literature provides strong evidence against the prevailing "belief that socioeconomic status . . . and various measures of academic achievement are strongly correlated" (p. 461).[56] Likewise, in theorizing about factors associated with students' academic achievement and intellectual performance, researchers would be prudent in their empirical investigations to use data in which the individual student is the unit of analysis.

In light of the substantial research we have examined in this chapter concerning the relation between measured SES and measured intelligence among minority students, what major conclusions may we draw? There are several:

1. Contrary to what is believed by many behavioral and social scientists, SES is only weakly associated with measured intelligence when the more appropriate unit of analysis (i.e., student) is used (White, 1982).

2. When SES is controlled, the difference in intellectual performance between White and minority means frequently is reduced and in some cases is negligible (Anastasi & D'Angelo, 1952; Christiansen & Livermore, 1970; Herrnstein & Murray, 1994; Nichols & Anderson, 1973; Valencia, 1979; Valencia et al., 1995).

3. Sampling bias might be one explanation of the residual mean intellectual difference (favoring Whites) when Whites and minorities are matched or equated on SES. It might be that the caste-like status of involuntary minorities (as defined by Ogbu, 1991) is uncontrolled in many investigations (see, e.g., Dreger & Miller, 1960, 1968).

4. SES indexes (e.g., occupational status, educational attainment) have in some studies (e.g., Blau, 1981; Valencia, Rankin, & Henderson, 1986) been found to predict differentially the intellectual performance of White and minority students. In light of this, researchers need to be aware that conventional SES indicators, when used to control the social class environments of Whites and minorities, do not necessarily lead to equalization. Blau (1981) has admonished that educational attainment and occupational status

are, at best, crude indexes of Whites' social class position. "But, in comparisons of white and black samples, it is a mistake to assume that the use of these indices alone or in combination suffices to equalize the social environments of black and white respondents with the same scores" (p. 17).

5. The assertion that intellectual differences among children (White and minority alike) of different social classes are largely due to innate factors has waxed and waned for decades. The most recent example of such a position is seen in Herrnstein and Murray (1994). As we have discussed, such a contention is quite shaky on empirical grounds. We discuss this assertion further in Chapter 6 of this book.

Notwithstanding the typically observed weak correlation between SES and intelligence, a child's social class remains an important aspect in understanding its role in predicting performance on measures of intelligence. One must be aware, however, of what SES is. It is a summarizing, or distal, variable (Suzuki & Valencia, 1997; Valencia, Henderson, & Rankin, 1985). As such, SES often can conceal what families actually do in structuring learning environments for their children. This masking aspect of SES often gets lost in the discussion about the relation between SES and intelligence. Might it be that what parents *do* (i.e., their behavior toward learning in their children) is a better predictor of children's intellectual growth and performance than what parents *are* (e.g., occupational status, educational status)? We examine this issue in Chapter 4.

4

Home Environment

Numerous researchers have discussed the limitations of socioeconomic status (SES) as a predictor of children's intellectual performance (Bradley & Caldwell, 1978; Bradley, Caldwell, & Elardo, 1977; Henderson, 1981; Johnson et al., 1993; Marjoribanks, 1972a; Valencia, Henderson, & Rankin, 1985; Walberg & Marjoribanks, 1976; Wolf, 1966). Such concerns include criticisms that SES is a "status" variable rather than a "process" variable (Johnson et al., 1993), is less accurate a predictor than are indexes of home environment quality (Bradley et al., 1977), is relatively static (Bradley & Caldwell, 1978), and obscures the reality that families within a particular SES level may differ considerably in home intellectual climate (Bradley et al., 1977). Walberg and Marjoribanks (1976) compressed matters as follows: "Although socioeconomic status is a convenient construct to measure, it does not yield comprehensive assessment of the factors in the home that foster [cognitive] ability. Proximate, detailed assessments provide better predictions of children's [cognitive] ability" (pp. 532-533). Furthermore, Henderson (1981) commented, "Its [SES's] utility is limited . . . because it fails to designate the nature of the experience it subsumes" (p. 28).

What is meant by the "home environment"? Broadly speaking, we are conceptualizing home environment as consisting of proximal variables that represent "learning experiences provided in the home or under the direction of the family" (Valencia et al., 1985, p. 324). Examples of such variables are the ways in which intellectual stimulation is provided in the home (e.g., parent reading to child, play materials made available to child), parents' academic aspirations for children, and parents serving as language models. This conceptualization of the home environment, which stems from the field of developmental psychology, views parents as a major source of environmental influence on their children's development (in our focus here, cognitive development). There is, however, a radically different perspective that has emerged from behavioral genetics during the past 20 years. These two developments are (a) the construct of "nonshared environment" and (b) genetic influences on environmental measures. Research on these factors has been developing at a rapid pace, and such findings are gradually entering the academic discourse on individual differences in children's development, generating considerable interest as well as controversy. The authors of a major textbook on behavioral genetics recently commented, "Genetic research is changing the way we think about the environment" (Plomin, DeFries, McClearn, & Rutter, 1997, p. 245). Although our focus in this chapter is on the home environment research findings from developmental psychology, we do introduce the reader to the environmental perspectives from behavioral genetics.

To be sure, the research base on home environment as a predictor of children's intelligence is quite large.[1] It is not our goal in this chapter to review this body of research. Rather, our focus is on three aspects: the nature of the construct of home environment, a discussion of research studies in which U.S. racial/ethnic minority children served as participants, and the implications of this research area for understanding minority children's intellectual performance. In the remaining sections of this chapter, we cover the following: (a) early research (1920s); (b) influences of Piaget, Hunt, and Bloom; (c) origins of the "Chicago" school of family environment research; (d) influence of the Chicago school of family environment research; (e) home environment research by Caldwell, Bradley, and associates; (f) home environment research from behavioral genetics; (g) home environment research: African Americans; (h) home environment research: Mexican Americans; and (i) conclusions.

Early Research (1920s)

Although research on the home environment as a predictor of children's measured intelligence did not begin in earnest until the mid-1960s (Wolf, 1964), there were scattered investigations several decades earlier. It appears that Burks's Stanford University doctoral dissertation, presented and published in 1928, was the first systematic study in this area. Burks's (1928, cited in Marjoribanks, 1979) sample consisted of 214 foster children and their foster parents as well as a control group of 105 children and their natural parents. The children, who were White, ranged in age from 5 to 14 years. For its time, Burks's study was comprehensive, as it included (a) the administration of the 1916 Stanford-Binet intelligence test (given to children and their parents), (b) a questionnaire to measure emotional stability, (c) the Whittier Scale for Home Grading (a family environment questionnaire[2]), (d) a "cultural scale" (designed to gather data on, e.g., schooling attainment of parents, parents' vocabulary, and number of books in parents' library), and (e) a personal information inventory (e.g., parental occupation, home instruction received by child, storytelling to child).

For the foster children group, the strongest predictor of children's IQs was the cultural index ($r = .29$, corrected for attenuation [related to unreliability of some measures]). In descending order, the other correlations of children's IQs were family income (.26), mother's vocabulary (.25), Whittier Scale (.24), mother's mental age (.23), father's vocabulary (.14), and father's mental age (.09).[3]

In sum, Burks's (1928) study showed that parent IQ and family environment scores had low to low-moderate correlations with the foster children's IQs. For the control group, such correlations were of moderate to high magnitudes. Regarding the importance of this early investigation, Marjoribanks (1979) commented, "The measurement procedures and methodological and statistical techniques [used in Burks's study] continue to form the basis of many family investigations, often, it appears, without knowledge of Burks' pioneering efforts in relating family environments and children's outcomes" (pp. 10-11).

Another example of seminal research on the relation between home environment and children's intelligence is the doctoral dissertation by Van Alstyne (1929, cited in Wachs & Gruen, 1982). In this investigation, Van Alstyne correlated various measures of the home environment (e.g., amount of time parent reads to child, opportunities to engage in constructive play, number of books child has and sees) to 3-year-old children's intellectual performance on the Kuhlman-Binet scale. Race/ethnicity of the families was not described. The author reported that the home environment variables were positively and significantly correlated with children's mental age.[4]

Influences of Piaget, Hunt, and Bloom

It would be remiss not to briefly mention the works of Piaget, Hunt, and Bloom, who had considerable influence in shaping our knowledge about the roles of experience and the environment in children's intellectual development. The writings of these pioneers have laid a strong foundation for contemporary research on the relation between home environment and children's measured intelligence.

Piaget

The work of Jean Piaget (e.g., Piaget, 1926/1929), a noted Swiss child psychologist and professor of child psychology and history at the University of Geneva, "revolutionized and still dominates the study of human development" (Slavin, 1986, p. 34). Central to Piaget's theory of cognitive development are his notions of schemata, adaptation (assimilation and accommodation), and equilibration. Piaget, who incorporated both genetic and environmental influences in his theorizing as to how children's cognition develops, asserted that children's constant interaction with the environment leads to cognitive development. Part of this environment is the home and how it is structured to allow for learning opportunities. A major implication of Piaget's theory for teaching/learning in the home is what Hunt (1961) called "the problem of the match" or the problem of achieving equilibration. Here, equilibration refers to the "search for mental balance between cognitive schemes and information from the environment" (Woolfolk, 1995, p. 31). In short, children strive to maintain such balance by continuously fitting new information into existing schemata (assimilation) or by making new ones as children respond to new information from the environment (accommodation). Suffice it to say, for very young children, the home environment becomes a daily source of information and stimulation.

Hunt

In a classic volume, *Intelligence and Experience,* J. McV. Hunt meshed writings on infrahuman and human research and drew conclusions about the development of human intelligence (Hunt, 1961). Hunt, a professor of psychology at the University of Illinois, began his book by providing a discussion and critique of hereditarian theory (i.e., fixed intelligence, predetermined development). Drawing on Hebb's (1949) conceptual synthesis of development in neurophysiology that explains perceiving and problem solving via central processes as well as Piaget's theorizing on the development of intellectual functions in children, Hunt asserted that beliefs in fixed intelligence and predetermined development were untenable. In sharp contrast to hereditarian theory, Hunt contended that (a) experience was a critical factor in intellectual development and (b) early intervention was a valuable means of promoting cognitive growth in children. In sum, Hunt's seminal volume was extremely important in that it laid a strong case for the role of experience in shaping children's intelligence and the malleability of human cognitive development.

Bloom

Another major early contributor to our knowledge base about the relation between home environment and children's intellectual development/performance was Benjamin Bloom, a professor of education at the University of Chicago. Bloom (1964) gave us *Stability and Change in Human Characteristics,* a masterful book on environmental variation and stability and change in several aspects of human development (e.g., height, intelligence, academic achievement). His volume has been described as "a watershed in the history of environmental measurement" (Bradley & Caldwell, 1978, p. 117). Bloom's 1964 volume can be compressed by way of several major conclusions and propositions:

1. Correlations between measurement of a particular characteristic at certain ages are very similar across studies.

2. The environment can be defined. Bloom (1964) refers to "the conditions, forces, and external stimuli which impinge upon individuals. These may be physical, social, as well as intellectual forces and conditions" (p. 187). Bloom also suggested, "Somewhere between the total environment and the specific interaction or single experience is the view of the environment as a set of persisting forces which affect a particular human

characteristic" (p. 187). We shall see later that this concept of the home environment was developed in the doctoral dissertations of several of Bloom's students.

3. The home environment, as defined previously, can be measured systematically. We discuss this in more detail later.

4. Environmental variation is particularly salient for a trait (e.g., intelligence) during the trait's period of most rapid growth. For example, the most rapid growth of intelligence occurs before 5 years of age. Furthermore, Bloom found that the average correlation between intelligence at 4 and 17 years of age was .71, meaning that about 50% of the variability in intelligence at 17 years of age can be accounted for by measuring intelligence at 4 years of age.

5. One of Bloom's (1964) most basic propositions was that *"the environment is a determiner of the extent and kind of change taking place in a particular characteristic"* (p. 209). Regarding this proposition and variations in home environment and individual differences in general intelligence, Bloom asserted,

> Differences in general intelligence are likely related to the following:
>
> 1. Stimulation provided in the environment for verbal development
> 2. Extent to which affection and reward are related to verbal-reasoning accomplishments
> 3. Encouragement of active interaction with problems, exploration of the environment, and the learning of new skills. (p. 190)

Origins of the "Chicago" School of Family Environment Research

In light of Bloom's (1964) major conclusion that the preschool years were the most important period for children's intellectual stimulation, it was not surprising that researchers began active work on identifying and measuring those home environment variables that facilitated intellectual development (Henderson, 1981). The pioneering work of Bloom and his doctoral students has been so influential that Marjoribanks (1979), one of the preeminent scholars in family environment research, referred to it as the "Chicago" school of family environment research. Although Bloom was the architect of the family subenvironments model, it was several of his doctoral students who actually carried out the research on the relation between home environment and key criterion variables (i.e., Davé, 1963: academic achievement; Wolf, 1964: intellectual performance; Weiss, 1969: achievement motivation and self-esteem). Our focus is on the work of Wolf (1964, 1966).

Wolf's model of measuring the home environment is based on four principles (Wolf, 1966):

1. Rather than conceiving of the environment as a single entity, the physical environment is best viewed as a number of "subenvironments." The subenvironments operate "to influence the development of a specific characteristic" (p. 492), for example, one environment for the development of independence and one for general intelligence. As such, "the problem of measuring an environment was reduced to the identification and measurement of those aspects of the total environment that were likely to be related to the development of selected specific characteristics" (p. 492).

2. Previous research on familial influences on intellectual performance/development focused on indirect or surface manifestations of the environment (e.g., SES). Wolf (1966) was interested in identifying and measuring home environment variables that would *directly* shape intellectual development. Such a departure "resulted in our investigating what parents *do* in their interactions with their children [e.g., parent reading to child] rather than what parents are in terms of father's occupation, type of dwelling, source of income, and so forth" (p. 492).

3. There was an attempt "to summarize and treat environmental data through the use of psychometric procedures" (Wolf, 1966, p. 493). What is meant here is that Wolf drew from the science of testing. The environment was viewed as a large finite universe of possible influences on behavior. As such, multiple features of the environment—not just a single particular environmental variable—must be identified and measured. Wolf commented, "A summarization of a number of variables is as important in describing an environment as a number of test items is in describing a student's competence in a subject" (p. 493).

4. An extremely important feature of Wolf's work was to correlate measures of home environment with the criterion variable, the child's intellectual performance. Wolf commented that a number of investigations developed instruments to measure the home environment but did not correlate such data to individual test scores.

After doing a comprehensive review of the literature, and based on Bloom's (1964) family subenvironments model, Wolf (1964) identified three "press variables" that he postulated were predictors of children's measured intelligence:[5] (a) press for achievement, (b) press for language, and (c) provisions for general learning. Each press variable contained "process characteristics," which can be viewed as operational definitions of the three constructs. The three press variables and 13 process characteristics (as listed in Bloom, 1964, p. 78) are as follows:

A. Press for Achievement Motivation

   1. Nature of intellectual expectations of child

   2. Nature of intellectual aspirations for child

   3. Amount of information about child's intellectual development

4. Nature of rewards for intellectual development

B. Press for Language Development

5. Emphasis on the use of language in a variety of situations

6. Opportunities provided for enlarging vocabulary

7. Emphasis on correctness of usage

8. Quality of language models available

C. Provisions for General Learning

9. Opportunities provided for learning in the home

10. Opportunities provided for learning outside the home (excluding school)

11. Availability of learning supplies

12. Availability of books (including reference works), periodicals, and library facilities

13. Nature and amount of assistance provided to facilitate learning in a variety of situations

The participants used in Wolf's (1964) doctoral dissertation were 60 White 5th-grade children (selected from 19 schools) and their mothers. Participants resided in a midwestern community comprising rural, suburban, and urban areas. The SES backgrounds of the participants were diverse, representing proportions based on Department of Labor data. Based on his 13 process variables, Wolf devised a semistructured interview consisting of 63 questions. The interview, in which the child's mother served as the respondent, lasted about 1½ hours. The criterion variable was the Henmon-Nelson Test of Mental Ability (Lamke, Nelson, & Kelso, 1961); scores were obtained from school files. Wolf's major finding was a multiple correlation of .69 between his rating of the quality of the home environment and the child's measured general intelligence. Thus, about 48% of the variance in children's intelligence could be accounted for by knowing their ratings on the home environment measure. Wolf also noted that proximal measures of the home environment were considerably more powerful predictors of children's intelligence than were distal SES variables. Referring to his doctoral dissertation in a subsequent publication, Wolf (1966) commented,

> The correlation between social status and measured general intelligence . . . has been found to lie between +.20 and +.40. Accepting the correlation of +.40 as the correct estimate of the relationship between social status and intelligence, it would seem that the newer approach to the measurement of the environment accounts for about three times as much of the variance in general intelligence as a measure of social status. (p. 497)

The Chicago school of family environment research also is exemplified by the doctoral dissertations of Davé (1963) and Weiss (1969), two other students of Bloom. Davé (1963) found high multiple correlations between a home environment measure he devised and children's academic achievement.[6] Weiss (1969) found moderate to high multiple correlations between a home environment measure he devised and affective characteristics (achievement motivation and self-esteem).[7]

In sum, the Chicago school of family environment research was a major breakthrough in examining the relation between the home environment and children's behavior.[8] Although this body of research was based on White parents and their children, later we discuss its generalizability to African American and Mexican American families. In the next section, we briefly examine how the Chicago school of family environment research influenced the works of other researchers. Our major focus is on children's intellectual performance as the criterion variable.

Influence of the Chicago School of Family Environment Research

The Chicago school of family environment research spurred a number of investigations to examine the link between home environment and children's measured intelligence and, in some cases, academic achievement. As a whole, these studies have provided theoretical support for Bloom's (1964) basic model of family subenvironments as well as empirical support for the findings of Wolf (1964) and Davé (1963). The post-Chicago school of family environment research has been undertaken in a variety of national settings and with different cultural groups. Examples of investigations conducted in various countries include those in the United States (Henderson, 1966, 1972; Henderson & Merritt, 1968), Canada (Marjoribanks, 1972b, 1974; Mosychuk, 1969; Williams, 1976), Trinidad (Dyer, 1967), Ireland (Kellaghan, 1977), and Australia (Marjoribanks, 1977, 1978, 1979). As a whole, these international studies (e.g., Marjoribanks, 1972b, 1979) have demonstrated that ethnicity is a significant variable that should be accounted for in examining the relation between home environment variables and children's intelligence and academic achievement. However, as noted by Walberg and Marjoribanks (1976), "Correlational or causal relationships established for one group may not hold for other times, social classes, ethnic groups, or countries" (p. 527).

Home Environment Research by Caldwell, Bradley, and Associates

In addition to the Chicago school of family environment research, there are other researchers who have had impacts in this field.[9] One group that has had a particularly strong influence is Bettye Caldwell, Robert Bradley, and associates. Although it is clear that the works of Caldwell, Bradley, and associates were influenced by the Chicago school of research, their works differed in target population and methodology, that is, a focus on infants and very young children. The method of collecting data on the home environment has used *observations* of primary caregiver-child interactions as well as interviews with the primary caregiver (usually the mother). This is a major methodological break from the Chicago school of studies that obtained data only through interviews with the caregiver.

Drawing from a dozen propositions that are believed to foster the cognitive and language development of young children (e.g., play materials that promote learning, contact with adults who value achievement),[10] Caldwell and coinvestigators developed a home environment measure with two versions: one for families of infants (birth to 3 years of age) and one for families of preschoolers (3 to 6 years of age). The instrument, which was first tested during the mid-1960s (Caldwell, Heider, & Kaplan, 1966), has gone through a number of revisions and is now known as the Home Observation for Measurement of the Environment (HOME; Caldwell & Bradley, 1984). The infant version of HOME contains 45 items (scored in a binary *yes/no* manner clustered in six subscales (Emotional and Verbal Responsivity of Mother, Avoidance of Restriction and Punishment, Organization of Physical and Temporal Environment, Provision of Appropriate Play Materials, Maternal Involvement With the Child, and Opportunities for Variety in Daily Stimulation).[11] The preschool version contains 55 items clustered in eight subscales.[12]

HOME has proved to be the most widely used home environment measure, as evidenced by numerous studies (with White and minority samples) published over the years[13] including longitudinal research (Gottfried, 1984). Stoolmiller (1999) noted that, at the time of his writing, there were 443 published works listed on Bradley's Web page in which HOME was used as the measure of home environment. Many of these early investigations using HOME have been conducted at the Center for Child Development and Education in Little Rock, Arkansas. Bradley and Caldwell (1978) summarized the results of these early Arkansas studies as follows:[14]

1. Scores on HOME obtained during the 1st year of life correlated at low but significant magnitudes with the 6- and 12-month Mental Development Index of the Bayley Scales of Infant Development (Bayley, 1969).

2. These same HOME scores also correlated at moderate to strong magnitudes with 36- and 54-month Stanford Binet Intelligence Test IQs (Terman & Merrill, 1973) as well as 37-month scores on the Illinois Test of Psycholinguistic Abilities (ITPA; Kirk, McCarthy, & Kirk, 1968).

3. Moderate to quite high correlations were found between 24-month HOME scores and 36-month Stanford-Binet and ITPA scores.

Home Environment Research From Behavioral Genetics

Research from behavioral genetics has produced two major findings regarding the home environment and children's behavior pertaining to personality, psychopathology, and cognition (our focus is on the latter). One discovery has to do with "nonshared environment," which Plomin et al. (1997) defined as "environmental influences that contribute to differences between family members" (p. 315). "Shared environment," on the other hand, is defined as "environmental factors responsible for the resemblance between family members" (p. 316).[15] In a major target article published in *Behavioral and Brain Sciences,* Plomin and Daniels (1987), who titled their piece "Why Are Children in the Same Family So Different From One Another?," concluded that research findings from behavioral genetics indicated that individual differences in children's measured intelligence were negligibly "influenced" by shared environments (by the end of early adolescence). By sharp contrast, Plomin and Daniels commented that IQ (after childhood) was substantially influenced by nonshared environment, perhaps as much as 40% to 60% of the total variance in intelligence. In sum, Plomin and Daniels drew this major conclusion:

> One of the most important findings that has emerged from human behavioral genetics involves the environment rather than heredity, providing the best available evidence for the importance of environmental influences [i.e., nonshared] on personality, psychopathology, and cognition. The research also converges on the remarkable conclusion that these environmental influences make two children in the same family as different from one another as pairs of children selected randomly from the population. (p. 1)[16]

Plomin and Daniels's (1987) target article certainly elicited diverse reactions (see the commentaries that immediately followed their article) ranging from agreements to disagreements with the authors' theorizing. Many commentators, who agreed that nonshared environment is a valuable construct, offered constructive feedback that the next step is to identify the likely sources and processes of nonshared environment. Scholars who are strong proponents of nature-nurture interactions disagreed with Plomin and Daniels. More recently, one of the most potentially damaging blows to the nonshared environment construct is seen in an article by Stoolmiller (1999). As Stoolmiller pointed out, estimates of the contributions of heredity and environment (both nonshared and shared) to measured intelligence are derived from behavioral genetics research designs in which adoption studies are used in part. The gist of Stoolmiller's article is that families who adopt children are not "representative of the general population of families because of the adoption situation itself . . . or because adoptive parents are more highly educated, have higher occupational status, and are more affluent than the general population of parents" (p. 393). Stoolmiller concluded that these adoption studies suffer from considerable range restriction. It is well known in statistics that range restriction attenuates (i.e., weakens) correlations. As such, Stoolmiller corrected the correlations (between HOME scores and IQ) for range restriction and concluded, "Corrections for range restriction applied to IQ data from recent adoption studies indicate that if adoptive families were representative of the U.S. Census, SE [shared environment] could account for as much as 50% of the total phenotypic variance" (p. 405).[17]

In conclusion, further research is needed to provide confirmation of the hypothesis that nonshared environment is indeed a powerful explanation of why children in the same family are different in measured intelligence. If nonshared environment does eventually prove to be a major explanatory factor, then developmental psychologists will need to reconceptualize how the home environment helps to shape cognitive development. That is, the child, rather than the family, "must be considered the unit of socialization" (Plomin & Daniels, 1987, p. 15). As a final note on nonshared environment, it is important to underscore that minority families need to be participants in these research endeavors. To date, the adoption studies that have been conducted have overwhelmingly used White families.

A second major discovery regarding the environment that stems from behavioral genetics research also is quite surprising. Plomin et al. (1997) noted, "Many environmental measures [e.g., HOME] widely used in psychology show genetic influence" (p. 245).[18] Once again in a target article in *Behavioral and Brain Sciences* (this time titled "The Nature of Nurture: Genetic Influences on 'Environmental' Measures"), Plomin (this time with Bergeman) drew conclusions from behavioral genetic research findings that would result in unrest in developmental psychology. Plomin and Bergeman's (1991) main argument was that "*measures* of the environment—not the environment itself—should be conceptualized as phenotypes" (p. 374). When the notion of environment is viewed in this manner, the authors asserted, significant genetic contributions to links between family environment and measured intelligence exist. The typical design in this line of research is to obtain scores from a home environment measure (e.g., HOME) and intelligence test scores from children in both nonadoptive and adoptive families (see, e.g., Coon, Fulker, DeFries, & Plomin, 1990). The common pattern is to see higher correlations between HOME and IQ in nonadoptive families than in adoptive families. Plomin et al. (1997) did point out, however,

> Although the HOME is correlated genetically with children's cognitive development, it is not correlated genetically with parents' general cognitive ability (Bergeman & Plomin, 1988). In other words, parental cognitive ability does not appear to be responsible for this passive genotype-environment correlation. (pp. 258-259)[19]

Our inspection of the 30 commentaries that followed Plomin and Bergeman's (1991) target article showed, unsurprisingly, a wide range of reactions. Comments ranged from kudos for a valuable and thoughtful review of research to major criticisms along conceptual and methodological lines. Regarding the positive reactions, a common point was that much of behavioral science research has misattributed environmental influence (see, e.g., McGue, Bouchard, Lykken, & Finkel, 1991). With respect to criticisms, a fairly common point was that Plomin and Bergeman (1991) failed to adopt a clear interactionist position (see, e.g., Baumrind, 1991; Bradley & Caldwell, 1991). On this, Bradley and Caldwell (1991) noted, "Johnston (1987) presents a revealing account of how myths about genetic or environmental influence persist despite the overwhelming evidence in favor of the interactionist position" (p. 389).

In closing, what can one make of these advancements that behavioral genetics has proffered about the environment? Notwithstanding the criticisms leveled against this body of research, the field of developmental psychology has some issues with which to reckon. For example, does labeling an instrument a measure of home environment automatically imply that it only measures the environment? If nonshared environment is indeed a viable construct, then how can developmental psychology assist behavioral genetics in teasing out the intricate processes in the home that shape individual differences in intelligence among siblings? Of the many issues that arise from the environmental research in behavioral genetics, the one that we think is most critical is this: Does this body of research suggest that the role of parents and significant others in creating and fostering intellectually stimulating environments for children (e.g., reading to children) is now significantly diminished? This is no trivial issue. We opened this chapter asserting that what parents *do* with their young children, rather than who they are in terms of SES, is more important in facilitating cognitive development. This is a major finding stemming from the research efforts of developmental psychologists who have investigated the relation between home intellectual environment and children's measured intelligence and who have consistently reported positive correlations. That is, parents are the central unit of socialization. This is the case in White and minority homes.

In our view, much more research and theorizing on nonshared and shared environments, as well as on the contention that home environment measures are phenotypes, need to be conducted before the findings from developmental psychology on the connection between home environment and intelligence are dismissed as invalid. Our sense of the literature is that Plomin and associates were careful in not making arguments that parents and the home environment (as conventionally conceptualized by developmental psychologists) do not matter much. There are, however, some scholars who have taken extreme positions that alarm us. For example, Harris (1995) developed a "group socialization theory." She based her theory, in part, on the notions of nonshared environment and home environment measures as phenotypes. Harris's argument is that parenting styles are not very important: "What GS [group socialization] theory implies is that children would develop into the same sort of adults if we left them in their homes, their schools, their neighborhoods, and their cultural or subcultural groups but switched all the parents around" (p. 461). Such extremism also is seen in *The Limits of Family Influence: Genes, Experience, and Behavior,* a book by Rowe (1994), a professor of family studies with split appointments in genetics and psychology at the University of Arizona. Writing in a reductionist mode that sometimes is characteristic of behavioral geneticists, Rowe asserted his thesis that parent practices are unimportant:

> I share with other behavioral geneticists (Scarr, 1992) the position that parents in most working- to professional-class families may have little influence on what traits their children will eventually develop as adults. Moreover, I seriously doubt that good child-rearing practices can greatly reduce an undesirable trait's prevalence, whether it be low IQ, criminality, or any other trait of social concern. (p. 7)

Bradley (1994), a staunch proponent of the interactionist position in the nature-nurture controversy, reviewed Rowe's (1994) book and concluded that there is a potential danger in the author's theorizing. We strongly concur with Bradley's (1994) assertion, which is as follows:

> By implying little influence, Rowe may afford families an excuse to be less vigilant or to provide less nurturance—thereby falling below a threshold level necessary for felt security or effective action. As most cultures are currently structured, shared rearing experiences are probably "a weak source of trait variation." Such a conclusion is fine as far as it goes—so long as it is not confused with total lack of family influence. (p. 780)

Now that we have discussed the construct of the home environment; early influences by key child developmentalists; the origins and influences of the Chicago school of research; research by Caldwell, Bradley, and associates; and home environment research from behavioral genetics, we turn to the core of this chapter—a review of home environment research studies in which U.S. racial/ethnic minority groups served as participants. Our discussion is confined to African Americans and Mexican Americans because we did not identify any research investigations that used other minority groups.

Home Environment Research: African Americans

Our search of the existing literature on the relation between home environment and intellectual performance and/or academic achievement identified a small but informative set of eight published studies in which African American children served as participants, either singularly or in conjunction with White participants. These eight investigations are listed in Table 4.1. We also identified six other studies in which African American children were participants (along with White children). These investigations, however, are not useful for our analysis because the findings were not disaggregated by race/ethnicity.[20]

The investigations listed in Table 4.1 have, as a whole, several salient characteristics. First, although the ages of participants range from birth to about 14 years, the strong majority of participants were very young children not enrolled in elementary school (in six of the eight studies). Second, the HOME inventory was by far the home environment measure of choice (in seven of the eight studies). Third, the Stanford-Binet was the most frequently used criterion measure (six of the eight studies, although not exclusively).[21]

Regarding the findings of the eight studies listed in Table 4.1, it was observed, with some exception, that the home environment ratings were positively correlated with African American children's measured intelligence and/or academic achievement. To minimize redundancy, we have selected four of the eight studies for discussion. These four investigations appear to be representative of the corpus of studies in which African American children served as participants. In our discussion, we are guided by this ques-

TABLE 4.1 Home Environment Investigations: African American Participants

Study	Other Group	N^a	Age or Grade[b]	Home Environment Measure	Criterion Measure
Bradley, Caldwell, and Elardo (1977)	White	68 (37)	2 and 3 years (L)	HOME	Stanford-Binet Intelligence Scale
Trotman (1977)	White	50 (50)	9th grade	Wolf interview schedule[c]	Otis-Lennon Mental Ability Test, Metropolitan Achievement Tests, grade point average
Ramey, Farran, and Campbell (1979)	—	71	Birth to 3 years (L)	Lab observations, HOME	Bayley Scales of Infant Development, Stanford-Binet Intelligence Scale
Bradley and Caldwell (1981)	—	60	3 to 8 years (L)	HOME	SRA Achievement Battery
Bradley and Caldwell (1982)	White	36 (36)	6 months to 3 years (L)	HOME	Stanford-Binet Intelligence Scale
Bradley et al. (1989)	White, Mexican American	161 (497, 262)	1.0 to 3.5 years (L)	HOME	Bayley Scales of Infant Development, Stanford-Binet Intelligence Scale, McCarthy's Scales of Children's Abilities
Johnson et al. (1993)	White	56 (65)	2 to 3 years (L)	HOME	Stanford-Binet Intelligence Scale, Peabody Picture Vocabulary Test–Revised
Brooks-Gunn, Klebanov, and Duncan (1996)	White	483[d]	Birth to 5 years (L)	HOME	Stanford-Binet Intelligence Scale, Wechsler Preschool and Primary Scale of Intelligence–Revised

NOTE: HOME = Home Observation for Measurement of the Environment; SRA = Science Research Associates.
a. *n* in parentheses is sample size for "other group."
b. L in parentheses indicates that the study is longitudinal in design.
c. Interview schedule developed and used by Wolf (1964).
d. The *N* for total sample is 483, as *n* values for African American and White samples were not provided.

95

TABLE 4.2 Multiple Correlations Between Status, HOME, and Child IQ

Measure	African American (n = 68)	White (n = 37)	Total (N = 105)
Status	.35	.58	.56
HOME	.58	.62	.74
Status and HOME	.61	.68	.77
Status and HOME minus HOME	.04	.06	.02
Status and HOME minus Status	.27	.10	.21

SOURCE: From "Home Environment, Social Status, and Mental Test Performance," by R. H. Bradley, B. M. Caldwell, & R. Elardo, 1977, *Journal of Educational Psychology, 69,* p. 699. Copyright 1977 by the American Psychological Association. Adapted with permission of first author.
NOTE: HOME = Home Observation for Measurement of the Environment.

tion: Are there racial/ethnic differences or similarities in the pattern of relations between measures of the home environment and intellectual performance?

Bradley, Caldwell, and Elardo (1977)

The investigation by Bradley, Caldwell, and Elardo (1977) appears to be the first published study undertaken to examine the relation between home environment and measured intelligence with an African American school sample of young children.[22] The participants were 68 African American children (39 males, 29 females) and 37 White children (18 males, 19 females) and their respective mothers and fathers. In addition to the HOME inventory, which was administered when the children were 24 months old, status data (i.e., occupation and schooling attainment, father's absence) were collected. When the children were 3 years old, the Stanford-Binet (assumed to be the third revision [Terman & Merrill, 1973]) was administered.

The key findings reported by Bradley et al. (1977) are shown in Table 4.2. It can be seen that HOME was a slightly stronger predictor of children's measured intelligence among Whites (multiple correlation = .62) than among African Americans (multiple correlation = .58). Second, for both groups, HOME was a stronger predictor of IQ than was status. Third, for both groups, HOME predicted intellectual performance about as well as did the combination of HOME and status. Fourth, HOME was a considerably stronger predictor for the African American children; about three times as much variance in IQ could be accounted for by HOME as could be attributed to SES (34% vs. 12%, compared to respective percentages of 58% vs. 34% for the White children).

Regarding different patterns of how HOME subscales predicted IQ across the two groups, Bradley et al. (1977) reported, "Avoidance of Restriction and Punishment appeared less strongly associated with IQ for blacks than whites, while Variety in Daily Stimulation appeared less strongly associated with IQ for whites than blacks" (p. 699). Notwithstanding these patterns of relations, the "subtests of HOME are tapping aspects of the early socialization of intelligence in both black and white homes" (p. 699).

TABLE 4.3   Correlations Between Index of Social Characteristics, Home
Environment, and Student IQ

Measure	African American (n = 50)	White (n = 50)	Total (N = 100)
Index of Social Characteristics	.25	.18	.24
Home Environment	.68	.37	.63

SOURCE: From "Race, IQ, and the Middle Class," by F. K. Trotman, 1977, *Journal of Educational Psychology, 69*, p. 270. Copyright 1977 by the American Psychological Association. Adapted with permission of author.

This pioneering study by Bradley et al. (1977) is very important in that (a) process variables, as measured by HOME, appear to be more accurate predictors (compared to SES) of measured intelligence; and (b) home environment measures predict intellectual performance about equally well across African American and White groups.

Trotman (1977)

In this study, Trotman (1977) sought to examine the relation among SES, home environment, and several criterion variables. The participants were 50 African American and 50 White 9th-grade boys and girls. Trotman controlled for SES by using only middle-class participants. Based on data from Warner's (1949) Index of Status Characteristics (ISC), a measure of social class, all families were within the middle-class range (25 to 50) of the scale (*M* = 36.9 for African Americans, *M* = 36.1 for Whites).[23] For the measure of home environment, Trotman used the 63-question interview schedule developed by Wolf (1964). Criterion variables were scores on the Otis-Lennon Mental Ability Test (Otis & Lennon, 1967), the Metropolitan Achievement Tests (Durost, Bixler, Wrightstone, Prescott, & Balow, 1975), and grade point averages. We restrict our discussion to measured intelligence.

Major findings reported by Trotman (1977) are presented in Table 4.3. It can be seen that for both racial/ethnic groups, home environment, as hypothesized, was more strongly associated with students' intellectual performance than was ISC. Trotman acknowledged, however, that the different correlations between home environment and IQ for African Americans (.68) and Whites (.37) were unexpected. Trotman's observed correlation of .68 for her middle-class African American participant group was nearly identical to Wolf's (1964) observed correlation of .69 "between the scores on the same home environment scale and the IQs of a Midwestern socioeconomically heterogeneous white population" (p. 270). As to how to explain the different correlations between home environment and IQ across the two groups, Trotman called for future research to be undertaken. Nonetheless, her major conclusion lent support to the findings of the previously discussed study by Bradley et al. (1977) that used African American and White participants. Trotman concluded that, for both African Americans and Whites, "traditional indices of SES—such as income, occupation, and living conditions—represent insufficient assessments of important environmental variables related to intelligence test results" (p. 272).[24]

Bradley et al. (1989)

Bradley et al.'s (1989) comprehensive, collaborative, longitudinal investigation involved 11 researchers, six sites in North America, and three racial/ethnic groups (Whites, $n = 497$; African Americans, $n = 161$; Mexican Americans, $n = 262$).[25] The total sample was pooled from a number of investigations begun during the 1970s.[26] Predictor variables included SES (based on Hollingshead's [1975] Four-Factor Index of Social Status) and HOME. Criterion variables were the Mental Development Index (MDI) of the Bayley Scales of Infant Development (Bayley, 1969) and the Stanford-Binet (Terman & Merrill, 1973).

HOME was administered to each participating family when the child was 1 and 2 years old (Infant version) and at 3 years of age (Preschool version). The Bayley scales were administered at 1 and 2 years of age, and the Stanford-Binet was administered at about 3 years of age.

In light of the comprehensive analyses and reporting done by Bradley et al. (1989), space limitations lead us to a brief discussion of major findings. Also, results for the Mexican American sample are discussed in the next section. Table 4.4 presents bivariate correlations between total HOME and the two criterion variables for the three participant groups, matched on the 12-month total HOME scores.[27] Regarding comparisons between African Americans and Whites, the data in Table 4.4 indicate that in two of the six comparisons (12-month HOME and 24-month MDI; 24-month HOME and 24-month MDI), r values for the African American children are significantly lower than those for the White children. With respect to developmental levels, r values between Infant HOME scores and the MDI tend to be about .10 to .20 lower for African Americans than for Whites. By contrast, r values between Preschool HOME scores and IQ were of about the same general magnitude.

Bradley et al. (1989) also sought to examine whether HOME scores added to the prediction of IQ over and above SES. Abbreviated results of the authors' regression analyses, based on matched samples, are shown in Table 4.5. As the data show, for the White group, there was no increase in the adjusted R^2 when SES was added to HOME in predicting Stanford-Binet IQ measured at 3 years of age. By contrast, when HOME was added to SES, the adjusted R^2 increased from .11 to .26. For the African American group, SES did not add a significant proportion of variance to the regression model. When HOME was added to SES, however, the adjusted R^2 increased from −.01 to .32. As such, when SES was used alone to predict IQ, the regression equation was not significant. One of the authors' major conclusions was as follows:

> The finding that HOME scores were of much greater value in predicting IQ among Blacks than were SES indices seems rather easily explainable in terms of historical restrictions in educational and job opportunities among Blacks.[28] Occupational status, in

TABLE 4.4 Correlations Between Total HOME Scores and MDI and IQ

Measure	African American			White			Mexican American		
	MDI (12 months)	MDI (24 months)	IQ (36 months)	MDI (12 months)	MDI (24 months)	IQ (36 months)	MDI (12 months)	MDI (24 months)	IQ (36 months)
12 months HOME	.17	.19[a]	.41	.23	.52[b]	.47[c]	.03	.41	.22[d]
24 months HOME	.13	.28[a]	.49[e]	.22	.62[b,c]	.57[c]	−.08[d]	.24[d]	.14[d,f]
36 months HOME	.25	.33[e]	.50[e]	.09	.46[c]	.42[c]	.17	.05[d,f]	.10[d,f]

SOURCE: From "Home Environment and Cognitive Development in the First 3 Years of Life: A Collaborative Study Involving Six Sites and Three Ethnic Groups in North America," by R. H. Bradley, B. M. Caldwell, S. Rock, K. Barnard, C. Gray, M. Hammond, S. Mitchell, L. Siegel, C. Ramey, A. W. Gottfried, & D. L. Johnson, 1989, *Developmental Psychology, 28*, pp. 222-223. Copyright 1989 by the American Psychological Association. Adapted with permission of first author.

NOTE: Analysis based on matched samples. HOME = Home Observation for Measurement of the Environment; MDI = Mental Development Index.

a. Significantly lower than r for Whites.
b. Significantly higher than r for African Americans.
c. Significantly higher than r for Mexican Americans.
d. Significantly lower than r for Whites.
e. Significantly higher than r for Mexican Americans.
f. Significantly lower than r for African Americans.

99

TABLE 4.5 Multiple Regression Analysis: HOME and SES Predictors of
Children's IQ

Sample	Step and Variable Entered	R	R^2	Adjusted R^2	p
African American	1 / HOME	.60	.36	.33	.001
	2 / SES	.60	.36	.32	.001
	1 / SES	.03	.00	−.01	n.s.
	2 / HOME	.60	.36	.32	.001
White	1 / HOME	.58	.34	.28	.001
	2 / SES	.58	.34	.26	.001
	1 / SES	.37	.14	.11	.02
	2 / HOME	.58	.34	.26	.001
Mexican American	1 / HOME	.38	.14	.04	n.s.
	2 / SES	.39	.14	−.01	n.s.
	1 / SES	.13	.02	−.02	n.s.
	2 / HOME	.38	.14	−.01	n.s.

SOURCE: From "Home Environment and Cognitive Development in the First 3 Years of Life: A Collaborative Study Involving Six Sites and Three Ethnic Groups in North America," by R. H. Bradley, B. M. Caldwell, S. Rock, K. Barnard, C. Gray, M. Hammond, S. Mitchell, L. Siegel, C. Ramey, A. W. Gottfried, & D. L. Johnson, 1989, *Developmental Psychology, 28,* p. 227. Copyright 1989 by the American Psychological Association. Adapted with permission of first author.
NOTE: Analysis based on matched samples. HOME = Home Observation for Measurement of the Environment; SES = socioeconomic status; n.s. = not significant.

> particular, appears largely unrelated to the kinds of parenting the Black children receive. . . . For Whites, there appears to be a closer link between social status and the quality of parenting that a child receives. (p. 233)

Johnson et al. (1993)

In a relatively recent study, Johnson et al. (1993) examined "whether HOME adds to the predictability of child intelligence beyond that provided by SES" (p. 33). The participants were 56 African American and 65 White 3-year-old boys and girls.[29] SES was measured by Hollingshead's (1975) Four-Factor Index of Social Status (i.e., parental occupation and education, marital status). HOME, which served as the measure of home environment, was administered, and mother-child observations were conducted, when the children were 2 years of age. Within 1 month of their 3rd birthdays, children were administered the Stanford-Binet Intelligence Scale: Fourth Edition (SB:FE; Thorndike, Hagen, & Sattler, 1986a) and the Peabody Picture Vocabulary Test–Revised (PPVT-R; Dunn & Dunn, 1981). Here, we confine our discussion to the SB-FE because there is a controversy as to whether the PPVT-R is indeed a measure of general intelligence (see Chapter 6 of this book).

Johnson et al. (1993) reported significant correlations between the two predictor variables and the SB-FE composite. These correlations are presented in Table 4.6. In a break from the pattern seen in most other research studies, SES was more strongly asso-

TABLE 4.6 Correlations Between Socioeconomic Status, Home Observation for Measurement of the Environment, and Child IQ

Measure	African American (n = 56)	White (n = 65)	Total (N = 121)
Socioeconomic Status	.37	.50	.55
Home Observation for Measurement of the Environment	.28	.47	.52

SOURCE: Adapted from Johnson et al. (1993).

TABLE 4.7 Multiple Regressions of Fourth Revision of Stanford-Binet Intelligence: Fourth Edition Composite as Predicted by Socioeconomic Status and Home Observation for Measurement of the Environment

Measure	African American (n = 64)	White (n = 54)	Total (N = 118)
Socioeconomic Status (change in R^2)	.08*	.07*	.09*
Home Observation for Measurement of the Environment (change in R^2)	.00	.06*	.06*
Socioeconomic Status and Home Observation for Measurement of the Environment (R^2)	.15*	.22*	.36*

SOURCE: From "Does HOME Add to the Prediction of Child Intelligence Over and Above SES?" by D. L. Johnson, P. Swank, V. M. Howie, C. D. Baldwin, M. Owen, & D. Luttman, 1993, *Journal of Genetic Psychology, 154*, p. 37. Copyright 1993 by *Journal of Genetic Psychology*. Adapted with permission of first author.
*$p < .05$.

ciated (although only slightly) to measured intelligence for the total sample and the White group than were HOME scores. This pattern also was observed for the African American group, but the magnitudes of the correlations were lower and the SES-HOME discrepancy was more pronounced.

To examine their main research question—whether HOME scores add to the prediction of IQ over and above SES—Johnson et al. (1993) performed multiple regression analyses based on SES and HOME data from the African American, White, and total groups. The criterion variables were the SB-IV composite and the 13 subscales. In Table 4.7, we show the multiple regression of the SB-IV composite as predicted by SES and HOME scores. Several conclusions can be drawn from these data. First, when the range of scores was considerably wide, as seen in the total sample, HOME contributed additional power in predicting intelligence over and above SES. Second, for the White group, HOME also provided additional information, beyond that of SES, in predicting the children's intelligence. Third, for the African American group, HOME also contributed additional predictive power beyond that of SES but not to the same magnitude as seen in the White group.

Johnson et al. (1993) pointed out that their results were in contrast to those of some previous research. For example, in Bradley et al. (1989), which we discussed earlier, it was found that HOME was a good predictor of African American children's intelligence but that SES was not. Johnson et al. (1993) suggested that the difference in results might be related to methodological differences. The authors commented,

> (Bradley et al. (1989) predicted Stanford-Binet LM scores at age 3 [years] with HOME total scores obtained at 12, 24, and 36 months. The presence of the 36-month total score may have had a major effect on the outcome because environmental factors other than those measured by either SES or HOME may have been involved.
>
> The relationship between environmental variables and child intelligence may also be affected by child age. Children in the present sample were very young, and Gottfried and Gottfried (1984) reported that environmental influences appear to be greater later in development. (p. 39)

Home Environment Research: Mexican Americans

As is the case with research on African Americans, our search of existing literature on Mexican Americans resulted in a small set of six published studies.[30] These studies are listed in Table 4.8. We also identified one other published study (Beckwith & Cohen, 1984) in which Mexican American children might have been participants. The authors merely refer to the participants as "English-speaking" and "Spanish-speaking" samples. Given that the ethnic background of neither sample is described, we have decided not to discuss this study.

Regarding the studies listed in Table 4.8, there are some aspects that differ from the characteristics of the investigations in which African Americans served as participants. First, in the Mexican American studies, 50% of the investigations used young elementary school-age children and 50% used very young infants and children not yet enrolled in elementary school. In the African American studies, the strong majority (75%) of the investigations used very young children not yet enrolled in elementary school. Second, the home environment measure used in the Mexican American studies was quite varied, whereas in the African American studies the HOME inventory was used in nearly all of the investigations. Third, the criterion measure used in the Mexican American sample was diverse, whereas the measure used in the African American investigations was fairly limited to the Stanford-Binet.

As was seen in the case of the studies that had African American participants, the set of Mexican American studies also presented results that showed, with some exceptions, positive correlations between the measures of home environment and intellectual performance and/or academic achievement. To minimize redundancy, we discuss four of

TABLE 4.8 Home Environment Investigations: Mexican American Participants

Study	Other Group	N^a	Age or Grade[b]	Home Environment Measure	Criterion Measure
Henderson and Merritt (1968)	—	80	1st grade	Davé and Wolf interviews,[c] sociological factors	Goodenough-Harris Drawing Test, Van Alstyne Picture Vocabulary Test
Henderson (1972)	—	35	3rd grade (L)	Davé and Wolf interviews	California Reading Test
Henderson, Bergan, and Hurt (1972)	White	60 (60)	1st grade	Henderson Environmental Learning Process Scale	Stanford Early Achievement Test, Boehm Test of Basic Concepts
Johnson, Breckenridge, and McGowan (1984)	—	85	12, 24, and 36 months	Home Observation for Measurement of the Environment	Bayley Scales of Infant Development, Stanford-Binet Intelligence Scale
Valencia, Henderson, and Rankin (1985)	—	140	Preschool	Henderson Environmental Learning Process Scale	McCarthy's Scales of Children's Abilities
Bradley et al. (1989)	African American, White	262 (161, 497)	1.0 to 3.5 years (L)	Home Observation for Measurement of the Environment	Stanford-Binet Intelligence Scale, Peabody Picture Vocabulary Test–Revised

a. *n* in parentheses is sample size for "other group."
b. L in parentheses indicates that the study is longitudinal in design.
c. Includes process variables from interview schedules developed and used by Davé (1963) and Wolf (1964).

the six investigations. These four studies are fairly representative of the set of investigations in which Mexican American children served as participants.

Henderson and Merritt (1968)

The investigation by Henderson and Merritt (1968) appears to be the first published study that sought to investigate the relation between the home environment and cognitive performance with Mexican American children. The participants were 80 Mexican American 1st-grade children ("who spoke Spanish and who had Spanish surnames" [p. 101]) attending public schools in Tucson, Arizona. Based on weighted scores from 378 children who were administered two criterion measures, the Goodenough-Harris Drawing Test (Harris, 1963) and the Van Alstyne Picture Vocabulary Test (Van Alstyne, 1961), Henderson and Merritt (1968) designated one sample as the "high-potential group" ($n = 38$) and one sample as the "low-potential group" ($n = 42$). The samples were based on the highest and lowest weighted criterion scores, respectively, and had "relatively high and relatively low potential for success in school[, respectively]" (p. 101).

For selection of the home environment measure (an interview schedule), Henderson and Merritt (1968) built on the works of Davé (1963) and Wolf (1964). Specifically, Henderson and Merritt used six of the previous variables identified by Davé and Wolf: (a) achievement press, (b) language models, (c) academic guidance, (d) activeness of family, (e) intellectuality in the home, and (f) work habits in the family. The authors also used three additional process variables: (a) identification with models, (b) range of social interaction, and (c) perception of practical value of education. In all, the nine process variables contained 33 characteristics. In addition to the process variables, Henderson and Merritt gathered data (through interviews) on what they called "sociological factors" (i.e., family structure, national origin, occupational status of parents, educational status of parents, linkages and interpersonal relations, travel and diversion, value and achievement motive).[31]

Based on a multivariate analysis (Hotelling's T^2), Henderson and Merritt (1968) found that the means on all nine environmental process variables were significantly higher for the high-potential group than for the low-potential group. The authors concluded, "Children in the high potential group apparently came from backgrounds that offered a greater variety of stimulating experiences than were available to those children in the low potential group" (p. 103).

With respect to the findings based on data analyses of the sociological factors, 14 of 32 variables resulted in significant differences (high-potential group greater than low-potential group). These significant differences are presented in Table 4.9. Some of the variables that yielded the strongest significant differences between the high- and low-potential groups were "engage in weekend travel," "periodicals in home," "discrepancy between educational desires and expectations," and "engage in active diversions." It also should be noted that the high-scoring children had fathers with higher occupational status and mothers with higher schooling attainment than did their low-scoring peers. In all, Henderson and Merritt (1968) concluded, "The significant differences . . . are of general theoretical interest because, taken collectively, they support the hypothe-

TABLE 4.9 Comparison of High- and Low-Potential Groups on Sociological Factors

Factor	Chi-Square	Mann-Whitney U Test
Family structure		
Number of children in family		2.45**
Occupational status of parents		
Father employed full-time	7.84**	
Status level of father's work	4.40*	
Educational status of parents		
High- and low-status mothers		2.30*
Linkages and interpersonal relations		
Membership in sodalities	3.91*	
Periodicals in home	10.63**	
Travel and diversion		
Engage in weekend travel	16.24***	
Travel for educational purposes	5.01*	
Travel to gain experiences	4.85*	
Engage in active diversions	8.26**	
Values and achievement motive		
Future orientation in use of finances	6.76**	
Feel unskilled labor undesirable	4.09*	
Discrepancy between educational desires and expectations	10.48**	
High estimate of children's ability	6.10*	

SOURCE: From "Environmental Background of Mexican-American Children With Different Potentials for School Success," by R. W. Henderson & C. B. Merritt, 1968, *Journal of Social Psychology, 75,* p. 104. Copyright 1968 by Heldref Publications. Adapted with permission of first author.
All significant differences favor the high-potential group: *$p < .05$; **$p < .01$; ***$p < .001$.

sis that the low potential children were exposed to a more restricted range of experiences than were their high potential counterparts" (p. 103). The authors also commented,

> The data seem to refute the common assumption that children from families that are "most Mexican" in their behavior and outlook will have the most difficulty in school. Children from families recently arrived from Mexico were no more common in the low potential group than in the high potential group. High potential families also had as much contact with Mexico as families in the contrasting sample. Since high potential children significantly outscored low potential children on a test of Spanish vocabulary[32] . . . . , it appears that high potential families may participate more fully than the families of low potential children in both Anglo-American and Mexican culture. This again indicates, for the high potential group, the greater availability of experiences which facilitate intellectual development. (p. 105)

Henderson, Bergan, and Hurt (1972)

As we have seen, the interview schedules developed by Davé (1963) and Wolf (1964) have proved to be very good predictors of children's academic achievement and intelligence, respectively. The administration of these measures, however, is time-consuming and requires the services of highly trained interviewers. Given these concerns, Henderson, Bergan, and Hurt (1972) sought to develop, using the focused interview instruments developed by Davé (1963) and Wolf (1964), an interview instrument that had the advantages of requiring a short administration time (20 to 25 minutes), could be used by interviewers with a small amount of training, and could be adapted for local use. Thus, the study by Henderson et al. (1972) was designed to develop and validate such an instrument—the Henderson Environmental Learning Process Scale (HELPS).

Participants in Henderson et al.'s (1972) study were 60 low-SES Mexican American 1st-graders and 60 middle-SES White 1st-graders attending several public schools in Tucson. The children's mothers served as the respondents. Depending on their choice, the Mexican American mothers were interviewed in either English or Spanish.

All analyses were based on the combined sample. Via a principal components analysis of 25 HELPS items, five factors were yielded: Factor I (Extended Interests and Community Involvement), Factor II (Valuing Language and School-Related Behavior), Factor III (Intellectual Guidance), Factor IV (Providing a Supportive Environment for School Learning), and Factor V (Attention).[33] Using these five factors as concurrent predictors, Henderson et al. (1972) performed stepwise regressions. Criterion variables were the Stanford Early Achievement Test (SEAT; Madden & Gardner, 1969) and the Boehm Test of Basic Concepts (BTBC; Boehm, 1971).[34] Results of the regression analysis showed that all factors, except Factor III, were statistically significant predictors of scores on the SEAT and the BTBC. Factors II and IV accounted for the majority of the variance in the scores on the criterion measures, leading the authors to conclude that parents who demonstrated higher degrees of valuing language (e.g., reading to the child), valuing school-related behavior (e.g., reinforcing good work), and providing a supportive environment for school learning (e.g., helping the child recognize words or letters during the preschool stage) had children who tended to perform higher on the criterion measures.

Valencia, Henderson, and Rankin (1985)

In this study, Valencia et al. (1985) examined the relation between a number of predictor variables (e.g., family size, child and parent language, parental schooling attainment and location [Mexico or the United States], SES, home environment) and children's intellectual performance. Participants were 140 Mexican American boys and girls enrolled in a number of preschools in Southern California serving children of low-income families (e.g., Head Start). The HELPS (Henderson et al., 1972) served as the measure of the home environment.[35] The McCarthy Scales of Children's Abilities (MSCA; McCarthy, 1972) was the measure of intellectual performance.[36] The children's mothers were administered the HELPS in their preferred language (English or Spanish). Based on several indirect methods (e.g., preschool teacher judgment, child's preference), the children were administered the MSCA in either Spanish (*n* = 88) or English (*n* = 52).[37]

The main analyses performed by Valencia et al. (1985) consisted of (a) a principal components factor analysis (with varimax rotation) of all predictor variables (except HELPS scores) and (b) a simultaneous regression analysis. The factor analysis yielded three factors, which were labeled Factor I (Language/Schooling), Factor II (SES), and Factor III (Family Size). Following this analysis, a multiple regression analysis was computed in which the MSCA General Cognitive Index (GCI) was regressed on the three factor scores and the HELPS total score. The results of this analysis are presented in Table 4.10. It can be seen that the multiple correlation between the three factors and the HELPS and GCI was .42, which explained 17.6% of the variance in the GCI. The semipartial square correlations shown in Table 4.10 indicate that the largest amount of *unique* variance in the GCI was explained by the HELPS (6%), with the next largest proportion explained by the Language/Schooling factor (3%). Neither the SES nor the Family Size variable made a significant unique contribution to predicting the GCI.

The findings of Valencia et al. (1985) support the general pattern of home environment investigations that the variance in scores on measures of intelligence is best accounted for by the variance in scores on proximal measures of the home environment. Furthermore, the authors concluded,

> The findings showed that the nature of intellectual experiences provided by the [Mexican American] parents for their children was not completely controlled by parental language, schooling, SES, and family constellation variables. It was evident, for example, that at least some parents who had lower schooling attainments, who were schooled in Mexico, who spoke only Spanish, and who had Spanish-speaking children provided richer and more varied experiences for their children than did parents with contrasting, presumably more favorable, characteristics along these dimensions. (pp. 328-329)

Bradley et al. (1989)

Bradley et al.'s (1989) investigation, which we discussed earlier in the section titled "Home Environment Research: African Americans," also used Mexican American participants. As such, we refer the reader to that text of the chapter for descriptions of samples, instruments, data analyses, and findings (see Tables 4.4 and 4.5). The bivariate correlations between total HOME and the two criterion variables (MDI and Stanford-Binet IQ) for Mexican Americans revealed a pattern that was substantially different from that for African American and White participants. First, as a whole, there was little relation between HOME scores and mental test scores for the Mexican American group. Second, in six of nine correlations, *r* values were significantly lower than those for Whites. In three of nine correlations, *r* values were significantly lower than those for African Americans. Regarding whether HOME scores added to the prediction of IQ over and above SES for the Mexican American group, the findings in Table 4.5 showed that "none of the regression models provided significant prediction at 3 years" (p. 226).

TABLE 4.10 Zero-Order and Multiple Correlations With General Cognitive Index Regressed on Factor Scores and Henderson Environmental Learning Process Scale

Variable	r	Beta Weight	Semipartial-squared	F	p
Language/Schooling	.23	.17	.03	4.20	.042
Socioeconomic Status	−.23	−.14	.02	2.95	.088
Family Size	−.14	−.11	.01	1.90	.170
Henderson Environmental Learning Process Scale	−.35	.26	.06	9.17	.002

SOURCE: From "Family Status, Family Constellation, and Home Environmental Variables as Predictors of Cognitive Performance of Mexican American Children," by R. R. Valencia, R. W. Henderson, & R. J. Rankin, 1985, *Journal of Educational Psychology, 77*, p. 328. Copyright 1985 by the American Psychological Association. Adapted with permission of first author.
NOTE: $R = .42$; $R^2 = .18$; $F = 7.18$; $p < .001$.

As to why there was little relation between HOME scores and the MDI/IQ scores for the Mexican American sample, Bradley et al. (1989) commented,

> The failure to observe a relationship between home environmental measures and mental test scores in the Mexican American sample is . . . difficult to explain. Part of the difference may reflect the homogeneity of the Mexican American sample with respect to social status. Differential access to education may have contributed to the negligible correlations. It is also possible that the level of parental education below high school has a near uniform effect on children's mental development, thus allowing other factors (e.g., parental IQ) to exert a relatively greater degree of influence (see Laosa, 1987). Another part of the difference may be associated with yet to be explicated cultural factors. (p. 233)

The authors also noted that the failure to find a relation between HOME and mental test scores might be related to differential validity given that the Bayley Scales of Infant Development, the Stanford-Binet, and the HOME were not initially developed on a Mexican American sample. Bradley et al. suggested, "This possibility, coupled with the possibility that some undetected difference in test administration could have occurred from site to site, leaves the reason for the observed difference uncertain" (p. 233).

Conclusions

As noted by Henderson (1981), the research findings on the relation between various measures of the home environment and various measures of intelligence and academic achievement "have been remarkably consistent" (p. 19). Although this body of research has employed correlational analyses and thus causality cannot be assumed, it has led to some interesting theoretical considerations about how home experiences may shape children's intellectual performance and development. A major conclusion derived from home environment research is that an intellectually stimulating and supportive home environment tends to produce a bright child. One needs to be cautious, however, about assuming only a parent (∅) child directionality. It also might be likely that child (®) parent effects exist. That is, bright children might capture the attention of their parents, who in turn might respond to them in stimulating ways.[38] Furthermore, home environment researchers need to be aware of the advancements that behavioral genetics has made regarding the construct of nonshared environment and the contention that home environment measures are phenotypes. To study these important issues, perhaps we will see greater collaboration between developmental psychology and behavioral genetics (Wachs & Plomin, 1991). Nonetheless, numerous

> studies of relations between home environmental processes and intellectual development show that measures of specific characteristics of home environments account for a statistically and educationally significant portion of the variance in children's intellectual performance, and they provide stable predictions over time. (Laosa & Henderson, 1991, p. 183)

Another major finding stemming from the home environment research base is that measures of the home environment (e.g., HOME) are more accurate predictors of children's measured intelligence than is SES. This fairly consistent finding strongly suggests that families in each SES stratum differ considerably in the ways in which the home intellectual climate is structured and in the amounts of stimulation provided. To be sure, this is not to say that SES is not an important predictor variable of children's intellectual performance. As we have seen, SES accounts for a share of the variance in measured intelligence. In this chapter, those studies that have examined both SES and home environment as predictors of children's measured intelligence have used *parental* SES background as the measure of SES. It is important to note, however, what we discussed in Chapter 3. That is, an *individual's* schooling attainment—one index of SES—has been found to correlate quite strongly with one's IQ (Ceci, 1990).[39]

In light of the existing literature on the relation between home environment and African American and Mexican American children's measured intelligence, what implications can be made for understanding their intellectual performance? Although the number of studies amounts to a small corpus, several points can be made.

First, as can be seen in the case of White children and their families, most of the research on minority families has demonstrated significantly positive correlations between measures of the home environment and children's intelligence. The implication here is that the ways in which parents structure the home learning environment and are reciprocally influenced by their children probably are fairly common across racial/ethnic groups. One needs to be very cautious, however, in inferring that there is a "global early environmental action"—the belief (myth) that " 'good' environmental stimulation will uniformly enhance all aspects of cognitive development at all ages for all children and that 'bad' environmental stimulation will uniformly depress all aspects of development at all ages for children" (Wachs, 1984, p. 274). Although there probably are some common ways in which White and minority parents provide stimulating home environments for their children (e.g., parent reading to child), researchers must consider that there may be variations across racial/ethnic groups. Models to understand such possible relativism can be seen in the works of Wachs (1979, 1984), Wachs and Gruen (1982), and Laosa (1981, 1990).

A second implication of this body of research is that the results of those investigations that used SES (in addition to a home environment measure) have the potential to deconstruct the belief that low-SES minority families cannot produce bright children. During the 1960s (Pearl, 1997), and to a lesser extent today (Valencia & Solórzano, 1997), there have been (are) some researchers, educators, and policymakers who believe that low-SES minority families cannot and do not provide intellectually stimulating home environments for their children. The research studies on African American and Mexican American children that we have reviewed refute such "deficit-thinking" views.[40] Given that African Americans and Mexican Americans (and other Latinos, particularly Puerto Ricans) are overrepresented in low-SES brackets (Chapa & Valencia, 1993; Pérez & De La Rosa Salazar, 1993), it is particularly important in considering the home environment in understanding these children's measured intelligence. As Valencia and Solórzano (1997) commented, "The intellectual performance of poor and low-SES children, particularly of color, when examined in the 1960s and early 1970s was seldom [accurately] detected due to methodological shortcomings [i.e., not measuring the home environment]" (p. 194). More specifically, Ginsburg (1986) noted, "The research techniques employed were not sensitive enough to uncover the true extent of poor children's [intellectual] performance" (p. 172).

A third and final implication stemming from research on the relation between measures of the home environment and intelligence with respect to minority children (particularly low SES) has to do with intervention (see, e.g., Henderson, 1981; Laosa & Henderson, 1991; MacPhee, Ramey, & Yeates, 1984). For low-SES minority families (as well as any low-SES families), unless one wins the lottery, SES is difficult to change. On the other hand, it is possible for parents to modify their behavior by acquiring knowledge about how to structure an intellectually stimulating home environment for their children. For example, in a series of field experiments conducted by Henderson and colleagues that used low-SES Mexican American and American Indian (Papago) families (Henderson & Garcia, 1973; Henderson & Swanson, 1974; Swanson & Henderson, 1976), results "showed that parents can be trained effectively to teach specific intellectual skills to their children and to influence their motivation toward academic activities" (Laosa & Henderson, 1991, p. 186). It appears that interventions, based on the home environment research base, can be effective. Interventionists, however, need to be sensitive to cultural variations and to avoid a paternalistic posture.

In closing, research on the connection between the home environment and measures of intelligence provides profound insights into understanding minority children's intellectual performance and development. Paraphrasing Wolf (1964), a pioneer in this field of research, in understanding and predicting children's performance on measures of intelligence, it appears that what parents *do* is of greater importance than what parents *are* in terms of SES.

5

Test Bias

(written with Moises F. Salinas)

Thus far in Part II of this book, we have focused on two major factors (socioeconomic status [SES] and home environment) that have been identified, through research investigations, as partial explanations of White and minority average differences in intellectual performance. In this chapter, we concentrate on a third area of research—intelligence tests themselves. Can intellectual performance differences between Whites and minorities be partially accounted for by cultural bias in tests? The focus here is on the following concern: To what degree, and in what manner, are individually administered standardized tests of intelligence culturally biased against minority students?

This chapter is organized around several sections. The first section discusses the emergence of claims and counterclaims of cultural bias in intelligence tests. The second section explains the concepts of "cultural loading" and "cultural bias." The third section, the core of this chapter, presents a review we undertook of 62 empirical investigations of cultural bias. The chapter closes with a summary discussion of our literature review of cultural bias research findings and then conclusions.

AUTHORS' NOTE: Moises F. Salinas contributed substantially to this chapter and is acknowledged as a coauthor. He is a faculty member in the Department of Psychology at Central Connecticut State University.

The Emergence of Claims and Counterclaims of Cultural Bias in Intelligence Tests

Historically, there have been charges that standardized intelligence tests penalized minority children because the content (cultural and linguistic) favored the exclusively White standardization samples (Klineberg, 1935b; Sánchez, 1934; see also Chapman, 1988; Guthrie, 1976; Thomas, 1982; Valencia, 1997b). Notwithstanding these allegations of cultural bias during the 1920s and 1930s, it was difficult for early critics to sustain a concerted critique against intelligence tests given the entrenchment of hereditarianism and the reification of intelligence tests by influential scholars (Valencia, 1997b; see also Chapter 1 of the present book).

It was not until several decades later that the issue of cultural bias reappeared. As we discussed in Chapter 1, the civil rights movement of the late 1950s and 1960s brought forward national attention to the rights of minorities. Two aspects of this debate were the role of group-administered intelligence testing in shaping curriculum differentiation (e.g., tracking) in the educational mainstream and the role of individually administered intelligence testing in influencing the overrepresentation of minority students in special education (Valencia, 1999).

As allegations of discriminatory assessment escalated, a major influence that brought charges of cultural bias to the national limelight was litigation (Henderson & Valencia, 1985; San Miguel & Valencia, 1998; Valencia, 1999). As we noted in Chapter 1 of this book, *Hobson v. Hansen* (1967) was the *first* case that focused on the legality of using group-administered intelligence tests in the curricular assignment of minority (i.e., African American) students.

Following *Hobson*, there was a round of cases in which minority plaintiffs asserted that intelligence tests were being used discriminatorily, resulting in overrepresentation of minority students in classes for the educable mentally retarded. The most important of these post-*Hobson* cases are (a) *Diana v. State Board of Education* (1970), (b) *Covarrubias v. San Diego Unified School District* (1971), (c) *Guadalupe v. Tempe Elementary School District* (1972), and (d) *P. v. Riles* (*P.* is popularly known as *Larry P. v. Riles*, 1972, 1979).[1] It is particularly interesting that in these post-*Hobson* cases, plaintiffs' arguments of bias largely rested on "armchair analyses" of certain items as being culturally biased. This is important to underscore because these cases were brought forth at a time when a body of scientifically based research on test bias was simply nonexistent. Nonetheless, these four victorious post-*Hobson* cases, which were brought forth by African American, Mexican American, and American Indian plaintiffs, collectively were highly influential in bringing attention to the issue of minority student overrepresentation in educable mentally retarded classes (Henderson & Valencia, 1985; Valencia, 1999).

In addition to the issues of omission of minority students in the standardization samples, allegations of culturally biased items, and failure to appropriately assess language-minority students, critics during the late 1960s and early 1970s argued that the observed mean-level differences of test performance on intelligence tests between, for example, White and African American students were due to an inherent bias in the overall structure of the tests themselves. Proponents of this view argued that there is no a priori reason to believe that mean mental performance should differ across racial/ethnic groups (Alley & Foster, 1978; G. D. Jackson, 1975).

It was not until the mid-1970s and through the 1980s that empirical investigations of test bias were vigorously undertaken (we discuss these trends further later). A number of researchers (Jensen, 1980; Reynolds, 1982b; Reynolds & Kaiser, 1990) challenged the early intelligence test critics, asserting that (a) based on the available *empirical* research, intelligence tests are *not* biased against racial/ethnic minority students; (b) the "mean difference as bias" definition is a misconception (Jensen, 1980, referred to the definition as the "egalitarian fallacy," which is "the gratuitous assumption that all human populations are essentially identical or equal in whatever trait or ability the tests purport to measure" [p. 370]); and (c) the concern of inappropriate standardization sample is a nonissue. Perhaps the strongest statement on the nonbias nature of mental ability tests was voiced by Jensen (1980) in his book *Bias in Mental Testing*. With conviction, Jensen stated in his preface,

> Many widely used standardized tests of mental ability consistently show sizeable differences in the average scores obtained by various native-born racial and social subpopulations in the United States. Anyone who would claim that all such tests are therefore culturally biased will henceforth have this book to contend with.
>
> My exhaustive review of the empirical research bearing on this issue leads me to the conclusion that the currently most widely used standardized tests of mental ability—IQ, scholastic aptitude, and achievement tests—are, by and large, *not* biased against any of the native-born English-speaking minority groups on which the amount of research evidence is sufficient for any objective determination of bias, if the tests were in fact biased. For most nonverbal standardized tests, this generalization is not limited to English-speaking minorities. (p. ix)

Jensen made his bold claim 20 years ago. In this chapter, one of our objectives is to examine the veracity of this assertion in light of a considerably larger research base from which to draw conclusions. On a final note, it should be emphasized that Jensen's *Bias in Mental Testing* presented only one perspective on test bias. The interested reader might wish to examine *Perspectives on Bias in Mental Testing* (Reynolds & Brown, 1984) for diverse views on the subject of test bias.

The Concepts of "Cultural Loading" and "Cultural Bias"

One way in which to understand the concept of "test bias" is to first examine the notion of "cultural loading." Given that all intelligence tests are developed by men and women (i.e., cultural beings), all intelligence tests are inherently culturally loaded in varying degrees, reflecting their developers' experiences, knowledge, values, and conceptions of intelligence. But what does cultural loading mean?

Jensen (1980) contended that the term *cultural loading* should be viewed as a continuum rather than a dichotomy in that some intelligence tests may contain, in varying degrees, culturally loaded items. According to Jensen, "Items that make use of scholastic types of knowledge or skills (e.g., reading or arithmetic) or items in which the fundaments consist of artifacts peculiar to a period, locality, or culture are considered to be 'culture-loaded' " (p. 133). On the other hand, Jensen argued, tests that minimize or have no items of cultural artifacts but that focus on items involving, for example, lines or geometric shapes are culturally reduced.

Although we agree with Jensen's (1980) conception of cultural loading, it needs to be expanded to provide a fuller understanding of intellectual assessment in the cross-cultural context. For us, we find the notion of congruency and related terms (e.g., match, compatibility, fit) useful to conceptualize what a culturally loaded item, subtest, or test might be. If there is a fairly good congruence between the cultural content of an intelligence test and the cultural background of the examinee, then it is likely that cultural loading may be minimized for the individual member of a racial/ethnic group. On the other hand, if there is little or no congruence between the cultural content of the test and the cultural background of the examinee, then there is a good probability that cultural loading may be increased for the individual. Thus, central to our conception of cultural loading is the notion of congruence (or fit or match) between test content and examinee. We conceptualize this as an inverse relation between congruence and cultural loading. As congruence increases, cultural loading decreases.

For an intelligence test to be deemed culturally biased, it must be culturally loaded. A culturally loaded test does not, however, necessarily mean that such a test is culturally biased. In other words, cultural loading on an intelligence test is a necessary, but not a sufficient, condition for the existence of cultural bias. Failure to understand this point by judges in lawsuits that involve (in part) allegations of test bias, as well as by some scholars, has led to considerable confusion about the nature of cultural bias and its penalizing effects on minority students.

The measurement of cultural loading on an intelligence test remains elusive. This is so, Jensen (1980) claimed, because attempts to determine cultural loading on tests have relied on subjective judgments by individuals. The argument goes that, given the fallibility of judgmental analyses, it has been extremely difficult to derive objective determinations on the degree of cultural loading on an intelligence test. Jensen (1980) commented, "I have not found in the literature any defensible proposal for a purely objective set of criteria for determining the cultural-loadedness of individual test items, and perhaps none is possible" (p. 375). Jensen's conclusion, we assert, needs to be tempered with great caution. In reference to Jensen's claim, Hilliard (1984) commented,

> Although it may be true that there are few commonly used procedures for determining cultural bias in testing objectively, it is certainly not the case that descriptions of culture are merely subjective. . . . In order to come to such a conclusion, one must rule out years of systematic work by cultural linguists and cultural anthropologists. (p. 148)

In our view, the subject of cultural loading in intelligence tests frequently is downplayed in the psychometric tradition. Why this is so is not very clear. Perhaps it has to do with the psychometric paradigm itself, which views humans as fairly stable in cognitive functioning and measured intelligence as a culturally independent trait. By sharp contrast, the cultural-psychology perspective asserts that the role of culture in intelligence and its measurement are critical to understand (see, e.g., Miller, 1997).

Although it is quite difficult to operationalize and measure cultural loading in intelligence tests, there are objective and empirical methods to detect *bias* in such tests. When research on test bias began during the late 1970s, many scholars asserted that the construct of test bias can be explained in the context of validity theory and, as such, is an empirical, testable, quantifiable, and scientific matter. Novices to the subject of test bias typically are surprised that the notion of bias is derived from mathematical statistics. In this field, the term *bias* "refers to the *systematic* under- or overrepresentation of a population parameter by a statistic based on samples drawn from the population" (Jensen, 1980, p. 375). In traditional psychometrics, however, bias takes on a related but distinct meaning. It typically is conceived as the systematic (not random) error of some true value of test scores that are connected to group membership (see, e.g., Jensen, 1980; Reynolds, 1982b). Note that group membership is the general referent. It can refer to test bias in the contexts of race/ethnicity, sex, social class, age, and so on. Test bias in the context of race/ethnicity often is referred to as cultural bias, the subject of this chapter. As we discuss later, investigations of cultural bias in intelligence tests can, and sometimes do, employ empirically defined and testable hypotheses and complex statistical analyses (see, e.g., Reynolds, 1982b).

Another perspective on test bias was offered by Cole and Moss (1989). These authors commented that "many definitions of bias [e.g., Reynolds, 1982b; Shepard, 1981, 1982] have been closely tied to validity theory" (p. 205). Cole and Moss asserted that this approach might be limited (e.g., too narrow a focus on specific tests, test features, or methods of investigation). As such, they contended that test bias, which they defined as *"differential validity of a given interpretation of a test score for any definable, relevant subgroup of test takers"* (p. 205), should be investigated, based on several major features of "validation theory":

1. The appropriate unifying concept of test validation is construct validation.
2. Evidence traditionally associated with the terms *content validity* and *criterion validity* should be considered with the construct validity notion.
3. Construct validation should be context based; it should be guided by information about test content, the nature of the examinees, and the purposes the test is intended to serve.
4. Validity evidence should include logical and empirical evidence, divergent and convergent evidence collected within a hypothesis-generating orientation that requires the examination of plausible rival hypotheses. (p. 205)

We are well aware that the conception of test validation as a "unitary" concept has gained widespread consensus by measurement specialists (Cole & Moss, 1989; Cronbach, 1980; Messick, 1981, 1989) and that there have been a number of statistical advancements (e.g., item response theory, structural equation models) to investigate test bias. In our review of the cultural bias literature that we present in this chapter, however, we focus on how the various researchers investigated cultural bias given the theories and technologies with which they had to work. In the heyday of this research (from the mid-1970s to the late 1980s), cultural bias research typically was guided by "validity theory" (i.e., looking at specific aspects of validity [e.g., predictive validity] on specific intelligence tests) and not specifically by "validation theory" as conceptualized by Cole and Moss (1989). Furthermore, the advancements we now have (e.g., item response theory) were fields of statistical technology not yet fully developed during the early years of test bias research.

If test bias is conceptualized as being objective and technical, then what is meant by charges that a particular intelligence test is *unfair* toward, for example, Mexican American students? Some measurement experts assert that the notion of test unfairness (and its opposite, test fairness) denotes a subjective value judgment regarding how test results are used in the decision-making process (e.g., selection procedures). As such, the concepts of test fairness and its reciprocal, test unfairness, "belong more to moral philosophy than to psychometrics" (Jensen, 1980, p. 49). Other scholars have similar, but broader, ideas about values associated with tests and how test results are used. For example, Cole and Moss (1989) placed these concerns in what they referred to as the "extra-validity domain." These authors saw these issues (i.e., values and validity) as linked and not necessarily as flip sides of a coin (i.e., bias and unfairness as conceptualized by Jensen, 1980). Cole and Moss (1989) explained,

> Any investigation is guided by a variety of implicit values, and the validation process is no less subject to investigator values than any other investigation. Values guide the type of questions asked, the type of information collected, and the weight given [to] various types of evidence. In this sense, values are an important implicit part of the conception of validity.
>
> There [is] a host of other types of values held by those with an interest in testing situations. These are values about different types of outcomes from testing: instructional goals, equal opportunity, selection based on qualifications, diverse representation in instructional groups, labels attached to test scores. (p. 204)

Still, other scholars view test bias (i.e., an inherent feature of a test) and test unfairness (i.e., the use of test results) as inextricably linked and not converse constructs (see, e.g., Hilliard, 1984; Mercer, 1984). We return to this point later.

Two final terms pertinent to the study of test bias are the "major group" and "minor group" (Jensen, 1980). Although validity coefficients are helpful in detecting test bias, they are limited in the amount of information provided. As Jensen (1980) commented, validity and bias are separate notions. Validity can apply to a single group, whereas bias always involves a comparison of two or more groups.[2] In the case of a two-group comparison, one population is referred to as the major group and the other as the minor group. These terms are not intended as value judgments. More specifically, Jensen noted,

The major group can usually be thought of as (1) the larger of the two groups in the total population, (2) the group on which the test was primarily standardized, or (3) the group with the higher mean score on the test, assuming the major and minor groups differ in means. (p. 376)

To these distinctions discussed by Jensen, we offer a fourth feature. The major group is the group that the test is believed not to be biased against. On the other hand, the minor group is the group that the test, if determined to be biased, is biased (in most instances) *against*.[3] As such, test bias research can involve various groupings. For example, in the case of test bias investigations (e.g., mathematics) regarding gender, the major group is male and the minor group is female. In investigations of cultural bias—the subject of this chapter—the major group is made up of White participants and the minor group consists of minority participants.

Cultural Bias in Intelligence Tests: A Review of Research Findings

In this section, we present a comprehensive review of published empirical investigations that sought to examine cultural bias in intelligence tests. Our review of pertinent literature is organized and presented as follows: (a) procedures; (b) results: study characteristics; and (c) results: cultural bias findings.

Procedures

Search. Studies (journal articles) to be included in this review were identified using *Current Index to Journals in Education* and *Psychological Abstracts* via computerized searches (ERIC and PsycINFO databases, respectively). Furthermore, studies were identified by scrutinizing the bibliographies of studies already obtained by the authors as well as those studies obtained via the computerized searches. In all, 54 articles were identified for review. In some articles, more than one psychometric property (e.g., reliability, construct validity) of the intelligence test under investigation was examined for cultural bias. As such, a total of 62 investigations were available for review (see Appendix A for a numerical listing of the studies).

Criteria for Selection. The following criteria were used for selection of a germane study for review:

1. The investigation was empirical in design; participants were tested, and the results were analyzed for cultural bias. Only studies published in refereed scholarly journals were examined. In Chapter 9 of this book, we discuss test designers' attempts to study and eliminate cultural bias at the test development stage.

2. The measure under investigation was designed as an *individually* administered, standardized intelligence test. In some studies, more than one intelligence test was administered.

3. There was a major group (i.e., White participants) and a minor group (i.e., minority participants). In some investigations, more than one minor group served as participants.

4. Participants were limited to preschool children and to elementary, middle school, and high school students.

5. The author(s) of the study sought to investigate, for cultural bias, one or more of the following psychometric properties of the intelligence test(s) under examination: (a) reliability, (b) construct validity, (c) content validity, and (d) predictive validity.

Variables Coded for Each Study. The following variables were coded for each study if such data were available:

1. Year of study

2. Psychometric property examined for possible cultural bias

3. Intelligence test(s) under investigation

4. Race/ethnicity of major and minor group(s)

5. Location (state) of study

6. Total number of participants (N) as well as total number of participants for major group (n) and minor group (n)

7. Total number of female (N) and male (N) participants as well as total number of female (n) and male (n) participants for major and minor groups

8. Age of participants

9. SES of major and minor groups and, if reported, how SES was measured

10. Language status of participants and, if reported, whether language was assessed

11. Educational status (i.e., regular classroom, referred for possible special education placement, special education placement) of major and minor groups

12. Research finding of cultural bias in intelligence test(s) (i.e., bias, nonbias, mixed)

Results: Study Characteristics

In this first subsection of the results, we present a summary of the various study characteristics (Nos. 1 to 11) provided on the previous list of variables. In some instances, there is an accompanying table of data.

Year of Study. Table 5.1 presents data on the frequency of the 62 cultural bias investigations by decade and half decade. As noted, research on cultural bias began during the 1970s, a period when 24 (39%) of the 62 investigations were published. The decade of the 1980s accounted for the majority (53%) of publications (33 of 62 investigations). Table 5.1 also shows that during the 1990s only a handful (8%) of the total studies were published. In the discussion section later, we offer possible explanations as to why such research has waned during the 1990s.

TABLE 5.1   Frequency of Test Bias Investigations, by Decade and Half Decade
(*N* = 62)

Decade	*n*	*Percentage of Grand Total*	*Study Numbers*
1970s (*n* = 24, 38.7%)			
1970-1974	8	12.9	4, 12, 13, 14, 15, 16, 19, 50
1975-1979	16	25.8	7, 18, 21, 28, 30, 31, 32, 33, 37, 38, 39, 41, 46, 47, 52, 62
1980s (*n* = 33, 53.2%)			
1980-1984	19	30.7	2, 3, 5, 9, 10, 17, 20, 22, 23, 25, 29, 40, 42, 43, 44, 48, 49, 54, 55
1985-1989	14	22.6	1, 6, 8, 11, 24, 26, 27, 35, 36, 53, 56, 57, 58, 59
1990s (*n* = 5, 8.1%)			
1990-1994	3	4.8	34, 45, 51
1995-1998	2	3.2	60, 61
Grand total	62	100.0	

NOTE: The study numbers provided in Tables 5.1 to 5.6 refer to the investigations listed in Appendix A.

Psychometric Property Examined. Table 5.2 shows how the 62 investigations sorted out by psychometric properties examined for possible cultural bias. Construct validity was the property studied most frequently (37% of the 62 studies). In descending order, the other three psychometric categories examined for bias were predictive validity (29%), content validity (23%), and reliability (11%). In all, it appears that the observed proportions represent a fairly good psychometric mix of cultural bias studies.

Intelligence Test Examined. Table 5.3 contains information on the various intelligence tests that were examined for cultural bias. A total of 15 different intelligence measures, most of which are well known, are listed (author citations are provided in Appendix B).

The data in Table 5.3 show that the Wechsler Intelligence Scale for Children–Revised (WISC-R) was by far the instrument of choice by researchers. Alone, the WISC-R was examined in the *majority* (*n* = 34, 55%) of the 62 studies. When the WISC-R and its predecessor, the WISC, are combined, these two Wechsler scales were investigated in nearly two thirds (65%) of all studies. This finding is not surprising given the clinical popularity of the WISC and WISC-R over the years. In general, the Wechsler tests, compared to other measures of intelligence, are the most widely used tests by practitioners (Harrison, Flanagan, & Genshaft, 1997; Harrison, Kaufman, Hickman, & Kaufman, 1988; Wilson & Reschly, 1996). In addition, these intelligence tests had been heavily researched. Reynolds and Kaufman (1990) noted, "Since the original publication of the WISC [in 1949], there have been more than 1,100 articles published in the scholarly literature, and most of these pertain to the WISC-R" (p. 127). On a final note about the Wechsler instru-

TABLE 5.2   Test Bias Investigations, by Psychometric Property ($N = 62$)

Psychometric Property	n	Percentage of Grand Total	Study Numbers
Reliability	7	11.3	3, 13, 30, 44, 47, 55, 58
Construct validity	23	37.1	1, 4, 5, 8, 9, 10, 11, 17, 18, 19, 22, 27, 31, 37, 40, 45, 48, 50, 52, 53, 57, 61, 62
Content validity	14	22.6	12, 14, 15, 20, 21, 23, 24, 25, 32, 43, 46, 49, 56, 60
Predictive validity	18	29.0	2, 6, 7, 16, 26, 28, 29, 33, 34, 35, 36, 38, 39, 41, 42, 51, 54, 59
Grand total	62	100.0	

ments, we did not identify a single cultural bias investigation in which the third revision of the WISC (WISC-III; Wechsler, 1991) was examined for cultural bias.[4] Nor did we find a single cultural bias investigation of the Stanford-Binet Intelligence Scale: Fourth Edition (SB:FE) (Thorndike, Hagen, & Sattler, 1986a, 1986b).

Regarding the other intelligence tests listed in Table 5.3, the Kaufman Assessment Battery for Children (K-ABC), which was examined in 15% of the total studies, was a distant second in frequency to the WISC-R. The Raven Coloured Progressive Matrices (RCPM) was third in frequency (11%), and the WISC was fourth (10%). Each of the McCarthy Scales of Children's Abilities (MSCA), Raven Standard Progressive Matrices (RSPM), and Peabody Picture Vocabulary Test (PPVT) was examined in several studies. Each of the remaining intelligence tests was examined in one study. With respect to the PPVT, there is some concern that it is not truly a measure of intelligence, yet we included it as part of the present review. Although the PPVT (and its successor, the PPVT–Revised; Dunn & Dunn, 1981) is more of a measure of receptive vocabulary and not of general intelligence (see, e.g., Anastasi, 1988; Cohen, Swerdlik, & Smith, 1992; Sattler, 1992), we kept the PPVT in our corpus of instruments because texts on testing still list this test under "assessment of intelligence" (e.g., Salvia & Ysseldyke, 1988).

Race/Ethnicity of Minor Group. Table 5.4 presents data on the minor group participants by race/ethnicity. What we know about cultural bias in intelligence tests is overwhelmingly based on African American and Mexican American children. African Americans participated in a strong majority (76%) of the 62 investigations, whereas Mexican Americans were participants in a slight majority (58%) of the 62 studies. By sharp contrast, other minority groups were poorly represented—American Indians (10%), Puerto Ricans (3%), and Asian Americans (2%). The American Indian partici-

TABLE 5.3   Intelligence Tests Examined for Bias Investigations

Intelligence Test	n	Percentage of Total[a]	Study Number(s)
1. Wechsler Intelligence Scale for Children–Revised	34	54.8	5, 8, 9, 10, 11, 17, 23, 26, 27, 29, 30, 31, 32, 33, 34, 35, 36, 37, 38, 39, 40, 41, 42, 43, 44, 45, 46, 47, 48, 49, 52, 53, 54, 61
2. Kaufman Assessment Battery for Children	9	14.5	6, 26, 27, 34, 35, 57, 58, 59, 60
3. Raven Coloured Progressive Matrices	7	11.3	3, 12, 13, 14, 15, 16, 55
4. Wechsler Intelligence Scale for Children	6	9.7	7, 20, 21, 41, 50, 62
5. McCarthy Scales of Children's Abilities	4	6.5	18, 22, 25, 56
6. Raven Standard Progressive Matrices	4	6.5	1, 4, 14, 16
7. Peabody Picture Vocabulary Test	3	4.8	12, 13, 15
8. Stanford-Binet: Form L&M, 1960 norms	1	1.6	24
9. Stanford-Binet: Form L-M, 1972 norms	1	1.6	2
10. Differential Ability Scales	1	1.6	51
11. Slosson Intelligence Test	1	1.6	28
12. Wechsler Preschool and Primary Scale of Intelligence	1	1.6	19
13. Woodcock-Johnson Psycho-educational Battery	1	1.6	34
14. Other measures	2	3.2	16[b]

NOTE: Author citations for the intelligence tests are listed, in numerical order, in Appendix B.
a. Divisor for percentage total is 62 (total number of investigations).
b. In Study No. 16, Jensen (1974b) used two additional measures of intelligence: figure copying and memory for numbers.

TABLE 5.4   Minor Group Participants by Race/Ethnicity Based on 62
                 Investigations

Minor Group	n^a	Percentage[b]	Study Number(s)
African Americans	47	75.8	1, 2, 3, 6, 7, 8, 10, 11, 12, 13, 14, 15, 16, 17, 18, 19, 20, 21, 24, 26, 27, 28, 29, 30, 31, 32, 33, 34, 35, 36, 37, 38, 39, 40, 41, 42, 43, 44, 46, 47, 48, 49, 50, 52, 53, 61, 62
Mexican Americans	36	58.1	5, 7, 9, 12, 13, 14, 16, 22, 25, 28, 29, 30, 31, 32, 33, 34, 35, 37, 38, 39, 43, 44, 45, 46, 47, 48, 49, 50, 54, 55, 56,[c] 57, 58, 59, 60, 61
American Indians	6	9.7	23, 37, 38, 39, 43, 44
Hispanics/Latinos (unspecified)	3	4.8	3, 52, 53
Puerto Ricans	2	3.2	4, 6
Asian Americans	1	1.6	51

a. n refers to number of times members of a particular minority group participated in a study.
b. Divisor for percentage computation is 62 (total number of investigations).
c. In Valencia and Rankin (1985, No. 56), the major group was English-speaking Mexican Americans and the minor group was Spanish-speaking Mexican Americans.

pants were Papagos (Study Nos. 37, 38, 39, 43, 44 in Appendix A) and Navajo (No. 23). The Asian American participant group (No. 51) was unspecified regarding ethnic background.

Location of Study. We also gathered data on the states and regions in which the 62 studies were conducted. California was the leader, with 27% of all investigations being done in that state. Arizona was second in frequency (15%), followed by Ohio (7%). All other states accounted for rather small percentages. Surprisingly, more than one third (36%) of all studies failed to mention the states in which the investigations were conducted. Regarding region, nearly one half (47%) of all studies were conducted in the Southwest. The Midwest (13%), Southeast (5%), and Northeast (3%) were considerably lower in frequency.

Number of Participants. Based on the 62 investigations, our calculations show that there were nearly 46,000 total participants.[5] Of these, about 22,000 (48%) were major group children and nearly 24,000 (52%) were minor group children. Sample sizes varied greatly, ranging from studies having fewer than 100 participants (major plus minor groups) to studies having more than 2,000 participants. The most frequent interval of sample size was 201 to 300 ($n = 10$ studies), followed by intervals of 101 to 200 ($n = 9$ studies), 401 to 500 and 1,001 to 2,000 (each with $n = 8$ studies), and 901 to 1,000 ($n = 7$ studies).

In sum, it appears that sample sizes generally were large enough to provide adequate power for the various statistical analyses used to examine cultural bias.

Sex of Participants. Of the total 62 investigations, only 30 (48%) reported the sex of the children. Of these 30 studies, 6 (20%) reported sex data for the total sample and 24 (80%) provided sex data for the major and minor groups.

Age of Participants. About two thirds of the 62 studies reported the age of the participants. We found that the age of participants covered a wide range, spanning 2 years to 17 years. Children ages 5 to 14 years served extensively as participants. Compressed even further, children ages 7 to 11 years served as participants most frequently. The modal age was 10 years; that is, 53% ($n = 33$) of the 62 investigations used participants who were 10 years old (but not exclusively).

SES of Participants. Of the total 62 investigations, 36 (58%) reported data on the SES of participants. We identified those studies that actually measured SES with some index. For the 36 investigations that did report SES, 29 (81%) used some index to measure SES and 7 (19%) did not use an index.[6] Regarding the SES levels of participants for those 36 investigations that reported SES, low-SES children served (but not exclusively) in 21 (58%) of the 36 investigations. Lower middle- and middle-SES children were participants in 28% of the studies. In only 2 studies did upper middle-SES children serve as participants.

Language Status of Participants. Here, language status pertains to those studies in which participant groups had histories of *some* children who might be "language minorities" (i.e., English is a second language). In our analysis, these groups include Mexican Americans, Puerto Ricans, American Indians, and Asian Americans. There were 43 investigations that had participants who might or might not have been language-minority children. Of these 43 studies, the majority ($n = 24, 56\%$) failed to report the language status of the participants. For the 19 (44%) investigations that did report language status of the participants, the children were described as being English speaking in the majority of these studies ($n = 11, 58\%$). In the remaining 8 studies, the participants were said to be bilingual in 5 studies (26%) and limited English proficient in 3 studies (16%). We also identified those investigations that actually assessed language status with some measure. For the 19 studies that reported language status, 14 (74%) used an indirect or a direct measure of language status.

Educational Status of Participants. Children in regular classes were by far the most frequent type of participants, with 45 (73%) of the 62 investigations using participants who were enrolled in regular classes. Only 10 (16%) of the studies had participants who were referred for possible special education placement. Even less frequently studied were children in special education ($n = 1$ study, 2%). Finally, there was a mixed category (e.g., regular/referred participants) with very low frequencies.

Results: Cultural Bias Findings

In this subsection, we provide a summary of test bias findings based on the 62 investigations. Table 5.5 lists the investigations by psychometric property, indicating the numbers and percentages of studies of the grand total that resulted in "nonbias," "bias," and "mixed" findings. By mixed, we mean findings that were not unequivocally nonbiased or biased. As an example of how to interpret the data in Table 5.5, let us examine the content validity category. There are 14 investigations of test bias, and 7 (11%), 2 (3%), and 5 (8%) of these studies' results were deemed to be nonbiased, biased, and mixed, respectively. Another observation that can be gleaned from Table 5.5 is that 44 (71%) of the total 62 investigations were concluded to be nonbiased in interpretation of results, whereas 18 (29%) of the studies were deemed to be biased ($n = 8$, 13%) or mixed ($n = 10$, 16%) in findings.

Table 5.6 provides a closer examination of cultural bias findings within each category of the four psychometric properties. For those 7 investigations that examined possible cultural bias in reliability, all of them concluded that the various intelligence tests were nonbiased. Table 5.6 indicates that for the 23 construct validity studies, 21 (91%) resulted in conclusions of nonbias, 0 in bias, and 2 (9%) in mixed. Regarding the 14 content validity investigations, 7 (50%) found the intelligence tests under study to be nonbiased, whereas 2 (14%) and 5 (36%) were observed to contain biased and mixed findings, respectively. Finally, in the category of predictive validity, as in the case of content validity, the overall findings are fairly equivocal, compared to the findings for reliability and construct validity. For these predictive validity investigations, 9 (50%) of the 18 studies were considered nonbiased in findings. Half of the studies ($n = 9$) were deemed biased ($n = 6$, 33%) or mixed ($n = 3$, 17%) in findings.

In sum, the data in Table 5.6 strongly suggest that the existence of cultural bias in the 62 investigations appears to be a function of the type of psychometric property studied. In descending order, findings of nonbias in the various intelligence tests that constituted the corpus of the present study were observed in investigations of (a) reliability, (b) construct validity, and (c) content validity and (d) predictive validity (the last two were tied in rank).

Next, we report an even closer examination of test bias findings by describing, in detail, results for each of the four psychometric properties. For each property, we first describe the study characteristics, and then we report the findings of bias.

Reliability

Table 5.6 lists seven investigations that examined various intelligence tests for possible bias in reliability. The key study characteristics are as follows:

1. Mexican American students participated in six studies (Nos. 13, 30, 44, 47, 55, and 58), African American students in five studies (Nos. 3, 13, 30, 44, and 47), unspecified Hispanics in one study (No. 3), and American Indians (Papagos) in one study (No. 44).

TABLE 5.5 Test Bias Findings ($N = 62$)

Psychometric Property	Total		Nonbias			Bias			Mixed		
	n	Grand Total (%)	n	Grand Total (%)	Study Numbers	n	Grand Total (%)	Study Numbers	n	Grand Total (%)	Study Numbers
Reliability	7	11.3	7	11.3	3, 13, 30, 44, 47, 55, 58	0	0.0	—	0	0.0	—
Construct validity	23	37.1	21	33.9	4, 5, 8, 9, 10, 11, 17, 18, 19, 22, 27, 31, 37, 40, 45, 48, 50, 52, 53, 57, 62	0	0.0	—	2	3.2	1, 61
Content validity	14	22.6	7	11.3	14, 15, 20, 21, 25, 46, 49	2	3.2	56, 60	5	8.1	12, 23, 24, 32, 43
Predictive validity	18	29.0	9	14.5	2, 6, 16, 29, 36, 38, 39, 41, 42	6	9.7	7, 34, 35, 51, 54, 59	3	4.8	26, 28, 33
Grand total	62	100.0	44	71.0		8	12.9		10	16.1	

125

TABLE 5.6 Test Bias Findings: Within Psychometric Categories ($N = 62$)

Psychometric Category	Nonbias			Bias			Mixed		
	n	Category (%)	Study Numbers	n	Category (%)	Study Numbers	n	Category (%)	Study Numbers
Reliability (n = 7)	7	100.0	3, 13, 30, 44, 47, 55, 58	0	0.0	—	0	0.0	—
Construct validity (n = 23)	21	91.3	4, 5, 8, 9, 10, 11, 17, 18, 19, 22, 27, 31, 37, 40, 45, 48, 50, 52, 53, 57, 62	0	0.0	—	2	8.7	1, 61
Content validity (n = 14)	7	50.0	14, 15, 20, 21, 25, 46, 49	2	14.3	56, 60	5	35.7	12, 23, 24, 25, 32, 43
Predictive validity (n = 18)	9	50.0	2, 6, 16, 29, 36, 38, 39, 41, 42	6	33.3	7, 34, 35, 51, 54, 59	3	16.7	26, 28, 33

2. With the exception of Study Nos. 44 (which included regular and special education children) and 3 (which did not report educational status), all major and minor group participants were enrolled in regular classrooms.

3. In the seven studies, the WISC-R was the test of choice for bias examination in three instances (Nos. 30, 44, and 47), the RCPM also in three studies (Nos. 3, 44, and 48), and the PPVT (No. 13) and K-ABC (No. 58) in one study each.

Reliability is a critical psychometric property of intelligence tests. As such, it is not surprising that researchers have investigated whether reliability estimates of intelligence tests are acceptably high for major and minor groups. In general, "Reliability refers to the *consistency* of measurement—that is, how consistent test scores or other evaluation results are from one measurement to another" (Gronlund & Linn, 1990, p. 77). In cultural bias investigations of reliability, the research question is as follows: Is the intelligence test consistently accurate, at a comparable level, for both the major and minor groups? As seen in Table 5.6, all seven investigations resulted in conclusions of nonbias. This is not surprising. When major intelligence tests are developed (e.g., the WISC-R), test developers go to great detail in establishing sound reliability. Highly trained psychometricians expend considerable time and energy in ensuring that "true score" components are large and error components are slight. In establishing reliability estimates during test construction, it is critical that items intracorrelate at fairly high magnitudes (i.e., internal consistency is sought). In short, the finding of nonbias in reliability in the seven studies probably is connected to the tests' histories in which acceptable levels of reliability were established for the standardization samples during the test construction stage.

Notwithstanding the finding of nonbias in the seven reliability studies, it is important to examine each study individually because the magnitudes of the reliability estimates sometimes vary across major and minor groups. For example, Carlson and Jensen (1981, No. 3), in a study of the reliability of the RCPM with 783 White and Hispanic children (ages 5.5 to 8.5 years), reported that the reliability estimate (Cronbach's alpha) for the Anglo group was higher (r_{tt} = .83) than for the Hispanic group (r_{tt} = .76) and the Black group (r_{tt} = .76).[7] Similar patterns can be seen in several other studies (i.e., Nos. 13, 30, and 58).

Regarding how the authors of these seven reliability studies concluded that the intelligence tests were equally reliable (i.e., nonbiased) for the major and minor groups, the majority (Nos. 3, 13, 30, 44, and 47) merely visually inspected the observed reliability coefficients and drew conclusions of equivalence; that is, no statistical tests were run. These conclusions should be tempered with caution, however, because such differences in reliability estimates could be explained by differences in variability between the major and minor groups. Two studies (Nos. 55 and 58), however, used the preferred technique of testing for a significant difference between the observed reliability coefficients.[8]

On a final note, a caveat is in order. The reliability estimate of an intelligence test is a necessary, but not sufficient, condition for validity. As Jensen (1980) cogently commented,

A very high reliability coefficient in both groups [major and minor] tells us that in both groups the test is measuring whatever it measures with reasonable accuracy. In such a case, the reliability coefficients are not indicative of bias, but neither can they prove its nonexistence, as the test could possibly measure different things in the two groups with comparably high reliabilities. (p. 430)

As such, we now turn to the question of bias in validity.

Construct Validity

Table 5.6 lists 23 studies that investigated a number of different intelligence tests for possible bias in construct validity. Key study characteristics are as follows:

1. Of the 23 construct validity studies, African Americans were by far the most represented minority, serving as participants in 17 studies (74%). Mexican Americans were second in frequency, participating in 10 studies (44%), with unspecified Hispanics/Latinos participating in 2 studies (Nos. 52 and 53) and Puerto Ricans (No. 4) and American Indians (Papagos; No. 37) in only 1 study each.

2. The majority of the construct validity investigations focused on participants enrolled in regular classrooms ($n = 18, 78\%$). Participants in four studies (Nos. 5, 9, 17, and 52) were referred for assessment for possible special education placement, and participants in one study (No. 11) were enrolled in a special education program.

3. Overwhelmingly, the WISC-R was the test of choice for investigations that examined possible bias in construct validity. The WISC-R was used in a total of 15 of the 23 studies (65%). In addition, the WISC was used in 2 studies (Nos. 50 and 62), and the Wechsler Preschool and Primary Scale of Intelligence (WPPSI), a downward extended version of the WISC scale, was used in 1 study (No. 19). That brings the total of construct validity studies using a Wechsler test to 18 or 78% of the total investigations. Other tests examined were the RSPM in 2 studies (Nos. 1 and 4), the MSCA in 2 studies (Nos. 18 and 22), and the K-ABC in 2 studies (Nos. 27 and 57).

To be sure, reliability is an important psychometric property of a test. The notion of validity, however, is "the most central concept in the whole testing enterprise. It is the main goal toward which reliability and stability are aimed" (Jensen, 1980, p. 297). Validity can be defined in many ways. The concept we find most useful is that provided by Jensen (1980): "*A test's validity is the extent to which scientifically valuable or practically useful inferences can be drawn from the scores*" (p. 297). Jensen's point about "inferences" is very important. For the sake of convenience, we often refer to "the validity of a test." Actually, the correct reference is to the *interpretations* to be made from test scores. Thus, validity always is specific to a certain use or interpretation (Gronlund & Linn, 1990).

Two other points about validity should be borne in mind. First, as noted by Gronlund and Linn (1990), validity is a *matter of degree,* not an all-or-nothing element. It is preferable to specify the degree of validity (e.g., high, moderate, low). Second, as we introduced earlier, validity is a unitary concept. Although in this article we speak of several types of validity, we do so for the ease of reporting. Technically, validity is best viewed as a unitary concept that stems from various forms of evidence (Cole & Moss, 1989).

From a *scientific* standpoint, construct validity is the most important type of validity (Jensen, 1980). This is particularly true in the case of those developers of intelligence tests who begin with some theory of intelligence that guides their development of items and subtests. Such theoretical formulations provide us with psychological understanding of what the test attempts to measure. For example, in the development of the K-ABC, Kaufman and Kaufman (1983a) drew from clinical neuropsychology, cerebral specialization, and cognitive psychology to construct their model of sequential and simultaneous information processing (Kamphaus & Reynolds, 1987).

In cultural bias investigations of construct validity, the research question is as follows: Does the intelligence test measure similar constructs for both the major and minor groups? More specifically, "Bias exists in construct validity when a test is shown to measure different hypothetical traits (psychological constructs) for one group than another or to measure the same trait but with differing degrees of accuracy" (Reynolds, 1982b, p. 194). To investigate whether a particular intelligence test has structural similarity in constructs across populations, researchers often use one of the more popular and empirically necessary approaches, factor analysis (Reynolds, 1982a). If one finds that for the test has factorial similarity for the major and minor groups, then it can be concluded, to some degree, that the test is not biased in construct validity.[9] Although modern techniques related to the measurement of construct validity have been developed during recent years (Cole & Moss, 1989), nearly all the investigations of construct validity identified in the present review make use of traditional factor-analytic techniques. It must be underscored, however, that although the establishment of construct validity via factor analysis is useful, such an approach is not sufficient evidence of construct validation (Messick, 1989).

The construct validity studies identified for review overwhelmingly had conclusions of nonbias. In all, 21 of the 23 studies (91%) concluded that there was no bias, whereas only 2 studies (9%) found mixed results. As examples, it is useful to comment on a couple of studies that found nonbias in construct validity. Silverstein (1973, No. 50), in a study comparing the factor structure of the WISC for 1,310 White, African American, and Mexican American children (ages 6 to 11 years), examined the invariance of a two-factor solution (i.e., Verbal Comprehension and Perceptual Organization) for the test. He found factorial similarity in the WISC for the three groups and, therefore, concluded that "the test measures the same abilities in Anglo, Black, and Chicano children" (p. 410). In another study, Sandoval (1982, No. 48) compared the factor structure of the WISC-R for a group of 953 White, African American, and Mexican American children (ages 5 to 11 years). He also found very few factor differences among the three groups and concluded that "the WISC-R is measuring the same abilities in these three groups" (p. 200).

It is important to note, however, that although a large number of studies reported nonbias, some interesting differences in factor structures for White and minority participants were found. For example, Valencia and Rankin (1986, No. 57) found in their study that two factors on the K-ABC (Achievement and Simultaneous Processing) indicated factorial similarity for 200 White and Mexican American children (10.0 to 12.4 years of age). The authors reported, however, that the third factor (Sequential Processing) was substantially different across groups. In another study, Mishra (1981, No. 22) concluded that the factor structures of the MSCA were quite similar for 312 White and Mexican American children (5.6 to 7.6 years of age). Mishra pointed out, however, that the results of the factor analyses need to be interpreted with caution because of some ethnic group differences in the loading patterns of the scales. Also, Reschly (1978, No. 37), in a study comparing factor structures of the WISC-R for 950 White, African American, Mexican American, and American Indian (Papago) children (6.3 to 15.9 years of age), reported that the common two-factor solution was found for all groups (Verbal Comprehension and Perceptual Organization) and, given that the two factors were highly similar for all groups, "the construct validity of the WISC-R as an *intellectual* measure for different groups is supported" (p. 422). However, the commonly found third factor of the WISC-R (Freedom From Distractibility) did not show similarity given that "the three-factor solutions for Native American Papagos and Blacks were clearly different from the previous reported three-factor solutions and should, perhaps, *not* be interpreted at all" (p. 420).

Regarding the investigations that reported mixed findings, Bart, Rothen, and Read (1986, No. 1), in an investigation comparing item hierarchies on the RSPM for 808 White and African American children (10.0 to 12.4 years of age), found that even though item hierarchies for both groups were similar for complex items, they were significantly different for simpler items. Valencia, Rankin, and Oakland (1997, No. 61), in a mixed study exploring the factor structures of the WISC-R for 451 White, African American, and Mexican American children (6.8 to 14.6 years of age), reported that although a three-factor solution emerged for all groups, Factors 2 and 3 (Perceptual Organization and Freedom From Distractibility, respectively) were reversed in order for the minority participants, compared to those for Whites.

Overall, it can be concluded that even though the overwhelming majority of the construct validity studies reported findings of nonbias, some of the results raise concerns about claiming factorial invariance across ethnic groups.

Content Validity

Table 5.6 shows 14 studies that investigated various intelligence tests for possible bias in content validity. The key study characteristics are as follows:

1. African Americans were participants in 10 studies (Nos. 12, 14, 15, 20, 21, 24, 32, 43, 46, and 49). Mexican American students were a close second, participating in 9 investigations (Nos. 12, 14, 25, 32, 43, 46, 49, 56, and 60). American Indian (Navajo and Papago) students served as participants in 2 studies (Nos. 23 and 43, respectively).

2. With the exceptions of Study Nos. 20 (which used referred students), 24 (which included regular and low-birthweight children), and 43 (which included regular and special education students), all major and minor participants in the remaining 11 studies were enrolled in regular classes.

3. In the 14 studies, the WISC-R was the intelligence test of choice in 5 investigations (Nos. 23, 32, 43, 46, and 49), followed by the RCPM (Nos. 12, 14, and 15). Next in frequency (tied) were the PPVT (Nos. 12 and 15), the MSCA (Nos. 25 and 56), and the WISC (Nos. 20 and 21). Last in frequency (tied) were the RSPM (No. 14), Stanford-Binet (SB, 1960 edition, No. 24), and the K-ABC (No. 60).

In a general sense, content validity refers to "the extent that the items in the test are judged to constitute a representative sample of some clearly specified universe of knowledge or skills" (Jensen, 1980, p. 297). The content validity of a test can be established through two techniques. First, there is the use of judgmental methods in which experts visually scrutinize items and, through consensus, offer conclusions whether a particular item might be biased against, for example, females and racial/ethnic minorities (see, e.g., Tittle, 1982). These judgmental methods fall under the rubric of face validity. The second major type of test bias analysis is statistical (sometimes referred to as "empirical" [Shepard, 1982]). Such a method incorporates sophisticated item-by-group approaches (e.g., race/ethnicity) such as the delta-plot method (Angoff, 1982) and the chi-square and latent trait approaches (Ironson, 1982).

As noted by Berk (1982), the issue of bias in content validity is particularly important in test bias research. First, test items form the most basic level of content analysis; thus, such studies are necessary for all tests.[10] Second, it is the item level of tests at which most charges of bias against minorities are directed (Hilliard, 1979; G. D. Jackson, 1975; Jensen, 1980; Koh, Abbatiello, & McLoughlin, 1984, No. 20; Williams, 1971). Third, test developers, via bias investigations, can debias tests during the early stages of test construction.

In cultural bias studies of content validity, the research question is as follows: Are the items (or subtests) of the intelligence test at similar difficulty levels for both the major and minor groups? Such a broad conception can lead to a specific, empirically defined, and testable definition of content bias for intelligence tests. Reynolds (1982b) proffered the following:

> An item or subscale of a test is concluded to be biased in content when it is demonstrated to be relatively more difficult when the general ability level of the groups being compared is held constant and no reasonable theoretical rationale exists to explain group differences on the item (or subscale) in question. (p. 188)

Common statistical techniques to investigate possible item bias include analysis of variance (i.e., item-by-group interactions) and correlation of rank order of item difficulty across groups (e.g., Study Nos. 21 and 24). There has been some concern, however, that item-by-group interaction procedures can lead to inaccurate conclusions (e.g., the creation of false evidence of bias or even the masking of real bias [Camilli & Shepard, 1987; Cole & Moss, 1989]). Other investigations (e.g., Study Nos. 56 and 60) have employed the item-group (partial correlation) method in which ability is controlled.[11] A major problem in investigations of item bias lies in the interpretation of biased findings. That is, although it is possible to detect biased items using statistical techniques, it is very difficult "to deduce any logical [i.e., psychological] reason why this is true" (Burrill, 1982, p. 174; see also Valencia, Rankin, & Livingston, 1995, No. 60).

Table 5.6 lists 14 investigations of possible bias in content validity. Of this total, 50% ($n = 7$) resulted in findings of nonbias, whereas the remaining 50% had findings of bias ($n = 2$, 14%) and mixed results ($n = 5$, 36%). Compared to the bias investigations of reliability and construct validity, which overwhelmingly resulted in findings of nonbias, the category of content validity resulted in equivocal findings. This is not surprising. Although test developers sometimes attempt to debias tests for item bias at the construction stage (e.g., the case of the K-ABC [Kamphaus & Reynolds, 1987]; see also Chapter 9 of the present book), once a test becomes available for clinical use or research on its psychometric properties, the issue of item-by-group interactions still can arise. This is so because although cultural bias at the item level can be minimized during test construction, there is no guarantee that bias will not be an issue when empirical investigations are done in the field. In short, although it is possible to establish nonbias of items at the test development level, it is difficult to generalize beyond the norming process. The standardization sample is representative of only one population and cannot account for all the permutations and variables that exist in actual testing situations and research investigations in the "real world."

Regarding the seven content validity studies that resulted in findings of nonbias, African American participants were well represented (participants in all studies except No. 25). Mexican American children served as participants in four studies (Nos. 14, 25, 46, and 49). No Puerto Ricans, American Indians, or Asian Americans participated in this set of investigations. As expected, the popular Wechsler instrument was the most prominent intelligence test examined for bias in content validity (the WISC in Study Nos. 20 and 21 and the WISC-R in Study Nos. 46 and 49). The RCPM and the PPVT were examined in two studies (Nos. 14 and 15), and the MSCA was examined in one study (No. 25). Although these seven investigations found the various tests to be nonbiased at the item level, such a general conclusion needs to be interpreted with caution. For example, Murray and Mishra (1983, No. 25), in their study of item bias in the MSCA (using 118 White and Mexican American children, 5.6 to 8.6 years of age) found three items (7%) of the total Verbal Scale items to be biased against the Mexican American participants. Likewise, Sandoval, Zimmerman, and Woo-Sam (1983, No. 49), in their study of the WISC-R in which 385 White, African American, and Mexican American children (7.5 to 10.5 years of age) served as participants, reported that several items on the Vocabulary subtest were more difficult for African American and Mexican American children and that several items on the Information subtest were more difficult for Mexican American children.

One of the most interesting investigations in the content validity's nonbias set is the study by Koh et al. (1984, No. 20). The authors sought to investigate empirically whether seven items on the WISC, which were singled out by Judge John F. Grady in the *Parents in Action on Special Education v. Joseph P. Hannon* (*PASE*; 1980) case, were actually biased.[12] Using a percentage passing analysis, Koh et al. (1984, No. 20) found no significant differences between 360 White and Black children's (7.0 to 15.7 years of age) pass rates on any of the seven contentious items.[13] The findings of the Koh et al. study are a reminder of the weakness of strictly relying on judgmental methods to identify biased items. Commenting on the need to examine empirically all of the items on the WISC for possible cultural bias, the authors concluded, "Such a study would put to rest any doubts reflected in 'armchair' inspection of the intelligence test items" (p. 94).

With respect to the five investigations listed in Table 5.6 that had mixed findings of bias in content validity, African Americans participated in all studies except No. 24, Mexican Americans were participants in three studies (Nos. 12, 32, and 43), and American Indians were participants in two studies (Navajos in No. 23 and Papagos in No. 43). As is the case throughout this review, the Wechsler instrument was the most frequently examined test for the category of content validity (WISC-R in Study Nos. 23, 32, and 43). The PPVT and RCPM were the instruments of choice in Study No. 12, and the SB: L&M was examined in Study No. 24.

The five investigations of possible bias in content validity that made up the mixed set are the most complex to understand. In some instances, bias was detected on items in one subtest but not in others, or bias was detected for one minor group but not another. For example, Jensen (1974a-1, No. 12), in a study of item bias in the PPVT in which 1,663 White, African American, and Mexican American children (6 to 12 years of age) were participants, reported that there were grounds for suspected cultural bias in the instrument against Mexican American boys and girls but not against African American boys and girls. Mishra (1982, No. 23), who examined the WISC-R with 80 White and Navajo participants (9.1 to 11.4 years of age) for possible item bias, analyzed the 79 items that make up the Information, Similarities, and Vocabulary subtests. Based on a percentage passing analysis, Mishra reported that for the Navajo children, 15 (19%) of the 79 items were significantly more difficult and, hence, biased.[14] Mishra did not advance a possible cultural or psychological explanation as to why the 15 items were biased. He did, however, proffer a conjecture about limited opportunity to learn: "The findings may lend some support to the notion that the experiential background of the Navajo subjects may not have provided them adequate opportunity for them to learn the content of certain vocabulary items" (p. 463). In their bias investigation of the WISC-R, Ross-Reynolds and Reschly (1983-1, No. 43) used three minor groups—698 African American, Mexican American, and American Indian (Papago) children (6.3 to 15.9 years of age)—and, of course, the White major group ($n = 252$). Based on transformed item difficulties and outlier analysis, the authors reported that no items were biased against the African American sample and that only one item was biased against the Mexican American sample. By sharp contrast, nearly one third of the items on the Verbal Scale subtests were found to be biased against the Papago group. Ross-Reynolds and Reschly suggested that, for the most part, the greater difficulty of the identified items was due to ceiling effects. The authors did note, however, that several biased items were in the middle of the sequence of items. "These items . . . clearly reflect an items-by-groups interaction and would be considered by most observers, both critics and proponents of tests, to be biased against NAP [Native American Papago] children" (p. 145). The authors offered no cultural or psychological explanation as to why these several items were biased.

Table 5.6 lists two studies that reported biased findings in content validity. Valencia and Rankin (1985, No. 56), using 304 Mexican American preschool children (2.7 to 6.3 years of age), modified the conventional test bias investigation approach (i.e., White is the major group, racial/ethnic minority is the minor group) by using Mexican American English-speaking and Spanish-speaking samples as the major and minor groups, respectively. The authors sought to investigate possible item bias on the conventional MSCA and a carefully constructed Spanish version. Using the item-group correlation technique and controlling for cognitive ability,[15] the authors reported that 23 (15%) of 157 MSCA items were biased—17 against the Spanish-speaking group and 6 against the English-speaking group. The pattern of item bias for the latter group was not discernible because the biased items cut across dissimilar subtests.[16] Also, all but one of the subtests that had items biased against the English-speaking group also contained items biased against the Spanish-speaking group. This resulted in a canceling-out effect for both language groups. By contrast, of the 17 items biased against the Spanish-speaking group, 12 (71%) clustered in two subtests: Verbal Memory I (9 biased items) and Numerical Memory I (3 biased items). Valencia and Rankin reported that this was a surprising finding given that the 12 items were rather simple in cognitive demands. That is, the items measured short-term serial-order memory. The authors suggested that these 12 items were biased against the Spanish-speaking children because of linguistic differences between English and Spanish. The 12 Spanish word and number items were longer in syllable length and contained more acoustic similarity, compared to the 12 English word and number items. As such, the Spanish-speaking children might have experienced information overload on short-term memory and, hence, a reduction in recall.[17] This test bias study by Valencia and Rankin is particularly noteworthy in that it appears to be the only investigation that has proffered a cultural/psychological explanation (here, a linguistic one) for the observed test bias.

The other study listed in Table 5.6 that found bias in content validity was by Valencia et al. (1995, No. 60). The K-ABC, which was administered to 200 White and Mexican American children (10.0 to 12.5 years of age), was the instrument examined (both the intelligence and achievement tests). Of the 120 items comprised by the Mental Processing Scales (the intelligence test of the K-ABC), 17 (14.2%) were identified as biased; the same bias detection method used in Valencia and Rankin (1985, No. 56) was used here. The strong majority of items (13 of 17, 76.5%) were biased against the Mexican American group, whereas 4 of 17 (23.5%) of the items were biased against the White group. Three major findings were reported. First, in the majority (75%) of the subtests, the number of biased items was very small (mode of 1 item).[18] Second, a pattern of biased items on six of eight subtests was not discernible, thus suggesting a random dispersion of bias. Third, of the 17 biased items, 10 (58.8%) clustered in two subtests (Word Order and Matrix Analogies), and the vast majority (8 of 10, 80%) were biased against the Mexican American sample. The authors were unable to offer a cultural or psychological explanation for the observed bias.

In addition to their bias examination of the K-ABC Mental Processing Scale, Valencia et al. (1995, No. 60) investigated possible bias in content validity of the K-ABC individually administered Achievement scale, a totally separate scale. The scale contains items that center on the children's knowledge of acquired facts and skills learned in the school and home environments. Although our focus in this review is on individually administered intelligence tests, we think that it is important to discuss the K-ABC Achievement scale bias investigation by Valencia et al. in that the authors, based on their findings, discussed the need to reconceptualize the notion of test bias. Of the 92 Achievement scale items examined for bias, a very large number of items ($n = 58$, 63%) were found to be biased—all against the Mexican American group.[19] In comparison to the other item bias investigations, the percentage of biased items identified by Valencia et al. is extraordinarily high. The authors contended that, given the extreme difficulty in offering a psychological explanation (i.e., item-by-group interactions) for cultural bias, it might be necessary to look elsewhere, particularly in learning opportunities experienced by White and minority children. In Valencia et al., the authors suggested that possible differences in learning opportunities (language based, SES based, school segregation) experienced by the White and Mexican American children might help to explain the observed item bias. For example, the Mexican American and White participants in the Valencia et al. study resided in a central California city of moderate size (population 80,000 in 1980). Yet, all the Mexican American children ($n = 100$) attended segregated schools. Of the White children, about one third ($n = 37$) also attended the Mexican American schools, whereas the remainder ($n = 63$) were enrolled in predominantly White schools in a contiguous school district. As part of their differential learning opportunities thesis, Valencia et al. suggested that, given the strong correlation between Mexican American school segregation and poor academic achievement (e.g., low test scores [Donato, Menchaca, & Valencia, 1991; San Miguel & Valencia, 1998; Valencia, 1991]), it also might be that the Mexican American participants in their study had inferior schooling due, in part, to reduced opportunities to learn. As such, "The construct of 'opportunity to learn' (OTL) may have some heuristic value in theory generation with respect to explaining culturally biased items" (Valencia et al., 1995, p. 165). Or, in more direct terms, what frequently has been termed *item bias* might better be labeled *instructional bias* (Linn & Harnisch, 1981). Valencia et al. (1995) summarized matters as follows:

> It must be underscored that the use of the OTL construct as an explanatory base for the observed item bias findings in achievement (particularly reading and arithmetic) is speculative on our part. No data were collected that investigated possible between-ethnic group differences (i.e., the 100 Mexican American students attending the segregated Mexican American schools vs. the 63 White students attending the segregated White schools) with respect to engaged time in learning, teachers' expectations, instructional content of the curricula, and the match between instruction and school learning. It can be stated, however, that given the ubiquitous association between school segregation of Mexican American students and their diminished academic achievement, reduced opportunity structures in the classroom may play a part in understanding the notion of bias. In the present study, perhaps the White students in the White segregated schools were being provided with greater opportunities to learn curricular material of the type that is tested on the K-ABC Achievement scale.[20] (p. 166)

Predictive Validity

Predictive validity is perhaps the most significant psychometric property when it comes to the actual use of intelligence tests in education. To a certain degree, when a child's intelligence is assessed in school, the implicit or explicit purpose is to predict future performance. Therefore, the question of bias in predictive validity is critical given that educational interventions (e.g., special education) for some minority students will depend, in part, on the accurate predictive validity of intelligence tests.

According to Reynolds (1982a), predictive validity bias can occur in three different forms. First, the predictor variable (e.g., intelligence test score) consistently under- or overpredicts the criterion variable (e.g., achievement test score) for the minor group. This type of bias can be observed when we compare the regression lines for the major and minor groups and we find significantly different *intercepts*. That is, the minor group regression line is parallel and significantly below (overprediction) or above (underprediction) the major group regression line.

In the second type of predictive validity bias, the regression line for the minor group is closer to the regression line of the major group at one end of the line but gets progressively farther apart at the other end. In this case, we find significantly different *slopes*, and that represents a different predictive relation between the predictor and the criterion for the two groups. The group for which the regression line has the steeper slope has a stronger predictive relation between the predictor and criterion variables; therefore, the group represented by the regression line with the flatter slope has a less valid measure of prediction.

The third way in which bias in predictive validity can occur is when we find that the test overpredicts for the minor group at low ability levels but underpredicts at the other extreme. Here we find both *slope* and *intercept* bias, which is graphically represented by regression lines that intersect with each other.

Table 5.6 lists 18 studies that examined various intelligence tests for possible bias in predictive validity. The key study characteristics are as follows:

1. Of the 18 predictive validity studies, African Americans once again were the most represented minority, participating in 15 studies (83%). Mexican Americans were second in frequency, participating in 11 studies (61%), followed by American Indians (Papagos) in 2 studies (Nos. 38 and 39) and Puerto Ricans (No. 6) and Asian Americans (No. 51) in 1 study each.

2. The majority of the predictive validity studies ($n = 11$ of 18, 61%) used participants enrolled in regular classrooms. Participants in 5 studies (Nos. 2, 36, 41, 42, and 54) were referred for assessment for possible special education placement, and participants in 2 studies (Nos. 34 and 35) were selected from both regular classrooms and referred groups.

3. The WISC-R once again clearly was the test of choice for bias examination. The WISC-R was used in 11 (61%) of the total 18 studies. The second most common test was the K-ABC, which was used in 5 studies (Nos. 6, 26, 34, 35, and 59). In addition, the WISC was used in 2 studies (Nos. 7 and 41), and a number of tests were used in only 1 study, namely the Differential Ability Scales (DAS, No. 51), the RCPM and the RSPM (No. 16), the SB (1972 edition,

No. 2), the Slosson Intelligence Test (SIT, No. 28), and the Woodcock-Johnson Psychoeducational Battery (WJ-PB, No. 34).

Earlier, we discussed several ways in which bias in predictive validity can occur. But how can the notion of such bias be conceptualized? In cultural bias studies of predictive validity, the research question is as follows: Does the test predict scores on a criterion measure equally well for both the major and minor groups? More specifically, Reynolds (1982a) suggested that a test is biased with respect to predictive validity "when the inference drawn from the test score is not made with the smallest feasible random error or if there is constant error in an inference or prediction as a function of membership in a particular group" (p. 216). The most common statistical techniques used to establish predictive validity bias include (a) the use of a test of differences between the regression slope and the intercept for both groups (Reynolds, 1982a; Palmer, Olivárez, Wilson, & Fordyce, 1989, No. 35); (b) the *simultaneous* analysis of regression slopes and *y* intercepts, also known as the Potthoff technique (Bossard, Reynolds, & Gutkin, 1980, No. 2; Jensen, 1980); and (c) the analysis of mean discrepancies between groups (Valdez & Valdez, 1983, No. 54). In this technique, the difference in means between the predictor and the criterion for the majority and minority groups are calculated and then compared. If the difference is significant, then the predictor test is assumed to be biased. Valdez and Valdez (1983, No. 54) suggested that this method is more sensitive to bias than are the other techniques discussed. In the investigations that we reviewed, however, biased and/or mixed findings were reported in studies regardless of the statistical technique used to explore bias in predictive validity.

Contrary to the cultural bias studies of reliability and construct validity reviewed in this chapter, considerably more investigations of bias in predictive validity have reported biased or mixed findings. As can be seen in Table 5.6, of the 18 predictive validity studies reviewed, 6 (33%) reported biased findings (Nos. 7, 33, 34, 35, 51, 54, and 59), and an additional 3 studies (17%) reported mixed findings (Nos. 26, 28, and 33). Yet, it is important to note that 50% of the studies failed to find any bias in predictive validity. African Americans were represented in all of the nonbias investigations, whereas Mexican Americans were represented in only 4 (Nos. 16, 29, 38, and 39), American Indians (Papagos) in 2 (Nos. 38 and 39), and Puerto Ricans in 1 (No. 6). As expected, the popular WISC-R was the most prominent test examined in these investigations of predictive validity. The WISC-R was used as the predictor in 6 studies (Nos. 29, 36, 38, 39, 41, and 42), whereas the WISC, the K-ABC, the RSPM, the RCPM, and the SB (1972 edition) were represented in 1 investigation each. Regarding the criterion variable, a wide variety of achievement measures were represented, the most common ones being the Wide Range Achievement Test (WRAT) in three studies (Nos. 2, 41, and 42) and the Metropolitan Achievement Test (MAT) in two studies (Nos. 38 and 39).

One example of a study that found nonbias in predictive validity is that of Reschly and Reschly (1979, No. 38), whose investigation examined the predictive validity of the WISC-R. The MAT (Reading and Mathematics sections) served as the criterion. The authors tested 787 White, African American, Mexican American, and American Indian (Papago) children (Grades 1 to 9). To investigate bias in predictive validity, Reschly and

Reschly performed correlations and partial correlation analyses between the predictor and criterion variables for each racial/ethnic group. Similarities in the correlation patterns for all factors led the investigators to conclude, "These data also again confirm the relatively strong relationship of WISC-R scores to achievement for most non-Anglo as well as Anglo groups. An exception to this trend was observed for a Native American group" (p. 359). More recently, Poteat, Wuensch, and Gregg (1988, No. 36) conducted a study in which they examined the predictive validity of the WISC-R. The California Achievement Test (CAT) and grade point average were the criterion variables. Poteat et al. tested 168 African American and White students (6.0 to 16.0 years of age). The results from simultaneously comparing the slopes and intercepts of the regression lines yielded no significant differences in prediction between African Americans and Whites. The authors concluded that for their samples differential prediction is not a meaningful problem in the use of the WISC-R.

Regarding the three mixed results studies, African Americans were represented in all investigations, whereas Mexican Americans were also represented in all three. No other minority groups were represented in the mixed results studies. As expected, the popular WISC-R was the most prominent test examined in these investigations of predictive validity; it was used in three studies (Nos. 26, 28, and 33) as the predictor variable. The other tests used were the K-ABC in one study (No. 26) and the SIT and WJ-PB in one study each (No. 28). Regarding the criterion variable, Study No. 26 used the Comprehensive Test of Basic Skills (CTBS), Study No. 28 used the MAT, and Study No. 33 used the CAT.

An example of a mixed finding study is that of Naglieri and Hill (1986, No. 26), who conducted an investigation comparing 172 White and African American children (9.3 to 12.4 years of age). The WISC-R and the K-ABC Mental Processing Composite (MPC) were used as predictors of the CTBS and the K-ABC Achievement scale. Naglieri and Hill reported no significant difference for the WISC-R as predictor of the two achievement tests. The K-ABC MPC and Simultaneous Scales, however, presented significant differences in predictive validity between the White and African American participants. The authors noted,

> The results suggest caution when using the K-ABC Mental Processing Composite scores to predict K-ABC achievement. The significant differences in the regression line intercepts suggest that using the Mental Processing Composite resulted in error in prediction of black student performance on the K-ABC Achievement scale. (p. 354)

In another example of a mixed finding study, Oakland and Feigenbaum (1979-4, No. 33) investigated the predictive validity of the WISC-R for the CAT with 467 White, African American, and Mexican American children (7.0 to 14.0 years of age). Using intercorrelation analysis, the authors reported "a number of differences in the magnitude of correlations for [the WISC-R and] Reading and Mathematics" (p. 973). Oakland and Feigenbaum concluded that the WISC-R predicted mathematics scores about equally well for all three groups but was a much better predictor of reading achievement for the White participants.

Finally, 6 of the 18 investigations (33%) found clear evidence of bias in predictive validity. Mexican Americans were the most represented group in these studies, being present in all but 1 (No. 51). African Americans were represented in 2 (Nos. 7 and 35) and Asian Americans in 1 (No. 51). American Indians were not represented in any of the biased results studies. Regarding the predictor tests, the WISC-R (Nos. 35 and 54) and the K-ABC (Nos. 35 and 59) were the most common tests used, with 2 studies each. Other tests used included the WISC (No. 7) and the DAS (No. 51) in 1 study each. The criterion measures varied widely, and no single test was used in more than 1 study. Some examples are the K-ABC Achievement scale (No. 35), the WRAT (No. 54), and the CTBS (No. 59).

One example of a predictive validity study that found bias is Palmer et al. (1989, No. 35). The participants were 236 White, African American, and Mexican American children (Grades 2 to 4). The authors focused on the issue of overidentification of minority children with learning disabilities (LD) and reported significant predictive validity bias in the WISC-R using the K-ABC Achievement Scale as the criterion. There was evidence of bias across minority groups for the K-ABC Achievement Scale (total) and the K-ABC Achievement Scale Arithmetic subtest. Predictive validity bias was found for both referred and nonreferred participants. In addition, the authors found language dominance-related bias for the Mexican American group in both the WISC-R and the K-ABC composite scales. The authors explained,

> Each case [of bias] involved overprediction of the Hispanic LEP [limited English proficiency] pupils in contrast to the Hispanic non-LEP sample. These findings suggest that . . . Hispanic LEP pupils may be overrepresented in special education classes due to biased assessment. (p. 272)

In another study that detected bias in predictive validity, Valencia and Rankin (1988, No. 59), who used 166 White and Mexican American children (10.0 to 12.4 years of age), conducted an investigation using several different statistical techniques to examine predictive validity bias in the K-ABC, with the CTBS and the K-ABC Achievement Scale serving as the criterion tests. The authors first computed Pearson product-moment correlations between the K-ABC and the CTBS for both the major and minor groups. Second, they used an interaction model between ethnicity and the K-ABC to predict the criterion. In this case, finding an interaction effect with race/ethnicity would suggest different slopes for the White and the Mexican American participants. Finally, they calculated a test of homogeneity of slopes of the regression lines. Overall, Valencia and Rankin found strong evidence of predictive validity bias. They reported, "By and large, the K-ABC [MPC] has less power to predict CTBS scores with the Mexican American group than with the Anglo group" (p. 262). The authors concluded,

> The K-ABC appears to be flawed or biased when used with Mexican American students because the test does not have the same predictive efficiency with the majority and minority students. . . . Caution is urged in using the K-ABC MPC to predict CTBS scores for upper primary-grade Mexican American children. (p. 263)

Evidence of predictive validity bias also is supported by the study of Valdez and Valdez (1983, No. 54). Their investigation contained 250 White and Mexican American children and used the WISC-R (predictor) and the Peabody Individual Achievement Test and the WRAT (criterion measures). The authors examined bias by using several different statistical techniques. Using the mean discrepancy of prediction technique, they found evidence that the WISC-R was a biased predictor of the Mexican American children's achievement. That is, the WISC-R showed a significant underprediction of achievement for these participants. Valdez and Valdez concluded,

> In this study, conventional methods of analysis failed to detect bias against Mexican-American students. For example, the methods of analysis used by Jensen (1980) and Reynolds (1980 [No. 40]) are insensitive to the detection of predictive bias because of: (a) evaluation of differences outside the range of interest (IQ = 0); (b) attenuation and reversal of differences because of small and unstable slopes associated with moderate regression coefficients and restriction of range; [and] (c) inappropriate application of deviation from commonality. Continued use of the conventional model, therefore, is likely to mask the evidence of predictive bias in IQ testing as it has in the past. (p. 446)

A final example of a study that detected bias in predictive validity is that of Olivárez, Palmer, and Guillemard (1992, No. 34), who conducted a bias investigation of the predictive validity of several intelligence tests (i.e., K-ABC, WISC-R, WJ-PB). The achievement measures (subtests) of the WJ-PB served as criterion variables. Participants were 236 White, Mexican American, and African American children (Grades 2 to 4, referred and nonreferred). The authors concluded not only that these tests systematically overpredicted the achievement scores of minority students but also that the effect was amplified for limited English proficiency Mexican American students. It is important to note that Olivárez et al. mostly found intercept rather than slope effects, meaning that the intelligence tests examined overpredicted achievement for minorities. This overprediction, Olivárez et al. asserted, can lead to misidentification of students with LD because one of the criteria for learning disability identification is a severe discrepancy between measured IQ and achievement scores. The overprediction of achievement scores produces misidentification in the following manner. Using the common (i.e., White) regression line and assuming standard means of 100 on both the predictor and criterion tests, an IQ score of 100 might predict an equivalent achievement score of 100. The regression line for minority children, however, has a lower intercept and, therefore, would predict a lower score. Assuming a standard deviation of 15 points, a minority intercept line that is 1 standard deviation below the common regression line would predict that a minority student with an IQ of 100 would, in fact, score 85 on an achievement scale. But if we use the common regression line to predict a minority student's achievement score, then we would find a *severe discrepancy* between the predicted achievement score (100) and the actual achievement score (85). This difference could qualify the student for special education, although in reality the difference is the result of bias in predictive validity and not of a learning disability. In conclusion, Olivárez et al. asserted, "The findings of this and the Palmer et al. (1989 [No. 35]) investigation suggest that reliance on these statistical approaches to identify severe discrepancy in ethnically and linguistically diverse populations may result in misidentification of these students" (p. 185).

Discussion

This review of research on cultural bias in intelligence tests has identified a number of study characteristics as well as a reporting of extant findings of cultural bias in four psychometric categories: reliability, construct validity, content validity, and predictive validity. In this section, we discuss in more general terms what has been learned. We also comment critically on this body of research:

1. Empirical research on cultural bias in intelligence tests did not begin in earnest until 1973. It appears that such research endeavors were a reaction, in part, to early critics' allegations that intelligence tests were inherently biased against minority students (particularly African Americans and Mexican Americans). Likewise, the armchair analyses used to detect item bias by expert witnesses for plaintiffs in some of the early 1970s litigation (e.g., *Larry P. v. Riles*, 1972) helped to fuel investigations by researchers who believed that cultural bias should best be examined using empirical statistical methods.

Of the total 62 cultural bias investigations identified in our study, 24 (39%) were published during the 1970s, 33 (53%) during the 1980s, and only 5 (8%) during the 1990s. Beginning around 1989, why has there been such a tremendous waning in cultural bias research on standardized, individually administered intelligence tests? Such abatement can likely be attributed to the rather consistent finding during the 1970s and 1980s that prominent intelligence tests (e.g., WISC, WISC-R) were concluded to be nonbiased; thus, interest in pursuing research in this area rapidly declined. Later, we offer discussion on why such a diminishment of research activity during the 1990s is not warranted.

2. African American children served as participants in 76% of the total cultural bias investigations, and Mexican American children participated in 58% of the total studies. These high participation rates are not surprising given that (a) these groups were prominent as plaintiffs during the 1970s litigation involving alleged bias in intelligence tests and (b) proportionately, they are the two largest minority groups in the national school-age population.

In sharp contrast to African American and Mexican American participation in the total investigations, other minority groups have been vastly understudied. American Indian children have served as participants in only 10% of the investigations, Puerto Ricans in 3%, and Asian Americans in 2%. It is ironic that children of these minority groups have not been selected more frequently as participants in cultural bias investigations given their growing presence in the national school-age population.

3. Most of what we know about cultural bias in intelligence tests comes from the southwestern region of the United States. Nearly 1 of 2 investigations studied were conducted in the Southwest, with California ($n = 17$ of 62 studies, 27%) and Arizona ($n = 9$, 15%) taking the lead in frequency. Of the total 50 states, only 12 (24%) are represented as research sites. Surprisingly, only 1 study was conducted in Texas, the nation's second largest state in total population and a state with a 55% minority school-age population during 1997-1998 (Texas Education Agency, 1998). New York, the nation's third largest state in population and having large percentages of African American and Puerto Rican students, did not serve as a research site in any of the cultural bias investigations.

4. Another irony about the present findings is that although children referred for special education diagnoses and possible placement, as well as special education students, are the primary subjects of assessment using individually administered intelligence tests, they were not the primary participants in the 62 investigations. Referred children constituted 16% of all participants, about 3% were mixed (regular and referred), and less than 2% were children actually enrolled in special education. By dramatic contrast, about 73% of all participants were enrolled in regular classes. It is unclear why researchers have sought to have regular education children as the modal group for participation. Perhaps it is because researchers find it easier to gain permission from school authorities and parents to administer intelligence tests to children in regular classes. Whatever the reasons, the overall findings of the 62 investigations appear to create a problem with external validity. That is, there remains the question as to what degree the conclusions drawn about nonbiased/biased/mixed findings with children in regular classes can be generalized to children in special education.

5. As we have found, the Wechsler scales (the WISC-R and its predecessor, the WISC) have been the dominant choice of instruments by researchers to examine for cultural bias. Combined, the WISC-R (the most frequently examined intelligence test) and the WISC (the fourth most frequently examined intelligence test) were the instruments of choice in the vast majority (65%) of all 62 investigations. Interestingly, the WISC-III, which successfully replaced its predecessor (the WISC-R) in 1991, was not identified as an instrument of examination in a single study (but see Note 4 of this chapter). This omission probably is related to the general decline in cultural bias research seen during the 1990s.

Overall, the finding that the WISC and WISC-R combined were analyzed for cultural bias in two thirds of all studies, compared to distant followers such as the K-ABC, the RCPM, and the MSCA, does not bode well for the study of cultural bias in intelligence tests as a whole. It can be concluded that what we know most about cultural bias in intelligence tests is mostly confined to two sister instruments: the WISC and WISC-R. Furthermore, given that these two tests no longer are used in clinical assessments and have been replaced by the nonexamined WISC-III, much of what is known about cultural bias in intelligence tests is based on two obsolete instruments. It must be recognized, however, that given the large overlap between the WISC-R and the WISC-III (80%

of the former's content remains in the latter test [Wechsler, 1991; see also Sattler, 1992]), one could conclude that much of what is known about cultural bias in the WISC-R may be useful in understanding the existence of bias in the WISC-III.

6. As we have described in the "Results: Study Characteristics" section, a substantial proportion of the investigations failed to control for SES background of participants, language dominance and proficiency of participants who might have been language minorities, and sex of participants in the bias analyses. These control problems do not fare well for this body of research.

The failure to report SES, and not to control for it in those studies that did report it, represents potential problems of confoundment (see Chapter 3 of this book). Although SES is only weakly correlated with measured intelligence, it still is an important variable to control for in test bias research.[21] As we discussed in Chapter 3, when SES is controlled in general studies that compare the intellectual performance of White and minority children (Anastasi & D'Angelo, 1952; Christiansen & Livermore, 1970; Herrnstein & Murray, 1994; Nichols & Anderson, 1973), as well as in test bias studies (Valencia, 1984, No. 55; Valencia et al., 1995, No. 60), the mean difference in intellectual performance between Whites and minorities frequently is reduced and in some cases is negligible. In sum, given that only a handful of studies reviewed in this chapter attempted to control for SES, the role of this variable in cultural bias research remains largely unknown.

With respect to studies that failed to consider the possibility that some of their Mexican American, Puerto Rican, American Indian, and Asian American participants might have been language minorities, another likely confoundment exists. It has been known for decades that the children who are most likely to be penalized on intelligence tests during the time of administration are those who have been raised in environments where English is not the mother tongue, have limited second-language skills, and/or are developing bilingually (Klineberg, 1935b; Sánchez, 1932, 1934). Nonetheless, many researchers over the years have simply ignored this issue (see Valencia, 1997b). Such methodological disregard in research, as well as the inappropriateness of administering tests in English to language minorities during clinical assessments, likely prompted the American Educational Research Association, the American Psychological Association, and the National Council on Measurement in Education (1985) to note in their *Standards for Educational and Psychological Testing* that for non-native English speakers every test, in part, becomes a test of language or literacy. For those investigations in the present review that failed to report language, their findings of nonbias or bias are unclear.

Regarding sex of participants, only a few studies controlled for sex during the bias analyses (e.g., Valencia & Rankin, 1985, No. 56; Valencia et al., 1995, No. 60). The failure to control for sex represents another possible confoundment in this corpus of investigations. This is indeed a reasonable conclusion. Halpern (1997) noted that the available literature suggests that females, in general, "score higher [than males] on tests that require rapid access to and use of phonological and semantic information in long-term memory, production and comprehension of complex prose, fine motor skills, and perceptual speed" (p. 1091). Halpern also reported that males, in general, "score higher [than females] on tasks that require transformation in visual-spatial working memory, motor skills involved in aiming, spatiotemporal responding, and fluid reasoning, especially in abstract mathematical and scientific domains" (p. 1091).[22] In sum, given the widespread practice of researchers in these 62 investigations not to control for sex in their analyses, the role of this variable in possibly shaping the findings of bias or nonbias is not known.

7. Several studies reported predictive validity bias in a variety of intelligence tests that are commonly used in educational placement including the WISC-R and the K-ABC. In these studies (e.g., Olivárez et al., 1992, No. 34; Palmer et al., 1989, No. 35), the differences in slope and intercept between the major and minor groups' regression lines appear to lead to the possibility of unfair decisions regarding student placement into special programs. A serious consequence of this type of bias is the misidentification of minority students as learning disabled and subsequent placement in special education programs. Olivárez et al. (1992) pointed to the *Standards for Educational and Psychological Testing* (American Educational Research Association et al., 1985) for guidance. Although the *Standards* assert that special attention should be placed on the construction and use of nonbiased instruments that are administered to culturally and ethnically diverse populations, few studies of predictive validity bias currently are being conducted. This was evident in the present review, for example, by the lack of cultural bias studies on the newest version of the Wechsler instrument, the WISC-III, which replaced the WISC-R. Olivárez et al. went on to assert that "the large majority of the bias cases occurred with the nonreferred group, suggesting that test developers should conduct predictive bias studies with their newly developed or revised instruments on their standardization samples" (p. 185).

8. The whole issue of cultural bias presence in intelligence tests is related, as we have found, to three major aspects. One is that the identification of bias appears to be psychometric specific. Investigations of bias in reliability demonstrated the existence of nonbias, and studies of construct validity also showed nonbias (with a few exceptions). On the other hand, about 50% of all content and predictive validity investigations had mixed/biased findings. The second aspect is that what we know about cultural bias in intelligence tests is confined, for the most part, to two instruments: the WISC and WISC-R. Thus, to generalize about cultural bias in intelligence tests beyond these two instruments appears difficult. The third aspect is that African American and Mexican American children overwhelmingly served as participants in the 62 cultural bias investigations. Very little is known about cultural bias vis-à-vis other minority groups (e.g., Asian Americans).

9. Our final point of discussion goes full circle to an issue raised earlier in this chapter. Jensen (1980), in the preface to his *Bias in Mental Testing*, asserted that, based on the available cultural bias literature, it can be concluded that the most frequently used tests of mental ability (including intelligence measures) were "*not* biased against any of the native-born English-speaking minority groups" (p. ix). Given that Jensen's claim was made two decades ago, does his conclusion have veracity today in light of the extant lit-

erature we have reviewed? In the most general sense, the evidence tends to weigh on the side of a nonbiased conclusion. That is, the strong majority (71%) of the 62 investigations had conclusions of nonbias. Yet, given that 29% of the total investigations had conclusions of mixed/biased findings, the sweeping inference of Jensen's claim that the subject of cultural bias in intelligence is a closed issue is not accurate today. Our review of the available literature on cultural bias in intelligence tests leaves us with the conclusion that *some* of the observed mean differences in White and minority performance on such tests is related to test bias. This point truly needs to be underscored because it appears that a number of scholars have been influenced greatly by Jensen's book in which he presents his major conclusion of nonbias in tests including measures of intelligence. For example, Brody (1992), author of a major text on the subject of intelligence, referred to Jensen's book on test bias as the "definitive discussion" (p. 287). Drawing from Jensen's findings, Brody concluded, "The differences in intelligence test performance [between Blacks and Whites] are not attributable in any obvious way to bias in the tests" (p. 309). It appears that *Bias in Mental Testing* (Jensen, 1980) has led some scholars to take the position that possible cultural bias in intelligence tests is a closed issue. It is our position that it is an open issue.

Conclusions

By way of closure, we provide the reader with two final thoughts on the subject of cultural bias in intelligence tests. First, as we mentioned earlier in this chapter, test bias refers to an empirical, testable, quantifiable, scientific matter. Bias, a statistical notion, can be detected through empirical approaches. Any discussion of test bias is incomplete, we contend, without commenting on "test use." Test use explicitly involves decision making. We often make judgments (e.g., identification of a student who is diagnosed as having LD, identification of a student for admission to a gifted and talented program). These decisions can, at times, lead to claims of unfairness by examinees or their parents. The point here is that test bias research addresses only one aspect of testing—the psychometric integrity of the instrument. How we use test results is the other half of the whole. This interrelatedness is particularly important in school-based interventions. Ross-Reynolds and Reschly (1983-1, No. 43) cogently capture this concern:

> Fairness in assessment with diverse groups depends on events prior to and following the use of formal tests. The effectiveness of special programs or other interventions that result, in part, from the use of tests with individuals is the crucial factor in fairness. *Tests that are unbiased according to the statistical criteria . . . are necessary, but not sufficient, conditions to ensure fairness.* (p. 146, italics added)

Our second point has to do with the waning of research on cultural bias in intelligence tests. Compared to the flurry of research activity during the 1970s and 1980s, very little research was undertaken during the 1990s. It appears that such a precipitous decline is related to the high percentage of earlier studies that had findings of nonbias. The current state of the art on cultural bias research informs us, however, that this line of research needs to be renewed with vigor. Our review of the literature has identified a number of shortcomings in this body of research. We have found that members of some minority groups (e.g., Puerto Ricans) have been vastly understudied, that participants have overwhelmingly been students in regular classes, that most of what is known about cultural bias is based on the obsolete WISC and WISC-R, and that possible problems of confoundment (i.e., related to SES, language status, and sex) exist. In light of these issues, a forceful case can be made that test publishers and independent researchers have an obligation to engage in expanded research on cultural bias in intelligence tests. In such endeavors, psychometricians need to be more sensitive to the concerns of Helms (1997), who asserted that test developers need to have better informed models about the influences of race, culture, and social class on cognitive test performance. Furthermore, this call for research can be justified by demographic reality. Currently, the racial/ethnic makeup of the nation is changing dramatically due to the phenomenal growth of culturally/linguistically diverse school-age populations, especially Latinos and Asian Americans (Valencia & Chapa, 1993). It is important when intelligence tests are administered to minority children and youths that such tests be free of cultural bias and that the test results be applied fairly.

APPENDIX A: 62 Cultural Bias Studies Examined and Used in the Review

1. Bart, W., Rothen, W., & Read, S. (1986). An ordering-analytic approach to the study of group differences in intelligence. *Educational and Psychological Measurement, 46,* 799-810. **Construct Validity**

2. Bossard, M. D., Reynolds, C. R., & Gutkin, T. B. (1980). A regression analysis of test bias on the Stanford-Binet Intelligence Scale for Black and White children referred for psychological services. *Journal of Clinical Child Psychology, 9,* 52-54. **Predictive Validity**

3. Carlson, J. S., & Jensen, C. M. (1981). Reliability of the Raven Coloured Progressive Matrices Test: Age and ethnic group comparisons. *Journal of Consulting and Clinical Psychology, 49,* 320-322. **Reliability**

4. Corman, L., & Budoff, M. (1974). Factor structures of Spanish-speaking and non-Spanish-speaking children on Raven's Progressive Matrices. *Educational and Psychological Measurement, 34,* 977-981. **Construct Validity**

5. Dean, R. S. (1980). Factor structure of the WISC-R with Anglos and Mexican Americans. *Journal of School Psychology, 18,* 234-239. **Construct Validity**

6. Glutting, J. J. (1986). Potthoff bias analyses of K-ABC MPC and Nonverbal Scale IQs among Anglo, Black, and Puerto Rican kindergarten children. *Professional School Psychology, 4,* 225-234. **Predictive Validity**

7. Goldman, R. D., & Hartig, L. K. (1976). The WISC may not be a valid predictor of school performance for primary-grade minority children. *American Journal of Mental Deficiency, 80,* 583-587. **Predictive Validity**

8. Greenberg, R. D., Stewart, K. J., & Hansche, W. J. (1986). Factor analysis of the WISC-R for White and Black children evaluated for gifted placement. *Journal of Psychoeducational Assessment, 4*, 123-130. **Construct Validity**

9. Gutkin, T. B., & Reynolds, C. R. (1980). Factorial similarity of the WISC-R for Anglos and Chicanos referred for psychological services. *Journal of School Psychology, 18*, 34-39. **Construct Validity**

10. Gutkin, T. B., & Reynolds, C. R. (1981). Factorial similarity of the WISC-R for White and Black children from the standardization sample. *Journal of Educational Psychology, 73*, 227-231. **Construct Validity**

11. Juliano, J. M., Haddad, F. A., & Carroll, J. L. (1988). Three-year stability of WISC-R factor scores for Black and White, female and male children classified as learning-disabled. *Journal of School Psychology, 26*, 317-325. **Construct Validity**

12. Jensen, A. R. (1974a-1). How biased are culture-loaded tests? *Genetic Psychology Monographs, 90*, 185-244. **Content Validity**

13. Jensen, A. R. (1974a-2). How biased are culture-loaded tests? *Genetic Psychology Monographs, 90*, 185-244. **Reliability**

14. Jensen, A. R. (1974a-3). How biased are culture-loaded tests? *Genetic Psychology Monographs, 90*, 185-244. **Content Validity**

15. Jensen, A. R. (1974a-4). How biased are culture-loaded tests? *Genetic Psychology Monographs, 90*, 185-244. **Content Validity**

16. Jensen, A. R. (1974b). Ethnicity and scholastic achievement. *Psychological Reports, 34*, 659-668. **Predictive Validity**

17. Johnston, W. T., & Bolen, L. M. (1984). A comparison of the factor structure of the WISC-R for Blacks and Whites. *Psychology in the Schools, 21*, 42-44. **Construct Validity**

18. Kaufman, A. S., & DiCuio, R. (1975). Separate factor analyses of the McCarthy Scales for groups of Black and White children. *Journal of School Psychology, 13*, 10-18. **Construct Validity**

19. Kaufman, A. S., & Hollenbeck, G. P. (1974). Comparative structure of the WPPSI for Blacks and Whites. *Journal of Clinical Psychology, 30*, 316-319. **Construct Validity**

20. Koh, T., Abbatiello, A., & McLoughlin, C. S. (1984). Cultural bias in WISC subtest items: A response to Judge Grady's suggestion in relation to the *PASE* case. *School Psychology Review, 13*, 89-94. **Content Validity**

21. Miele, F. (1979). Cultural bias in the WISC. *Intelligence, 3*, 149-164. **Content Validity**

22. Mishra, S. P. (1981). Factor analysis of the McCarthy scales for groups of White and Mexican-American children. *Journal of School Psychology, 19*, 178-182. **Construct Validity**

23. Mishra, S. P. (1982). The WISC-R and evidence of item bias for Native American Navajos. *Psychology in the Schools, 19*, 458-464. **Content Validity**

24. Montie, J. E., & Fagan, J. F. I. (1988). Racial differences in IQ: Item analysis of the Stanford-Binet at three years. *Intelligence, 12*, 315-332. **Content Validity**

25. Murray, A. M., & Mishra, S. P. (1983). Interactive effects of item content and ethnic group membership on performance on the McCarthy scales. *Journal of School Psychology, 21*, 263-270. **Content Validity**

26. Naglieri, J. A., & Hill, D. S. (1986). Comparison of WISC-R and K-ABC regression lines for academic prediction with Black and White children. *Journal of Clinical Child Psychology, 15,* 352-355. **Predictive Validity**

27. Naglieri, J. A., & Jensen, A. R. (1987). Comparison of Black-White differences on the WISC-R and the K-ABC: Spearman's hypothesis. *Intelligence, 11,* 21-43. **Construct Validity**

28. Oakland, T. (1978). Predictive validity of readiness tests for middle and lower socioeconomic status Anglo, Black, and Mexican American children. *Journal of Educational Psychology, 70,* 574-582. **Predictive Validity**

29. Oakland, T. (1980). An evaluation of the ABIC, pluralistic norms, and estimated learning potential. *Journal of School Psychology, 18,* 3-11. **Predictive Validity**

30. Oakland, T., & Feigenbaum, D. (1979-1). Multiple sources of test bias on the WISC-R and Bender-Gestalt test. *Journal of Consulting and Clinical Psychology, 47,* 968-974. **Reliability**

31. Oakland, T., & Feigenbaum, D. (1979-2). Multiple sources of test bias on the WISC-R and Bender-Gestalt test. *Journal of Consulting and Clinical Psychology, 47,* 968-974. **Construct Validity**

32. Oakland, T., & Feigenbaum, D. (1979-3). Multiple sources of test bias on the WISC-R and Bender-Gestalt test. *Journal of Consulting and Clinical Psychology, 47,* 968-974. **Content Validity**

33. Oakland, T., & Feigenbaum, D. (1979-4). Multiple sources of test bias on the WISC-R and Bender-Gestalt test. *Journal of Consulting and Clinical Psychology, 47,* 968-974. **Predictive Validity**

34. Olivárez, A., Jr., Palmer, D. J., & Guillemard, L. (1992). Predictive bias with referred and nonreferred Black, Hispanic, and White pupils. *Learning Disability Quarterly, 15,* 175-187. **Predictive Validity**

35. Palmer, D. J., Olivárez, A., Wilson, V. L., & Fordyce, T. (1989). Ethnicity and language dominance: Influence on the prediction of achievement based on intelligence test scores in nonreferred and referred samples. *Learning Disability Quarterly, 12,* 261-274. **Predictive Validity**

36. Poteat, G. M., Wuensch, K. L., & Gregg, N. B. (1988). An investigation of differential prediction with the WISC-R. *Journal of School Psychology, 26,* 59-68. **Predictive Validity**

37. Reschly, D. J. (1978). WISC-R factor structures among Anglos, Blacks, Chicanos, and Native American Papagos. *Journal of Consulting and Clinical Psychology, 46,* 417-422. **Construct Validity**

38. Reschly, D. J., & Reschly, J. E. (1979). Validity of WISC-R factor scores in predicting achievement and attention for four sociocultural groups. *Journal of School Psychology, 17,* 355-361. **Predictive Validity**

39. Reschly, D. J., & Sabers, D. (1979). Analysis of test bias in four groups with the regression definition. *Journal of Educational Measurement, 16,* 1-9. **Predictive Validity**

40. Reynolds, C. R. (1980). Differential construct validity of intelligence as popularly measured: Correlations of age with raw scores on the WISC-R for Blacks, Whites, males and females. *Intelligence, 4,* 371-379. **Construct Validity**

41. Reynolds, C. R., & Hartlage, C. C. (1979). Comparison of WISC and WISC-R regression lines for academic prediction with Black and White referred children. *Journal of Consulting and Clinical Psychology, 47,* 589-591. **Predictive Validity**

42. Reynolds, C. R., & Nigl, A. J. (1981). A regression analysis of differential validity in intellectual assessment for Black and White inner city children. *Journal of Clinical Child Psychology, 10*, 176-179. **Predictive Validity**

43. Ross-Reynolds, J., & Reschly, D. J. (1983-1). An investigation of item bias on the WISC-R with four sociocultural groups. *Journal of Consulting and Clinical Psychology, 51*, 144-146. **Content Validity**

44. Ross-Reynolds, J., & Reschly, D.J. (1983-2). An investigation of item bias on the WISC-R with four sociocultural groups. *Journal of Consulting and Clinical Psychology, 51*, 144-146. **Reliability**

45. Rousey, A. (1990). Factor structure of the WISC-R Mexicano. *Educational and Psychological Measurement, 90*, 351-357. **Construct Validity**

46. Sandoval, J. (1979-1). The WISC-R and internal evidence of test bias with minority groups. *Journal of Consulting and Clinical Psychology, 47*, 919-927. **Content Validity**

47. Sandoval, J. (1979-2). The WISC-R and internal evidence of test bias with minority groups. *Journal of Consulting and Clinical Psychology, 47*, 919-927. **Reliability**

48. Sandoval, J. (1982). The WISC-R factorial validity for minority groups and Spearman's hypothesis. *Journal of School Psychology, 20*, 198-204. **Construct Validity**

49. Sandoval, J., Zimmerman, I. L., & Woo-Sam, J. M. (1983). Cultural differences on WISC-R verbal items. *Journal of School Psychology, 21*, 49-55. **Content Validity**

50. Silverstein, A. B. (1973). Factor structure of the Wechsler Intelligence Scale for Children for three ethnic groups. *Journal of Educational Psychology, 65*, 408-410. **Construct Validity**

51. Stone, B. J. (1992). Prediction of achievement by Asian-American and White children. *Journal of School Psychology, 30*, 91-99. **Predictive Validity**

52. Swerdlik, M. E., & Schweitzer, J. (1978). A comparison of factor structures of the WISC and WISC-R. *Psychology in the Schools, 16*, 166-172. **Construct Validity**

53. Taylor, R. L., & Zeigler, E. W. (1987). Comparison of the first principal factor on the WISC-R across ethnic groups. *Educational and Psychological Measurement, 47*, 691-694. **Construct Validity**

54. Valdez, R. S., & Valdez, C. (1983). Detecting predictive bias: The WISC-R vs. achievement scores of Mexican-American and nonminority students. *Learning Disability Quarterly, 6*, 440-447. **Predictive Validity**

55. Valencia, R. R. (1984). Reliability of the Raven's Coloured Progressive Matrices for Anglo and for Mexican-American children. *Psychology in the Schools, 21*, 49-52. **Reliability**

56. Valencia, R. R., & Rankin, R. J. (1985). Evidence of content bias on the McCarthy Scales with Mexican American children: Implications for test translation and nonbiased assessment. *Journal of Educational Psychology, 77*, 197-207. **Content Validity**

57. Valencia, R. R., & Rankin, R. J. (1986-1). Factor analysis of the K-ABC for groups of Anglo and Mexican American children. *Journal of Educational Measurement, 23*, 209-219. **Construct Validity**

58. Valencia, R. R., & Rankin, R. J. (1986-2). Factor analysis of the K-ABC for groups of Anglo and Mexican American children. *Journal of Educational Measurement, 23*, 209-219. **Reliability**

59. Valencia, R. R., & Rankin, R. J. (1988). Evidence of bias in predictive validity on the Kaufman Assessment Battery for Children in samples of Anglo and Mexican American children. *Psychology in the Schools, 25*, 257-263. **Predictive Validity**

60. Valencia, R. R., Rankin, R. J., & Livingston, R. (1995). K-ABC content bias: Comparisons between Mexican American and White children. *Psychology in the Schools, 32,* 153-169. **Content Validity**

61. Valencia, R. R., Rankin, R. J., & Oakland, T. (1997). WISC-R factor structures for White, Mexican American, and African American children: A research note. *Psychology in the Schools, 34,* 11-16. **Construct Validity**

62. Vance, H. B., Huelsman, C. B., & Wherry, R. J. (1976). The hierarchical factor structure of the Wechsler Intelligence Scale for Children as it relates to disadvantaged White and Black children. *Journal of General Psychology, 95,* 287-293. **Construct Validity**

APPENDIX B: Intelligence Tests Examined for Cultural Bias

1. Dunn, L. M. (1959). *Peabody Picture Vocabulary Test.* Minneapolis, MN: American Guidance Service.

2. Elliot, C. D. (1990). *The manual for the Differential Ability Scales.* San Antonio, TX: Psychological Corporation.

3. Ilg, F. L., & Ames, L. B. *School readiness: Behavior tests used at the Gesell Institute.* New York: Harper & Row.

4. Kaufman, A. S., & Kaufman, N. L. (1983). *Kaufman Assessment Battery for Children.* Circle Pines, MN: American Guidance Service.

5. McCarthy, D. (1972). *Manual for the McCarthy Scales of Children's Abilities.* New York: Psychological Corporation.

6. Raven, J. C. (1960). *Guide to the Standard Progressive Matrices.* London: H. K. Lewis.

7. Raven, J. C. (1962). *Coloured Progressive Matrices, Sets A, A$_B$, B.* London: H. K. Lewis.

8. Slosson, R. L. (1963). *Slosson Intelligence Test and Slosson Reading Test.* New York: Slosson Educational Publications.

9. Terman, L. M., & Merrill, M. A. (1960). *Stanford-Binet Intelligence Scale.* Boston: Houghton Mifflin.

10. Terman, L. M., & Merrill, M. A. (1973). *Stanford-Binet Intelligence Scale: 1972 norms edition.* Boston: Houghton Mifflin.

11. Wechsler, D. (1949). *Wechsler Intelligence Scale for Children.* New York: Psychological Corporation.

12. Wechsler, D. (1967). *Wechsler Preschool and Primary Scale of Intelligence.* New York: Psychological Corporation.

13. Wechsler, D. (1974). *Wechsler Intelligence Scale for Children–Revised.* New York: Psychological Corporation.

14. Woodcock, R. W., & Johnson, B. J. (1977). *Woodcock-Johnson Psychoeducational Battery.* Allen, TX: DLM Teaching Resources.

6

Heredity

The debate concerning the relative contributions of "nature" and "nurture" to measured intelligence remains prominent today. The term *nature* sometimes is used interchangeably by scholars with the terms *heredity, genetics,* and *biology*. The term *nurture* sometimes is referred to in the literature as *environment, socialization,* and *culture*. Regarding the role of genetics, Block and Dworkin (1976) commented, "There is perhaps no issue in the history of science that presents such a complex mingling of conceptual, methodological, psychological, ethical, political, and sociological questions as the controversy over whether intelligence has a substantial genetic component" (p. xi). Furthermore, the question of whether genetics plays a significant role in explaining racial/ethnic group mean differences in intelligence has been a particularly explosive area of scholarly contention over the years (for discussions of this early history, see Valencia, 1997b, and Chapter 1 of the present book).

To examine the literature on the possibility of genetic differences in average measured intelligence between racial/ethnic groups in the United States (which, by the way, predominantly focuses on African American and White samples), we have organized this chapter in the following manner: (a) issues in human behavioral genetics; (b) going beyond the nature-nurture debate; (c) racial/ethnic group mean differences in measured intelligence: the role of genetics; and (d) conclusions.

Issues in Human Behavioral Genetics

In general, behavioral genetics is an interdisciplinary field, drawing from genetics and the behavioral sciences. Research on genetic contributions to human intellectual performance has taken three forms over the decades (McArdle & Prescott, 1997). First, there has been research on cognitive deficits that has shown demonstrative genetic bases (e.g., Down's syndrome, a type of mental retardation). This search for distinct genetic abnormalities linked to impaired intellectual functioning continues at a swift tempo (McKusick, 1994, cited in McArdle & Prescott, 1997). Second, there is a broad research stream that has addressed the question of "whether genetic factors might regulate aspects of behavior within the *normal range of variation*" (McArdle & Prescott, 1997, p. 403, italics added). This research question, which is considerably different from genetically based intellectual abnormalities, requires its own methodology. Research in this area has led to the emergence of the field called behavioral genetics. As noted by McArdle and Prescott, "One of the most popular topics in the history of BG [behavioral genetics] is the biometric genetic analysis of intellectual abilities (BGIA)" (p. 403).[1] The third area of research that has attempted to investigate links between heredity and measured intelligence is a subfield within the BGIA domain, that is, studies of group differences in mean intellectual performance (e.g., race/ethnicity, socioeconomic status [SES]).

There is a considerable amount of literature on how investigations of the genetic basis of measured intelligence have generated substantial debate in both scientific and social contexts.[2] "Galton's (1865) initial report, for example, was interpreted by some as evidence for genetic control over economic and social achievement, with little attention given to the striking group differences in educational opportunity" (McArdle & Prescott, 1997, p. 405). Earlier controversies in genetic studies of intellectual abilities also involved the eugenics movement in the United States and England during the 1920s and 1930s. Eugenics, the belief that improvement in the human race could and should be encouraged through selective breeding, has engendered much attention in the United States in the realm of social issues that were particularly germane to the status of people of color, the poor, immigrants from southeastern Europe, and the "feeble-minded" (see, e.g., Haller, 1963; Kelves, 1985; Valencia, 1997b).

Recent technological advances in gene location and gene cloning (Lander & Botstein, 1989; McKusick, 1994; both cited in McArdle & Prescott, 1997) also likely have added to the contemporary controversies in BGIA studies (Lee, 1993, cited in McArdle & Prescott, 1997).[3] And, of course, there is the BGIA issue of whether racial/ethnic group differences in measured intelligence—the core subject of this chapter—have a predominantly genetic basis.

In addition to the preceding issues regarding BGIA research, there are two other concerns that are germane to our discussion. One, the "norm of reaction," has to do with questions of how genotype, environment, and phenotype are related.[4] The second concern pertains to issues of how "heritability" is estimated.

Norm of Reaction

Any general conclusions drawn about the relative contributions of heredity and environment in shaping individual differences in measured intelligence are meaningless unless a person's norm of reaction is considered. Norm of reaction, first proposed by Woltereck (1909, cited in Wahlsten & Gottlieb, 1997), refers to the interaction between an individual's environment and his or her genetic endowments. More specifically, the notion of norm of reaction (sometimes referred to as *reaction norm*) "asserts that each genotype is associated with a characteristic pattern of phenotypic changes in response to alteration to the environment, but it does not require parallel response profiles for all individuals over a wide range" (Wahlsten & Gottlieb, 1997, p. 172).[5] Norm of reaction is an important construct to understand because there appears to be a myth among those who are not knowledgeable about behavioral genetics that *genetic* connotes intractableness in human behavior (Plomin, 1983; Scarr, 1993; Weinberg, 1989).[6] Plomin (1983) underscored that there is some difficulty in "shaking the mistaken notion that genetic differences begin prior to birth and remain immutable ever after. . . . *Genetic does not mean immutable*" (pp. 253-254, italics added). And, as stated by Weinberg (1989), "Genes do not fix behavior; rather, they establish a range of possible reactions to the range of possible experiences that environments can provide. Environments can also affect whether the full range of gene reactivity is expressed" (p. 101). In a related vein, Hirsch (1976) commented that the ontogeny of a person's phenotype with respect to a specific observable developmental outcome "has a norm of reaction not predictable in advance" (p. 163).[7] Hirsch continued,

> In most cases the norm of reaction remains largely unknown. . . . Even in the most favorable materials, only an approximate estimate can be obtained for the norm of reaction when, as in plants and some animals, an individual genotype can be replicated many times and its development studied over a range of environmental conditions. The more varied the conditions, the more diverse might be the phenotypes developed from any one genotype. Of course, different genotypes should not be expected to have the same norm of reaction. . . . Therefore, those limits set by heredity can never be specified. They are plastic within each individual but differ between individuals. Extreme environmentalists were wrong to hope that one law or set of laws described universal features of modifiability. Extreme hereditarians were wrong to ignore the norm of reaction. (p. 163)

In sum, the norm of reaction underscores the complexity of study regarding the nature-nurture debate, and its importance has been acknowledged by those adhering to both the environmental and behavioral genetics perspectives in BGIA research.[8]

Heritability

The cornerstone of the field of BGIA research is the construct of "heritability" (typi-cally symbolized as h^2). Heritability typically is defined as "the proportion of phen-otypic differences among individuals in a particular population. *Broad-sense heritability* involves all additive and nonadditive sources of genetic variance, whereas *narrow-sense heritability* is limited to additive genetic variance" (Plomin, DeFries, McClearn, & Rutter, 1997, p. 313).[9] Our discussion here pertains to broad-sense heritability, which is the per-centage of phenotypic differences due to all sources that are genetic.

It is important to comment on two major misconceptions about heritability that sometimes are seen in the nature-nurture discourse. First, in the most general sense, heritability "refers to the contribution of genetic differences to observed differences among individuals for a particular trait in a particular population at a particular time . . . , *not* to the phenotype of a single individual" (Plomin et al., 1997, pp. 82-83). For exam-ple, the heritability of intelligence (a particular trait) might be .50 for White, middle-SES, American adults 25 to 50 years of age (a particular population) estimated between 1980 and 1985 (a particular time period). As such, heritability estimates are protean, depend-ing on how environmental or genetic factors might differ in diverse populations or at different times. Numerous scholars have underscored the point that heritability is a characteristic of a population, not an individual (Bouchard, 1997; Brody, 1992; Brody & Brody, 1976; Jensen, 1969; Plomin et al., 1997; Taylor, 1980; Tucker, 1994). Unfortunately, this major point that heritability is a population and not an individual trait is not always understood. For example, Tucker (1994) pointed out that it is not unusual to find misin-terpretations of heritability in biology texts. A case in point is seen in *Biology Today and Tomorrow* by Ward and Hetzel (1980, p. 302), who stated that "80 percent of our basic in-telligence is inherited and . . . the remaining 20 percent determined by our environment" (quoted in Tucker, 1994, p. 221). Even Charles Murray, coauthor of *The Bell Curve* (Herrnstein & Murray, 1994), has demonstrated ignorance about heritability

> as shown by a recent CNN interview reported in *The New Republic* (Wright, 1995). Murray
> declared, "When I—when we—say 60 percent heritability, it's not 60 percent of the varia-
> tion. It is 60 percent of the IQ in any given person." Later, he repeated that for the average
> person "60 percent of the intelligence comes from heredity" and added that this was true
> of the "human species," missing the point that heritability makes no sense for an individ-
> ual and that heritability statistics are population relative. (Block, 1995, p. 108)

Second, given that heritability refers to a characteristic of a particular population (a *within*-group notion), *it is not generalizable from one group to another* (see, e.g., Dorfman, 1995; Lewontin, 1975, 1976; Mackenzie, 1984; Mercer & Brown, 1973; Neisser et al., 1996; Taylor, 1980, 1992). Stated more directly, knowing within-population heritability for in-telligence only for a White population in the United States, for example, does not allow one to draw conclusions about between-population differences in measured intelli-gence (e.g., Whites compared to African Americans, Whites compared to Mexican Americans). Put more simply, "The fact is . . . that the high heritability of a trait [i.e., intelligence] within a given group has no necessary implications for the source of a difference between groups" (Neisser et al., 1996, p. 95). This is a principle that frequently is discussed in the behavioral genetics literature but often is ignored or misunderstood by scholars. As we shall see later, violation of this major principle has led to unwarranted conclusions about genetic bases of racial/ethnic differences in intelligence (e.g., as seen in Herrnstein & Murray, 1994, and Jensen, 1969).

Heritability estimates of intelligence are derived on the basis of correlations on intelligence test scores from individuals of various relationships. Such data have been gathered from numerous sources over the years, using thousands of pairings. As described in Plomin and DeFries (1980), these investigations can be clustered as follows: (a) studies of genetically identical individuals (e.g., identical twins reared together or apart); (b) studies of genetically related individuals: first degree (e.g., fraternal twins reared together, non-twin siblings reared together or apart); and (c) studies of genetically unrelated individuals (e.g., unrelated children reared together, adoptive parents-adoptive children). The rationales for the various research designs appear to be quite logical:

> In general, the study of separated identical twins is intended to approximate pairs of individuals with identical genes but random (uncorrelated) environments. The study of adopted pairs in the same families is intended to approximate the converse case—pairs with identical environments but random (uncorrelated) genes. The other techniques, involving the study of pairs of varying degrees of kinship, fall between these two. (Taylor, 1980, p. 78)

Of the present-day procedures used to estimate the heritability of measured intelligence, "the one with the greatest conceptual appeal" (Taylor, 1980, p. 75) uses correlations between the IQs of identical (monozygotic [MZ]) twins who have—it is assumed—been separated at birth and reared in separate and unrelated environmental settings.[10] Notwithstanding the attraction that this research design carries, studies on separated MZ twins have been heavily criticized on methodological grounds (Farber, 1981; Kamin, 1974; Steen, 1996; Taylor, 1980) as well as strongly defended (Bouchard, 1997).[11] One critic has asserted, "In the final analysis, what we are left with is a mass of faulty methods and data, which do not permit one to conclude in favor of a significant genetic effect on IQ score" (Taylor, 1980, p. 216).

One of the most sustained critiques of the research on investigations of MZ twins reared apart (MZA twins) is that proffered by Taylor (1980) in *The IQ Game*, a book devoted to a methodological inquiry into the heredity-environment debate. In his chapter "The Myth of the Separated Identical Twins," Taylor described three requirements that need to be met in MZA twin studies to conclude that twins' measured intelligence (i.e., IQ) is very likely related to their genes. These stringent conditions are (a) the twins were separated at birth, (b) the twins were raised in completely different and unrelated families after separation (but before any intelligence testing was done), and (c) all MZA twin pairs were separated over a wide range of different environmental settings after separation.[12]

To investigate whether these three requirements were met, Taylor (1980) examined three well-known MZA twin studies forming the bedrock of research that has concluded that intelligence is highly genetic in basis (Juel-Nielsen, 1965; Newman, Freeman, & Holzinger, 1937; Shields, 1962). Based on his reanalysis of data from the three studies, Taylor found "four ways in which inadvertent similarity between the presumably separated identical twins came about" (p. 77). These four sources of environmental similarity (which, according to Taylor, were not analyzed by Juel-Nielsen [1965], Newman et al. [1937], or Shields [1962]), are (a) late separation, (b) reunion prior to testing, (c) relatedness of adoptive families, and (d) similarity in social environment after separation. Although not all MZA twin pairs experienced these sources of environmental similarity, a large number of twins did. For example, Taylor (1980) reported that although all 68 twin pairs across the three investigations were separated in some way and for at least some period of time, about 44 pairs (65%) were united for some period prior to being tested for IQ. According to Taylor, the overall picture that arose from his reanalysis of the three studies led him to conclude that the inadvertent environmental similarity experienced by many of the MZA twin pairs resulted in artificially inflated IQ correlations, thus producing questionable estimates of heritability. Taylor concluded, "There is no hard and convincing evidence that the heritability of IQ is anywhere near substantial" (p. 206).[13] However, as we mentioned earlier, Taylor's reanalysis of the MZA twin data has not gone unchallenged (Bouchard, 1983, 1997).

The final issue we need to raise about the heritability of intelligence is that more recent research has produced lower heritability estimates, compared to those of previous studies. Erlenmeyer-Kimling and Jarvik (1963) published the first comprehensive summary of world literature on IQ correlations between various pairings of relatives (52 studies and 30,000 pairings). Although Erlenmeyer-Kimling and Jarvik did not provide an overall heritability estimate, their findings led them to conclude, "The composite data are compatible with the polygenic hypothesis which is generally favored in accounting for inherited differences in mental ability. . . . We do not imply[, however,] that environment is without effect upon intellectual functioning" (p. 1478).[14] In a subsequent survey of worldwide literature on familial resemblances in measured intelligence (111 studies and 113,942 pairings), Bouchard and McGue (1981) concluded, "As in the earlier review [Erlenmeyer-Kimling & Jarvik, 1963], the pattern of averaged correlations is remarkably consistent with polygenetic theory. This is not to discount the importance of environmental factors" (p. 1058). Although Bouchard and McGue (1981) did not proffer a heritability estimate, they did comment that the data they reviewed supported the inference that partial genetic determination for measured intelligence is incontestable but that "the precise strength of this effect is dubious" (p. 1058). In sum, the review by Bouchard and McGue suggested that the role of genetic influences on intelligence was smaller than that reported previously. Or, conversely, "These data suggest a greater role for family environment [in influencing intelligence] than evidenced previously" (McArdle & Prescott, 1997, p. 404).

As to why the newer data point to less genetic influence on intelligence than do the older and widely cited data, McArdle and Prescott (1997) commented that such differences appear to be related to the use of more representative samples,[15] a greater stringency in criteria for study inclusion in recent reviews, environmental and genetic secular changes in the populations studies, and advanced statistical models that have been applied to BGIA data.[16]

Perhaps the most significant implication stemming from the more recent research on the genetics of intelligence is that "heritability ratios for intelligence of 70% to 80% reported in the past (e.g., Burt, 1966, 1969, 1972; Jensen, 1969) no longer appear tenable" (McArdle & Prescott, 1997, p. 405). In light of current reviews and investigations, a heritability of about 50% for measured intelligence frequently is cited in the literature (Chipeur, Rovine, & Plomin, 1990; Loehlin, 1989; Plomin & DeFries, 1980; Plomin et al., 1997; Plomin & Rende, 1990; Scarr & Carter-Saltzman, 1982; Steen, 1996).[17]

Going Beyond the Nature-Nurture Debate

Students and scholars of human development frequently are reminded of the long-standing controversy regarding the relative contributions of heredity and environment to human behavior. The debate is indeed historically rooted, dating "at least to the beginnings of the modern philosophical era: Contemporary hereditarians . . . can date back their pedigree to Descartes and Leibnitz, while contemporary environmentalists . . . can trace their intellectual heritage to the British empiricists such as Locke and Hume" (Gardner, Hatch, & Torff, 1997, p. 243).

In the United States, the nature-nurture debate was unyielding during the first half of the 20th century. Apparently, the intractability of the controversy prompted Anne Anastasi, in her 1958 presidential address to the American Psychological Association, to challenge psychologists to go beyond the debate (Anastasi, 1958b). In her address, Anastasi (1958b) exhorted her colleagues to abandon the question of "how much" variance in measured intelligence can be accounted for by heredity and environment. She pressed psychologists to devote their energy to the question of "how," that is, how genotypes influence phenotypes.

Although Anastasi (1958b) clearly brought the "how" question to the center stage of the nature-nurture controversy, it appears that her call for a new direction of research has met with little consensus (Bidell & Fischer, 1997). An assessment of contemporary literature pertaining to the heredity-environment debate informs us that the division actually has intensified. Although there have been new developments in behavioral genetics and developmental behavioral genetics providing insights into how both genetics and environmental events shape measured intelligence (Brody, 1992), and although research has expanded on how the home environment is an important statistical predictor of children's intellectual performance (see Chapter 4 of this book), there remains considerable discord between the heredity and environment perspectives. This can be seen clearly, for example, in Lewontin, Rose, and Kamin's (1984) *Not in Our Genes: Biology, Ideology, and Human Nature.*[18] In our opinion, *Not in Our Genes* represents one of the most vigorous defenses for the environmental position and sustained critiques of behavioral genetics, what Lewontin et al. called "biological determinism."[19] Another example of a diverse perspective on the nature-nurture debate is seen in Sandra Scarr's 1991 presidential address to the Society for Research in Child Development (Scarr, 1992). Proffering a theory of genotype (→) environment effects, in which "*genotypes drive experience*" (p. 9), Scarr elaborated,

> In this model, parental genes determine their phenotypes, the child's genes determine his or her phenotype, and the child's environment is merely a reflection of both parents and child. Here, differences among children's common home environments, *within the normal species range,* have no effect on differences among children's outcomes. The obvious challenge posed by this model is the proposition that differences among normal child environments are a product of parental and child characteristics and not a causal path in the determination of differences among children's behavioral phenotypes. (p. 9)[20]

The recent edited book, *Intelligence, Heredity, and Environment* (Sternberg & Grigorenko, 1997), attests to the divergence of opinions on the contemporary nature-nurture debate. Sternberg and Grigorenko (1997) deliberately sought contributors who adhere to behavioral genetic perspectives as well as those having perspectives that focus on environmental, socialization, and cultural aspects of intellectual development. What resulted is a book that is comprehensive, up-to-date, and fairly balanced.[21] New advances in behavioral genetic research are presented (e.g., Bouchard, 1997; Plomin, 1997; Scarr, 1997), as are "novel theoretical perspectives on the genes and culture controversy." Here we see, for example, (a) the "symbol systems approach" advanced by Gardner et al. (1997);[22] (b) the "bioecological model" of Ceci, Rosenblum, de Bruyn, and Lee (1997), in which "proximal processes are the engines of intellectual development, with higher levels of proximal processes with increasing levels of intellectual competence" (p. 313);[23] and (c) a "cultural psychology" perspective presented by Miller (1997).[24] Also presented are the views of scholars asserting that attempting to understand the effects of nature and nurture by partitioning them is an invalid approach (Bidell & Fischer, 1997; Wahlsten & Gottlieb, 1997).

Probably few scholars would disagree with the need to move beyond the nature-nurture debate regarding "how much" variance in measured intelligence can be accounted for by differences in heredity and environment. If it is likely that few will disagree with this aphorism, then why has there been so little progress in meeting Anastasi's (1958b) challenge of more than four decades ago? Bidell and Fischer (1997) asserted that it is the very reductionistic framework of the nature-nurture debate itself that has obviated a satisfying answer. Specifically, the intractability of the controversy may be traced to an elemental fallacy:

> By artificially partitioning all variability in intelligent behavior into just two mutually exclusive sources, the traditional nature-nurture framework defines the problem in a way that excludes the integrative role of constructive activity, leading implausibly to a predeterministic model of epigenetic mechanisms. Despite the inability of the

predeterministic model to account for the ever-growing volume of findings in molecular genetics, developmental biology, and developmental psychology, the dominance of the nature-nurture framework has sustained the idea of a preset maturational program with environmental inputs as the default model of the epigenesis of human intelligence. (p. 235).[25]

In a growing alternative framework for comprehending the interactive role of genes and environment in the development of measured intelligence, Bidell and Fischer (1997) offered a treatise on the important role of human agency in the epigenesis of intelligence.[26] In a nutshell, Bidell and Fischer contended that genetics and environment are intrinsically related, active, self-organizing systems. Furthermore, the authors stated, "Change in such systems is a product of self-organizing activity; for this reason, the epigenesis of intelligence is a constructive, not predetermined, process" (p. 236).

Drawing from the works of Piaget, Vygotsky, and others, Bidell and Fischer (1997) asserted that genes and environment do not, either singularly or jointly, determine cognitive outcomes. As human agents, we shape such outcomes as we go through the complex process of making sense of our world and building skills so as to participate in it. In their concluding remarks, the authors offered a challenge to scholars that resonates with the timbre of Anastasi's call of four decades ago:

> Between nature and nurture stands the human agent whose unique integrative capacities drive the epigenesis of intelligence and organize biological and environmental contributions to the process. For too long, the role of human agency in the origins of human intellectual abilities has been obscured by the dominance of the nature-nurture framework in the scientific discourse on this topic. Progress in understanding mechanisms in the epigenesis of intelligence now depends on our ability and willingness to move beyond the nature-nurture framework and take advantage of the growing repertoire of constructs and methods for the study of cognitive epigenesis as a constructive process. (Bidell & Fischer, 1997, pp. 236-237)

The words of Bidell and Fischer (1997) make much sense to us. There is a serious need for scholars to give much more attention to the central role that human agency plays in the nature *and* nurture framework regarding the origins and development of human intelligence. There is simply too much research and writings from cognitive developmentalists to ignore. As Bidell and Fischer clearly pointed out, numerous researchers since the late 1950s from practically every theoretical tradition have made contributions "to the now nearly universal consensus that our intellectual skills and abilities are in one way or another the products of our own self-governed activity in relation to the world" (p. 193). A similar position is held by Lewontin et al. (1984) in their critique of biological determinism, a special case of reductionism. In closing their book, Lewontin et al. reminded us about the central importance of human agency:

Our brains, hands, and tongues have made us independent of many single major features of the external world. Our biology has made us into creatures who are constantly re-creating our own psychic and material environments and whose individual lives are the outcomes of an extraordinary multiplicity of intersecting causal pathways. Thus, it is our biology that sets us free. (p. 290)

Racial/Ethnic Group Mean Differences in Measured Intelligence: The Role of Genetics

The ground that we have just covered focused on several issues germane to the broader context of the nature-nurture controversy, an area in which issues abound. At this juncture, we move into a terrain that is even more disputatious, with the question of whether *between-racial/ethnic group average differences* in measured intelligence (e.g., between Whites and African Americans) can significantly be accounted for by genetic factors. To be sure, this is not an issue restricted to modern times. As we discussed in Chapter 1 of this book, more than a century ago, Galton (1870), in *Hereditary Genius*, made a racial—and racist—pronouncement in his chapter "The Comparative Worth of Different Races," that differences existed in intelligence between the "Anglo-Saxons" and the "African negro" race and that such differences were "hereditary" in basis. Since the time in which Galton opined that racial differences in intelligence were hereditary in root, such hereditarianism has waxed and waned. The vicissitudinous nature of the genetic hypothesis has been shaped by changing Zeitgeists.

In the present chapter, our major objective is to explore the highly charged issue that racial/ethnic group mean differences in measured intelligence might be strongly related to genetics. In so doing, we have organized our discussion around two major subsections. First, there are the writings of Audrey M. Shuey, Henry E. Garrett (Shuey's mentor), Arthur R. Jensen, Lloyd M. Dunn, and the team of Richard J. Herrnstein and Charles Murray. This set of writings is largely opinionated in nature, suggesting or hypothesizing that racial/ethnic differences in intelligence are largely attributed to genetics. Although these publications are fairly devoid of scientific and empirical groundings to support the authors' conclusions, they are important to incorporate into the contemporary debate about alleged genetically based racial/ethnic group differences in intelligence. Each of these authors' works have generated, in varying degrees, controversial discourse in academia. In the following subsections, we provide synopses, as well as critiques, of the works of these authors. Much has been written on the *science* and *ideology* of this debate (Blum, 1978; Kamin, 1974; Marks, 1981; Tucker, 1994; Valencia, 1997b). Given the highly politicized and ideological configuration of the controversy, we would be remiss not to include these polemics as part of the current disputation about race/ethnicity and measured intelligence.

The second cluster of publications that we discuss here are more scientifically oriented investigations that have sought, for example, to examine the heritability of intelligence in African American samples. We also discuss some of the research on transracial adoptions and racial admixture. These studies were designed to examine the veracity of the genetic hypothesis of racial differences in intelligence.[27]

Audrey M. Shuey

We daresay that most scholars familiar with the evolution of the nature-nurture debate regarding racial/ethnic differences in intelligence would agree that the modern version of this controversy was launched with Jensen's (1969) provocative treatise in which he suggested that the lower performance of Black children on intelligence tests, compared to that of White children, was strongly due to genetic influences. Although Jensen's treatise often is cited as the source that rekindled the coals leading to conjectures about genetic factors as the major attributor for Black-White average differences in intelligence, we believe that it is necessary to go back a decade further in time to get a better sense of what transpired.

In 1958, Shuey published her controversial book, *The Testing of Negro Intelligence* (Shuey, 1958). Her second edition, which we shall use as our source of discussion, was published in 1966 (Shuey, 1966). Shuey's massive review (578 pages) spanned 50 years of research and included 380 original investigations of Black intelligence (singularly studied or in comparison to White participants). The review included research studies in which more than 81,000 Black schoolchildren and 48,200 Black high school and college students were tested on various intelligence measures. Furthermore, military officers, enlisted men, veterans, and other adults (e.g., criminals, homeless men) also were participants in some of the reviewed studies. In all, 80 different intelligence tests (excluding different test editions or forms) were employed in the various investigations.

Of Shuey's (1966) numerous conclusions, the ones most germane to our discussion are the following:

1. There was a striking consistency in intelligence test results. Combining all studies, Blacks, as a group, had a mean IQ of 1 standard deviation (15 IQ points) below the mean IQ for Whites.

2. Small average differences (1 to 6 points) were found between the IQs of northern-born and southern-born Black children (favoring the former group) living in the same northern cities and attending the same public schools. Shuey concluded, "About half of this difference may be accounted for by environmental factors and half by selective migration" (p. 495).

3. There was a "tendency for racial hybrids to score higher than those groups described as, or inferred to be, unmixed Negro" (p. 521).

4. In general, when SES was controlled, the observed Black-White differences in IQ still were present.

5. Blacks, in comparison to Whites, did better on intelligence test items requiring concrete and practical solutions and did poorer on those items demanding logical analysis and abstract reasoning.

Regarding her major explanation for the mean differences in intellectual performance between Blacks and Whites, Shuey (1966) was absolutely silent throughout her lengthy review. Her conclusion did appear, however, in the *final* sentence on the *final* page. Atavistically, Shuey posited a genetic hypothesis reminiscent of 1920s hereditarian thought: "[The test results,] all taken together, *inevitably point to the presence of native differences between Negroes and Whites as determined by intelligence tests*" (p. 521, italics added).

Suffice it to say, Shuey's (1966) hereditarian conclusion of Black intellectual inferiority did not go unchallenged. At that time, a number of investigators (Bond, 1958; Hicks & Pellegrini, 1966; Klineberg, 1963; Pettigrew, 1964a) offered sharp critiques.[28] Following is a sampling:

1. Shuey (1966) failed to distinguish the comments and conclusions of the authors she reviewed from her own inferences. Bond (1958) cited examples of authors who concluded that the observed lower IQ scores of Blacks were largely due to environmental factors. Yet, Shuey apparently lumped together the results of a large number of such studies and concluded that the differences were largely due to heredity.

2. Bond (1958) took Shuey (1958) to task for not fully discussing her reporting that White southerners invariably scored lower on IQ tests than did White northerners.

3. Shuey (1966) cited a number of studies in which she claimed that Blacks still scored significantly lower even when environmental factors were "equated." Pettigrew (1964a) found fault with Shuey's analyses, however, critiquing her failure to carefully read the qualifications of the authors (e.g., the assumption of SES equality of Blacks and Whites was not entirely valid).

4. Pettigrew (1964a) reprehended Shuey's work in terms of a definitional issue of race, asserting, "Since Negro Americans do not even approach the status of a genetically pure 'race,' they are a singularly inappropriate group upon which to test racist theories of inherent intellectual inferiority of the Negroid subspecies" (p. 7).

5. The criticism presented by Hicks and Pellegrini (1966) was a novel, yet significant, rebuke of Shuey's (1966) book in that it dealt with the *meaningfulness* of racial differences regarding *policy*. That is, can any policy implications stemming from race studies in intelligence be empirically substantiated? For example, Garrett (1962) argued that Blacks were so constitutionally inferior that miscegenation and school integration should be prohibited. Hicks and Pellegrini (1966), arguing that nearly any study could be made to show significant differences if enough participants were used regardless of how nonsensical the variables might be, criticized Garrett for misconstruing the meaning of "statistical significance" and erroneously equating it with "practical significance." Hicks and Pellegrini reexamined 40 studies by 26 investigators that Shuey reviewed in her 1958 book (first edition) and computed an estimated ω^2 (omega-square) from the t value in each case.[29] The estimated ω^2 values ranged from .000 to .383; the median value was .061 (i.e., 6% explained variance). Of the 40 studies, 24 had estimated ω^2 values of less than .100, 11 of the values were between .100 and .199, 3 were between .200

and .299, and 2 were between .300 and .383. One study that Hicks and Pellegrini listed had an astronomical t of 149.05, but 93,955 Whites and 23,596 Blacks were used as participants; the estimated ω^2 was .159.

Hicks and Pellegrini (1966), in sum, argued that there was no established objective basis for Shuey's (1966) conclusion of innate intellectual inferiority of Blacks and any policy recommendation resultant of such a conclusion (e.g., Garrett's [1962] call for the segregation of Black schoolchildren). Rebuking both Shuey and Garrett, Hicks and Pellegrini (1966) commented,

> The results of this [Shuey's] study reflect directly on the conflicting interpretations of racial differences in IQ.
> The median [estimated] ω^2, .061, is thought to best represent the strength of association between skin color and intelligence. Six percent represents only a small reduction in uncertainty. When Garrett [1962, p. 2] claims that the differences in Negro and white IQ "are real and highly useful in guidance and prediction," he has greatly exaggerated the strength of the relationship between skin color and IQ.
> It is concluded that studies of racial intelligence have failed to establish the existence of meaningful ethnic differences in intelligence. Therefore, any interpretation of racial IQ data that stipulates differential treatment of Negroes and whites is unwarranted. (p. 45)

Henry E. Garrett

It is difficult to fathom why Shuey in 1966, a time when the civil rights movement was gaining ground and White America was becoming sensitized to the plight of African Americans and other people of color, would proffer a genetic interpretation of the Black-White gap in measured intelligence. After all, hereditarianism as a form of social thought had been, for the most part, silenced for more than three decades.[30] We speculate that Shuey's hereditarian conclusion in *The Testing of Negro Intelligence* was influenced, in part, by her mentor, Garrett.

Garrett received his Ph.D. from Columbia University and subsequently chaired Columbia's Department of Psychology (1940-1956). It appears that during Garrett's career, his colleagues held him in high esteem, as he was elected as presidents of the Eastern Psychological Association, the American Psychological Association, and the Psychometric Society. In addition, he was a fellow of the American Association for the Advancement of Science and a member of the National Research Council.

One would think that, in light of his prestigious appointments and leadership roles in a number of learned societies, Garrett would have been a man with an open mind and whose scholarly endeavors would have been guided by the basic principle of objectivity. In the area of racial comparisons, however, we know that Garrett approached his research with the vigor of a pseudoscientist. He was biased, dishonest, and frequently a proselytizer of his racial views about the alleged inferiority of Blacks and the alleged superiority of Whites (Tucker, 1994; Valencia & Solórzano, 1997). Although Garrett made some of his racial pronouncements in respectable scholarly journals (e.g., Garrett, 1962), his most egregious racial animus was voiced in a series of self-serving, pseudoscientific, nonpeer-reviewed pamphlets. For example, in *Breeding Down* (circa mid-1960s, cited in Chorover, 1979), Garrett (n.d.), a strident anti-miscegenationist, offered justification for race segregation on the grounds that Blacks were mentally inferior:

> You can no more mix the two races and maintain the standards of White civilization than you can add 80 (the average IQ of Negroes) and 100 (the average IQ of Whites), divide by two and get 100. What you would get would be a race of 90's, and it is that 10 percent differential that spells the difference between a spire and a mud hut; 10 percent—or less—is the margin of civilization's "profit"; it is the difference between a cultured society and savagery. Therefore, it follows, if miscegenation would be bad for White people, it would be bad for Negroes as well. For, if leadership is destroyed, all is destroyed. (quoted in Chorover, 1979, p. 47)

Garrett's racist views on race mixing through intermarriage also included a staunch posture against school desegregation. Tucker (1994) described Garrett as "the most eminent scientific spokesman for segregation" (p. 153). Such race mixing of Blacks and Whites in schools must be opposed, Garrett argued, because desegregation would likely lead to friendships and, subsequently, to intermarriage. It appears that his anti-miscegenationist and anti-desegregationist convictions were so fervently articulated that other racist ideologues embraced them with vigor: "During the 1960's, 500,000 copies of Henry Garrett's pamphlets on the evils of miscegenation were distributed free of charge to American teachers by opponents of integrated education" (Chorover, 1979, p. 48).

As late as the 1970s, Garrett still was spewing his racist diatribe. In 1975, during the middle of the emotionally charged school busing debate in Boston, an advertisement appeared in the *Boston Globe* announcing the publication of Garrett's (1973) "book" titled *IQ and Racial Differences* (Chorover, 1979). Our reading of Garrett's book informs us that it actually was a 57-page pamphlet. Writing, it appears, for the educated layperson, Garrett was not reticent about his views concerning the heritage, intelligence, and schooling of Blacks. Note the following quotes from his pamphlet:

- "In recent years it has become fashionable to depict in glowing terms the achievement of the Negro over the past 5,000 years, although the truth is that the history of the black African is largely a blank" (pp. 1-2).
- "Egalitarianism makes a bow to heredity but argues that almost all of the undeniable differences among mankind arise from environmental pressures, many of which are under man's control.... The author [Garrett] of this study holds to the thesis that egalitarianism is dead wrong. Black and white children do *not* have the same potential. They do *not* learn at the same rate" (pp. 10-11).
- "The American Negro is aided [intellectually] by his racial admixture with the American white" (p. 11).
- "The case for genetic differences in [Negro-white] intelligence is a solid one" (p. 47).

- "Since environmental theory has wrought havoc, why not try a 'new' set of premises based on genetic theory? For example . . ., institute separate and equally well-equipped schools for Negroes and whites, wherever feasible" (p. 51).

- "It is clear there cannot be complete desegregation of our classrooms on the one hand and first-rate education on the other. Under such conditions, there would only be second-rate education for the children of both races" (p. 53).

In the final analysis, it appears likely that Shuey (1966), in her genetic assertion of Black-White differences in IQ, was influenced by the racial views of her mentor, Garrett. Our conjecture is further supported by examination of the foreword, written by Garrett (1966), of Shuey's *The Testing of Negro Intelligence.* Several points that he made are noteworthy. First, on the subject of ethics, he commented, "The honest psychologist, like any true scientist, should have no preconceived racial bias" (p. viii). Apparently, Garrett felt that he was above this principle. Keep in mind that at the same time he made this statement, he was voicing, in his pamphlets, his repugnant and bigoted opinions about Blacks. Second, on the environmental hypothesis of racial differences in intelligence, he argued, "The American Negro is generally below the white in social and economic status, and his work opportunities are more limited. Many of these inequalities have been exaggerated" (p. viii). Notwithstanding all the empirical evidence at the time that Black-White differences in intelligence were greatly accounted for by environmental and SES factors, he shrugged off this body of data. Finally, with respect to Garrett's interpretation of the cause of Black-White differences in intelligence, he posited, as did Shuey 527 pages later, a genetic hypothesis:

> It [Shuey's book] is a careful and accurate survey which should command the attention of all serious students of the subject. Dr. Shuey finds that at each age level and under a variety of conditions, Negroes regularly score below whites [on intelligence tests]. . . . We are forced to conclude that the regularity and consistency of *these results strongly suggest a genetic basis for the differences.* I believe that the weight of evidence (biological, historical, and social) supports this judgment. (p. viii, italics added)[31]

Arthur R. Jensen

Although the debate over Jensen's (1969) controversial treatise has been covered ad infinitum, a good deal of this coverage has been an incomplete or inaccurate presentation of what Jensen actually stated and what led up to his conclusion regarding race differences in intelligence. As such, we offer the reader our analysis.

The 1960s was a decade of growing environmentalism regarding children's cognitive development. The works of Jean Piaget, J. McV. Hunt, and Benjamin Bloom had notable influences in shaping our understanding concerning the roles of experience and environment in the intellectual development of children (see Chapter 4 of the present book). Also, the 1960s was a time, as noted by Snyderman and Rothman (1988), of America's "growing disenchantment with intelligence tests as tools for achieving a more democratic society" (p. 25). Furthermore, the civil rights movement of the 1960s "awakened public consciousness to the deplorable social and economic circumstances of many minority groups, and equality became the watchword" (p. 26). This was the 1960s, a time of tremendous social unrest as well as a time—long overdue—for social justice and racial equality. Any conjectures that differences in intelligence between racial/ethnic groups were genetic in origin were guaranteed to result in vigorous denunciations. Then entered Jensen, a relatively unknown educational psychologist from the University of California, Berkeley.

In 1969, the *Harvard Educational Review* (*HER*) published Jensen's monograph, "How Much Can We Boost IQ and Scholastic Achievement?" (Jensen, 1969).[32] The title of the lengthy article is very significant. The wording, a double-barreled query, cleverly captured the core of Jensen's thesis. He asked two questions regarding Black and poor children: (a) How much can their IQ be raised? and (b) How much can their school achievement be raised? His answer to the first question: very little. His answer to the second: somewhat (but in a prescribed manner).

Jensen's (1969) opening line of his 123-page article was to the point: "Compensatory education has been tried and it apparently has failed" (p. 2). His conclusion was that compensatory education programs for "disadvantaged" children had been tried but that they failed to increase the children's IQs for any significant period of time. Given that compensatory education programs (e.g., Head Start) were theoretically based on the plasticity of human intelligence (Bloom, 1964), Jensen questioned whether the efforts of these programs to raise children's intelligence were being misdirected. A reasonable and rival interpretation, Jensen opined, is perhaps that these children lacked the cognitive capacity for higher level learning. To support this contention, he drew from his research on his "Level I-Level II theory of mental abilities." Interestingly, nearly 30 years later, Jensen (1998) clarified matters that his Level I-Level II theory *"is not really a theory* but rather a set of generalizations about the nature of the W-B [White-Black] differences on cognitive tests" (p. 404, italics added).

Level I, Jensen (1969) claimed, involves lower level skills (e.g., digit memory, serial rote learning, paired associate learning). Level II, by contrast, involves higher order skills (e.g., concept learning, problem solving). Jensen hypothesized, "Level I ability is distributed about the same [i.e., normally] in all social class groups, while Level II ability is distributed differently in lower and middle-SES groups [i.e., positively skewed in low-SES children, negatively skewed in middle-SES children]" (p. 115). Jensen went on to conclude,

> Heritability studies of Level II tests cause me to believe that Level II processes are not just the result of interaction between Level I learning ability and experientially acquired strategies and learning sets. That learning is necessary for Level II no one doubts, but certain neural structures must also be available for Level II abilities to develop, and these are conceived of as being different from the neural structures underlying Level I. *The genetic factors involved in each of these types of ability are presumed to have become differentially distributed in the population as a function of social class,* since Level II has been most important for scholastic performance under the traditional methods of instruction. . . . There can be little doubt that certain educational occupational attainments depend more upon *g* than

upon any other single ability. *But schools must also be able to find ways of utilizing other strengths in children whose major strength is not of the cognitive variety.* One of the great and relatively untapped reservoirs of mental ability in the disadvantaged, it appears from our research, is the basic ability to learn. We can do more to marshal this strength for educational purposes. (pp. 116-117, italics added)

In sum, Jensen (1969) hypothesized that compensatory education failed to boost, to any appreciable degree, the IQs of "disadvantaged" children in such programs because these children had limitations in Level II ability, which Jensen said is measured in cognitive tests with high loadings of *g*. Jensen then moved into an area that would prove to be incendiary in the eyes of many. He discussed evidence that "social class and racial variations in intelligence cannot be accounted for [almost entirely] by differences in environment but must be attributed partially to genetic influence" (p. 2). This is a reasonable hypothesis to raise, Jensen asserted. He came to this conclusion via the following route:

1. Based on his synthesis of the worldwide literature on various kinship correlations of measured intelligence (i.e., for White populations), Jensen (1969) concluded that .81 is "the best single overall estimate of the heritability of measured intelligence that we can make" (p. 51).[33]

2. Although Jensen (1969) acknowledged that (a) his heritability estimate is based on White European and North American populations, (b) no adequate investigations of heritability estimates existed on the Black population in the United States, and (c) heritability estimates do not necessarily apply to intellectual differences *between* populations, he nonetheless concluded,

So all we are left with are various lines of evidence, no one of which is definitive alone, but which, viewed all together, *make it a not unreasonable hypothesis that genetic factors are strongly implicated in the average Negro-white intelligence difference.* The preponderance of the evidence is, in my opinion, less consistent with a strictly environmental hypothesis than with a genetic hypothesis, which, of course, does not exclude the influence of environment or its interaction with genetic factors. (p. 82)

Responses to Jensen's (1969) monograph were swift and astonishing in quantity. Within a few years after the *HER* article was published, there were 117 articles and chapters about Jensen's article published in academic outlets (see bibliography in Jensen, 1972).[34] Furthermore, his article was highly publicized in major print and television news sources (Snyderman & Rothman, 1988). Kelves (1985) commented, "No single publication did more to precipitate the revival [i.e., the issue of race and intelligence]" (p. 269) than Jensen's (1969) *HER* article. Pearson (1991), a staunch supporter of Jensen, hailed him as "the foremost researcher responsible for the revival of 'hereditarian' thought in recent decades" (p. 41).[35] Tucker (1994) described Jensen's publication as "the most explosive article in the history of American psychology, triggering one of the most bitter scientific controversies since Darwin" (p. 199).

In light of the voluminous literature that was stimulated by Jensen's (1969) article—some positive, most negative—and because this debate has been covered extensively, we touch on only a few major criticisms.[36]

1. Jensen's (1969) conclusion that preschool compensatory education was ineffective in increasing the intellectual performance (as measured by IQ) of "disadvantaged" children was based, in part, on the massive Westinghouse-Ohio National Evaluation of Head Start study (Cicirelli, Evans, & Schiller, 1969).[37] This investigation, which was based on a national sample of 102 Head Start centers and nearly 4,000 children, concluded that Head Start failed to produce any significant and lasting cognitive and effective gains in the children. Smith and Bissell (1970), in a major critique of the Westinghouse-Ohio study, contended that there were serious methodological problems (e.g., the use of random rather than stratified sampling, an unrepresentative final sample in that more than half of the target centers refused to participate, Head Start and non-Head Start children ["controls"] not being equated adequately). It also has been noted that the "failure" of Head Start might have been related to gross misuse of expenditures and poorly organized and structured programs (Tucker, 1994).

2. Several scholars have raised criticisms regarding Jensen's (1969) contention that the heritability estimate of intelligence is about .81 for White populations (i.e., Europeans and U.S. Whites). For example, (a) the sample sizes in kinship/heritability of intelligence studies of twins and siblings reared together have been limited (Crow, 1969); (b) a number of assumptions of MZA twins reared apart were not met (Taylor, 1980); (c) Jensen chose the higher heritability estimate, although a range of estimates had been found in human populations (Lewontin, 1973); and (d) the original work on the heritability of intelligence by noted English psychologist Sir Cyril Burt (whom Jensen relied on not solely, but considerably, for deriving his estimate of heritability) is suspected of being fraudulent (Hearnshaw, 1979; Kamin, 1974; Wade, 1976).[38]

3. For his conclusion that study after study has shown that U.S. Blacks, as a group, perform on average about 1 standard deviation lower than the mean of U.S. Whites on IQ tests, Jensen relied heavily on Shuey's (1966) review, a work that was brought into question earlier in this chapter.

4. Jensen (1969) has been criticized for making an unwarranted leap from within-group to between-group variance regarding the estimated .81 heritability of intelligence. That is, heritability estimated for one particular population cannot be applied to another population (Bodner & Cavalli-Sforza, 1970; Cronin, Daniels, Hurley, Kroch, & Webber, 1975; Gage, 1972a, 1972b; Golden & Bridger, 1969; Wallace, 1975; all cited in Fischbein, 1980).

5. Jensen's (1969) implicit schooling recommendation—that educational attempts to boost low-SES children's IQs have been misdirected and that, therefore, schools should focus on teaching specific skills (i.e., Level I learning) that are commensurate to

the abilities of such children—has come under fire by several scholars. The major criticism has been directed toward Jensen's claim that his Level I-Level II theory is a hierarchical model of learning (i.e., Level I precedes Level II). Jensen viewed his two levels as being genetically and independently determined (i.e., factorially distinct). Phillips and Kelley (1975) pointed out, however, that Jensen made an assumption that confused his claim and contradicted his assertion that Level I-Level II abilities are independent.[39] The issues raised by Phillips and Kelley a quarter century ago appear to have merit regarding the validity of Jensen's "theory." As we noted earlier, Jensen (1998) recently recanted, saying that the Level I-Level II theory actually is *not* a theory.

Lloyd M. Dunn

In 1987, a controversial research monograph titled *Bilingual Hispanic Children on the U.S. Mainland: A Review of Research on Their Cognitive, Linguistic, and Scholastic Development*[40] was published by American Guidance Service (a well-known test publisher) and authored by Dunn, the senior author of the Peabody Picture Vocabulary Test (PPVT) series.[41]

Dunn's (1987) monograph, it appears, was a shock in particular to Latino scholars in that his treatise was the *first* research report (88 pages) ever published in which Latinos (Mexican American and Puerto Rican children) were the *prime focus* of a genetic interpretation (in part) regarding Latino-White mean differences in measured intelligence. As we have discussed in this chapter (and likewise covered in Chapter 1 of the present book), historical and contemporary studies of genetically driven explanations of group differences in intelligence have predominantly targeted Black-White comparisons. Such studies of Mexican American children were mostly confined to the 1920s (see Valencia, 1997b). Thus, it was to the dismay of many Mexican American and Puerto Rican behavioral and social scientists when Dunn, after reviewing Latino-White studies of intellectual performance with an emphasis on English/Spanish-language versions of the PPVT–Revised (PPVT-R), concluded,

> While many people are willing to blame the low scores of Puerto Ricans and Mexican-Americans on their poor environmental conditions, *few are prepared to face the probability that inherited genetic material is a contributing factor.* Yet, in making a scholarly, comprehensive examination of this issue, this factor must be included. (p. 63, italics added)

Dunn (1987), referencing the works of Vernon and Jensen (who have studied the nature-nurture issue), continued,

> It is argued that it would be simplistic and irresponsible to contend that the 10- to 12-point IQ differential is due exclusively, or even to all environmental influences combined, including cultural incompatibility. Such naive contentions continue to abound, showing a complete lack of knowledge of the scholarly works of Vernon (1979) and Jensen (1981), among others, who have presented strong cases for the important role of heredity. (p. 64)

Finally, in the absence of any research on heritability estimates of intelligence for Latino populations, Dunn concluded, "Therefore . . ., my best tentative estimate is that about half of the IQ difference between Puerto Rican or Mexican [American] school children and Anglos is due to genes that influence scholastic aptitude, the other half to environment" (p. 64).

Not only is the preceding statement totally unfounded, it demonstrates Dunn's (1987) ignorance about the meaning and interpretation of heritability of intelligence for a population. Suffice it to say, Dunn's monograph resulted in a rapid response by concerned scholars. A distinguished panel of experts, both Latino and White, participated in a symposium at the 1988 meeting of the American Educational Research Association (AERA). Furthermore, the papers presented at that AERA panel were published as critiques later in the year in a special issue of the *Hispanic Journal of Behavioral Sciences* (*HJBS;* see Fernandez, 1988a). Space limitations do not allow a detailed review of these critiques of Dunn's publication here (Berliner, 1988; Cummins, 1988; Prewitt Diaz, 1988; Fernandez, 1988b; Mercer, 1988; Trueba, 1988; Willig, 1988). What we present here is a brief list of representative criticisms:

1. Dunn (1987), although citing Jensen's (1981) hereditarian interpretations about Black-White differences in intelligence, failed "to explain or to defend Jensen's genetic conclusions. He [Dunn] does not review any of the voluminous literature which has criticized Jensen's methods and his conclusions" (Mercer, 1988, p. 200).

2. In his discussion of comparison of vocabulary scores (as measured by the PPVT-R) across Spanish, Puerto Rican, and Mexican American children, Dunn (1987) failed to control for SES (Willig, 1988).

3. "The Spanish version of the PPVT-R was a poor translation and is not appropriate to measure receptive language in Puerto Rican and Mexican children in the United States" (Prewitt Diaz, 1988, p. 249).

4. Dunn's (1987) assertions that Latino pupils "are inadequate bilinguals" (p. 49) and "do not have the scholastic aptitude or linguistic ability to master two languages well, or to handle the switching from one to the other, at school, as the language of instruction" (p. 7) are unsubstantiated and contrary to existing research (Prewitt Diaz, 1988; Willig, 1988).

5. In his contentions that schools are exculpatory in creating inferior education for Puerto Rican and Mexican American students, Dunn (1987) ignored the perspective that structural forces in schools have indeed been implicated in creating inequitable learning climates for Latinos (Fernandez, 1988b).

6. Finally, Valencia and Solórzano (1997) offered a criticism of Dunn (1987) that was not presented in the *HJBS* critiques. It has to do with the measure of "intelligence" emphasized in Dunn's analysis—the PPVT and PPVT-R. In the development of the PPVT, Dunn (1959) conceptualized his instrument as an intelligence test in that raw scores were converted into an IQ. This practice was abandoned in the PPVT-R (Dunn & Dunn, 1981). Yet, raw scores still are converted into standard scores (or age-equivalent scores) with a standardization mean of 100 and a standard deviation of 15 (similar to conventional scales of intelligence). Was the PPVT, and is the PPVT-R, a general measure of the construct of intelligence? Experts in testing say *no*, as shown in the following:

- Salvia and Ysseldyke (1988), on the nature of picture vocabulary tests in general, had these words of caution for test consumers: "It is important to state what these devices measure. The tests are *not* measures of intelligence per se; *they measure only one aspect of intelligence, receptive vocabulary*" (p. 179, italics added).

- Cohen, Swerdlik, and Smith (1992), noting that the practice of using IQ scores on the PPVT was misleading, said this of the revised version: "*The PPVT-R is not an intelligence test* but rather one that measures one facet of cognitive ability—receptive (hearing) vocabulary for standard American English" (p. 370, italics added). Cohen et al. also made the observation that concurrent validity research in which the PPVT-R has been compared to conventional intelligence measures (e.g., Stanford-Binet, Wechsler Intelligence Scale for Children–Revised) has yielded correlation coefficients of low to moderate magnitudes. As such, "these types of correlations should be expected once it is acknowledged that, unlike some of the tests to which it has been compared, *the PPVT-R is not a test of general intelligence*" (p. 371, italics added).

- Anastasi's (1988) comment has clear implications for the assessment of linguistically and culturally diverse children (e.g., Puerto Ricans, Mexican Americans): "Scores on the PPVT-R reflect in part the respondents' degree of cultural assimilation and exposure to Standard American English" (p. 296).

- Sattler's (1992) warning about the PPVT-R was directed specifically to the assessment of Hispanics: "The Peabody Picture Vocabulary Test–Revised should *never* be used to obtain an estimate of young Hispanic-American children's general intelligence" (p. 586).[42]

Even Dunn and Dunn (1981), in the prepublication copy of the PPVT-R manual, offered caveats about what the test is purported and not purported to measure:

> The PPVT-R is designed primarily to measure a subject's receptive (hearing) vocabulary for Standard American English. In this sense, it is an achievement test, since it shows the extent of English vocabulary acquisition.
>
> Another function is to provide a quick estimate of one major aspect of verbal ability for subjects who have grown up in a standard English-speaking environment. In this sense, it is a scholastic aptitude test. It is not, however, a comprehensive test of general intelligence. Instead, it measures only one important facet of general intelligence: vocabulary. (quoted in Salvia & Ysseldyke, 1988, p. 182)

It appears, then, that because the PPVT series does not measure general intelligence, any conclusions about general intellectual or cognitive differences that Dunn (1987) has raised about Latino-Anglo comparisons cannot be supported. At the very best, the PPVT series measures a very small sliver of the construct of intelligence—that being *receptive vocabulary*. Given that lexicon, in general, is very culturally bound, for Dunn to draw any genetic conclusions based on the administration of a 15-minute test in which children point to plates containing vocabulary words of Standard American English is psychometrically indefensible and professionally irresponsible.[43]

Fortunately, for the sake of discourse, Dunn was invited to present a rejoinder to his *HJBS* critics. In some of his comments, he was recalcitrant in addressing criticisms raised by the scholars who found fault in his monograph. For example, Dunn (1988) (a) continued to press his argument that bilingual education for Puerto Rican and Mexican American children does "more harm than good . . . and . . . most such children will have great difficulty even learning to speak Standard English well" (p. 319) and (b) held to his assertion that the educational problems and the low social and economic conditions that Latinos endure are, for the most part, due to their own limitations. Regarding the heavy criticism directed toward his suggestion that genetic factors are partially implicated in explaining the poor performance of U.S. Puerto Ricans and Mexican Americans on standardized scholastic aptitude tests, Dunn announced a shift in his thinking:

> Thanks to such critiques as have been presented in this journal, and further study and additional research on my part, I now view my monograph as only a "working paper" that is badly in need of extensive revision. It is now clear that many people were offended by certain of my comments, especially those suggesting that the poor performance of impoverished Hispanic children may be due, in part, to genetic factors. Now I see that the insertion of this element into my discussion was a tactical error and a distraction. From my point of view, it is not a central issue. Since it has aroused such strong reactions, I will downplay this point of discussion or eliminate it from any rewrite, so that the main body of my report may get more attention. Therefore, I wish to retract my statements in this area. Both AGS [American Guidance Service] and I now recognize that we showed [a] lack of sensitivity toward Mexican Americans and Puerto Ricans in introducing this point of contention. We apologize. I cannot resist saying, however, that none of us is consistently correct and that all of us end up with egg on our faces from time to time. (pp. 302-303)

It appears that Dunn's response on the genetic issue was more of an apology for offending people than a recantation of making scientifically unjustifiable conclusions about Latino-White differences in intelligence. In any event, the damage has been done.

Richard J. Herrnstein and Charles Murray

In the fall of 1994, *The Bell Curve: Intelligence and Class Structure in American Life* (Herrnstein & Murray, 1994) was published and immediately set off a maelstrom of disputation. The authors were the late Herrnstein (Ph.D. from Harvard University; psychology professor at Harvard from 1958 to 1994) and Murray (Ph.D. from the Massachusetts Institute of Technology; currently a Bradley fellow at the conservative research group, the American Enterprise Institute, in Washington, D.C.). Although Herrnstein and Murray sought to shed new light on the complex relations among social class, race, heredity, and intelligence, the two authors *never have published* any peer-reviewed scientific journal articles on the genetic basis of intelligence and its relation to race or poverty (Dorfman, 1995).

The reader might be asking why we are including a discussion of *The Bell Curve*, a book whose subtitle claims that it deals with "class structure," in our present chapter on the genetic hypothesis regarding racial/ethnic differences in intelligence. Giroux and

Searls (1996) commented that Murray, in his defense of *The Bell Curve*, has asserted that race is not a focal point of study in the book given that only 1 of 21 chapters focuses on ethnic differences in cognitive ability. By contrast, Murray has noted, 8 chapters (Part II of that book) are devoted to data analyses based exclusively on *White* samples. Giroux and Searls pointed out, however, that "this demonstrates that a third of the book is organized around questions of race—unless one is willing to make the argument that whiteness is not a racial category" (p. 73). Other scholars (Early, 1995; Nunley, 1995) contend that about one half of the book deals with race/ethnicity. Based on our own reading of *The Bell Curve*, the book indeed does have a central focus on race/ethnicity (i.e., White, minority). Given the strong colinearity between race/ethnicity and SES in the United States, it is disingenuous for Herrnstein and Murray to claim that *The Bell Curve* is predominantly about the "class structure" in our society.

Herrnstein and Murray's (1994) cognitive partitioning thesis contended that (a) having a high IQ greatly improves one's life chances of social mobility and possessing desirable behaviors and that (b) having a low IQ places one at substantial risk for possessing undesirable behaviors. The authors derived their thesis and larger model of social stratification as follows:

1. High IQ is an invaluable raw material for social and economic success in American society.

2. "Intelligence itself, not just its correlation with socioeconomic status, is responsible for these group differences" (Herrnstein & Murray, 1994, p. 117). That is, different levels of cognitive ability are linked with different patterns of social behavior; intelligence is endowed unequally across social classes.

3. Within the White population, Herrnstein and Murray noted that the heritability estimate for intelligence is between .40 and .80. Regarding racial/ethnic differences in intelligence, the authors stated that they are "resolutely agnostic" on the relative contributions of nature and nurture to "the mix" (p. 311). Notice that Herrnstein and Murray said that they are agnostic on the relative *proportions* of genetic and environmental influences on racial/ethnic differences in intelligence. They are not agnostic that such influences exist, however, as indicated by this assertion: "It seems *highly likely* that both genes and environment have something to do with racial differences [in intelligence]" (p. 311, italics added). Furthermore, Judis (1995) commented that the sources that Herrnstein and Murray marshal for this claim of agnosticism actually are weighted toward genetic causes.

4. The cognitive elite, through concentrated social pools and self-selection (i.e., assortative mating), has emerged. Members of this class (who are disproportionately White) have become the controllers of power, privilege, and status, "restructuring the rules of society so that it becomes harder and harder for them to lose" (Herrnstein & Murray, 1994, p. 509).

5. Concomitant with the increasing isolation of the cognitive elite and its growing influence over the control of America is the growth and perpetuation of the cognitive underclass (disproportionately Latinos and Blacks) and its accompanying intractable social problems. Furthermore, Herrnstein and Murray (1994) claimed, there is an alleged dysgenic effect presently occurring in which the intellectually disadvantaged (disproportionately Latinos and Blacks, according to the authors) are having the highest fertility rates.

6. Given that (a) the cognitive underclass allegedly is deficient in intellectual endowments and abilities and (b) attempts to raise the IQs of members of this class have been disappointing, a national policy agenda needs to be set in motion.

7. Such policy considerations that should be considered, for example, include evaluation of the immigrant situation ("legal" and "illegal") in that "immigration does indeed make a difference to the future of the national distribution of intelligence" (Herrnstein & Murray, 1994, p. 358); this is particularly seen in the cases of "Latino and black immigrants [who] are, at least in the short run, putting downward pressure on the distribution of intelligence" (pp. 360-361).

8. Regarding policy directed toward educational reform, Herrnstein and Murray (1994) tangentially mentioned the use of "national achievement tests, national curricula, school choice, vouchers, tuition tax credits, apprenticeship programs, restoration of the neighborhood school, minimum competency tests, [and] ability grouping" (p. 435).[44] One reform suggestion that they discussed in some detail, however, is a need for more attention and funding for the gifted, who they claimed were "out" when economically disadvantaged became "in."[45] To meet the needs of the neglected gifted, Herrnstein and Murray advocated that the federal government *"reallocate some portion of existing elementary and secondary school federal aid from programs for the disadvantaged to programs for the gifted"* (pp. 441-442).

9. Finally, Herrnstein and Murray (1994), in *The Bell Curve*'s penultimate chapter, offered a speculation of the future impact of cognitive stratification on American life and the workings of the government. Their prediction of the future, which can best be described as resembling a horrendous caste system, is frightening and has a calamitous tone for those people who will occupy the bottom rungs of the cognitive ability continuum. The authors commented, "Like other apocalyptic visions, this one is pessimistic, perhaps too much so. On the other hand, there is much to be pessimistic about" (p. 509).

Herrnstein and Murray (1994) contended that this deplorable future scenario, if realized, will be shaped by three tendencies currently in motion: (a) a cognitive elite that is increasing in its isolation, (b) a fusing of the cognitive elite with the affluent sector of society, and (c) a worsening of the quality of life for the cognitive underclass. The authors asserted that the merging of the cognitive elite and the affluent eventually will reach such a level of wealth that "the haves [will] begin to feel sympathy toward, if not guilt about, the condition of the have-nots" (p. 523). What will result, Herrnstein and Murray speculated, is the development and implementation of an "expanded welfare state." By "expanded," the authors suggested that this will be a welfare state in which the cognitive elite's motives are guided by sympathy *and* fear and hostility toward the underclass recipients. In that the new coalition of cognitive and affluent elites will have grown weary of spending money on "remedial social programs" that do little to advance the underclass, coupled with the elites' enmity toward the have-nots, the social program of choice will be the *custodial state*.[46]

Herrnstein and Murray (1994) commented that the coming of the custodial state will be influenced, over the next two decades, by a growing acceptance of the belief that members of the underclass are in dire conditions "through no fault of their own but because *of inherent shortcomings* about which little can be done" (p. 523, italics added). As such, there will be more open discussion among politicians and intellectuals that members of the underclass cannot, for example, fend for themselves (e.g., from violence, child abuse, and drug addiction) or be trusted to spend cash appropriately (thus, the custodial state will rely more on services than on cash for the underclass).

As was seen in the case of Jensen's (1969) article, there has been a titanic outpouring of critical writings and debate in response to *The Bell Curve*. Our literature search identified five books about *The Bell Curve* (Dickens, Kane, & Schultze, 1996; Fischer et al., 1996; Fraser, 1995; Jacoby & Glauberman, 1995; Kinchloe, Steinberg, & Gresson, 1996)[47] and nearly 70 published book reviews and commentaries.[48] Based on the literature we have reviewed, the responses to *The Bell Curve* have been, for the most part, negative, a point acknowledged even by Murray, coauthor of the book. The book was released in October 1994, and Murray commented in May 1995, "The initial reaction was encouraging. . . . Then came the avalanche. . . . Most of the comment has been virulently hostile" (Murray, 1995, p. 23). We agree with Murray up to a point. Indeed, some of the responses have been written in an ad hominem and invective manner, but a good number of the reactions have been thoughtful, scientifically based, and well informed.[49] Given the enormous reaction to *The Bell Curve*, it is well beyond the scope of this chapter to provide a comprehensive and integrative analysis of these responses. What we offer is a very brief encapsulation of this body of criticism.[50] For ease of discussion, we have organized these critiques around central themes that make most sense to us. Following is our analysis:

1. On what is new in *The Bell Curve*

Many critics assert that there is little new in *The Bell Curve* regarding argumentation germane to race/ethnicity, class, and intelligence and that the book rests on an old, disreputable, and debunked paradigm of genetic explanations of inequality (Duster, 1995; Fischer et al., 1996; Gould, 1995; Reed, 1994; Scott, 1994; Valencia & Solórzano, 1997). Duster (1995), for example, commented, "It should now be clear that the extraordinary success of this book is not a function of the presentation of new information, nor of the restructuring of a new line of argumentation by reassembling old data in a coherent and convincing manner" (p. 160). In a similar vein, Gould (1995) voiced this opinion: "*The Bell Curve*, with its claims and supposed documentation that race and class differences are largely caused by genetic factors and are therefore essentially immutable, contains no new arguments and presents no new compelling data to support its anachronistic social Darwinism" (p. 11).

2. On statistical data presented in *The Bell Curve*

There are comments by critics that the observed associations (correlations) between the main predictor variables (measured intelligence and SES) and the various social behaviors are weak in strength (Gardner, 1995; Gould, 1995; Hendricksen, 1996). On this issue, Gould (1995) noted,

> Indeed, almost all their [Herrnstein & Murray's (1994)] relationships are weak; very little of the variation in social factors is explained by either independent variable [IQ or SES]. ... Their own data indicate that IQ is not a major factor in determining variation in nearly all the social behaviors they study—and so their conclusions collapse. . . . Most of Herrnstein and Murray's correlations are very weak—often in the 0.2 to 0.4 range. (pp. 19-20)[51]

There also has been criticism that Herrnstein and Murray have approached the use of correlational analysis in an uncritical manner, that is, drawing conclusions of causality from correlations (see, e.g., Carspecken, 1996; Kamin, 1995; Sowell, 1995).

3. On the construct of intelligence embraced in *The Bell Curve*

Drawing from the classicist tradition, Herrnstein and Murray (1994) embraced and defended the construct of g as the central dominant perspective of what intelligence is. Some critics have contended that in so doing, Herrnstein and Murray shunned alternative perspectives of the intelligence construct (see, e.g., Kinchloe & Steinberg, 1996; Reed, 1994).

Another related criticism that has been leveled against *The Bell Curve* has to do with the basic measure of "intelligence" used by Herrnstein and Murray (1994) in their analyses, that is, the Armed Forces Qualifying Test (AFQT). Fischer et al. (1996), based on their own empirical analyses, provided an informed and insightful discussion that the AFQT (a) is more a measure of school tasks (high school-level mathematics and reading) than of general intelligence (e.g., g); (b) strongly correlates with years of schooling of the test taker; (c) was a better predictor of "past schooling than . . . [of] future schooling. That is, the AFQT measured what test takers had already learned, not their ability for future learning" (p. 64); and (d) was not administered under standardized conditions.

4. On the issue of raising cognitive ability in *The Bell Curve*

Based on their review of the literature, Herrnstein and Murray (1994) concluded, "No one yet knows how to raise low IQs substantially on a national level" (p. 416). A major criticism of Herrnstein and Murray is that they ignored "literally scores of studies that document the benefits of educational intervention" (Kinchloe & Steinberg, 1996, p. 36; for concurrences about the workability of interventions, see Duster, 1995, and Finn, 1995).

5. On the political nature of *The Bell Curve*

Of the different criticisms directed at *The Bell Curve*, it appears that the most frequently cited is that Herrnstein and Murray (1994), behind the smokescreen of science (some would say pseudoscience), used their book to tacitly outline and push their agendas on affirmative action bans, immigration restriction, and welfare reform (see, e.g., Carspecken, 1996; Cary, 1996; Cross, 1996; Gardner, 1995; Giroux & Searls, 1996; Gould, 1995; Hauser, 1995; Jones, 1995; Kinchloe & Steinberg, 1996; Lugg, 1996; Wolfe, 1995). For example, Kinchloe and Steinberg (1996) asserted that *The Bell Curve* is "the theoretical torch bearer for the right-wing insurgency of the 1990s" (p. 4). Gould (1995) opined that "*The Bell Curve* is scarcely an academic treatise in social theory and population genetics. It is a manifesto of conservative ideology; the book's inadequate and biased treatment of data displays its primary purpose—advocacy" (pp. 20-21).

6. On the nature-nurture debate in *The Bell Curve*

Surprisingly, few critics of *The Bell Curve* have drawn from the literature in behavioral genetics that has found no support for a genetic hypothesis in explaining racial/ethnic mean differences in intelligence. One scholar who discussed this research from behavioral genetics is Nisbett (1998), who raised, for example, the extremely important point that "estimates of heritability *within* a given population need not say anything about the degree to which differences *between* populations are genetically determined" (p. 87). In the next section, we discuss research findings from the behavioral genetics field that the genetic hypothesis is not supportable.

7. On the absence of rival interpretations in *The Bell Curve*

In our view, this is the most powerful criticism of Herrnstein and Murray (1994). Theories in the behavioral and social sciences describe, explain, and predict human behavior, and they often offer recommendations for behavioral change. If one's nomological net is not flexible enough to cast for competing interpretations, then dogma is likely to take form for what should be competitive discourse. Explicitly or implicitly stated in the many criticisms of *The Bell Curve* is the failure on the part of Herrnstein and Murray to entertain competing interpretations for their conclusions that individuals of low and high IQ end up situated in the "cognitive underclass" and "cognitive elite," respectively. In sum, what might be a rival and perhaps sounder way in which to explain America's system of inequality? To address this, Fischer et al. (1996)—six sociology professors at the University of California, Berkeley—reanalyzed the very same National Longitudinal Survey of Youth (NLSY) survey data used by Herrnstein and Murray (1994), who came to the conclusion that inherited differences in intelligence between social classes explain inequality. The results and conclusion of the reanalysis by Fischer et al. (1996) were published in *Inequality by Design: Cracking the Bell Curve Myth*. Through the use of a macro-level structural model involving caste status, social circumstances, social policies, inequal distribution of wealth, and challenges to deficit thinking, Fischer et al. offered a compelling study whose findings rebut the deterministic model presented in *The Bell Curve*. The authors offered an explanation of societal inequality that is antithetical to the one generated by Herrnstein and Murray. Fischer et al. (1996) argued, *"A racial or ethnic group's position in society determines its measured intelligence rather than vice versa"* (p. 173). Or, stated even more directly, *"Groups score unequally on tests because they are unequal in society"* (p. 172).

As some critics have opined, *The Bell Curve* offers little new information as to understanding contemporary social stratification in the United States and is based on an old refuted model. Valencia (1997b) identified a number of ideological and "scientific" streams that helped to shape 1920 hereditarianism and are pertinent to this particular criticism leveled against *The Bell Curve*. The major forces were (a) Galton's belief that one's social status was genetically predetermined, as were socially desirable and undesirable behaviors; (b) Terman's (and others') view that intelligence was largely innately based and predicted, fairly accurately, one's eventual social, economic, and occupational status; and (c) McDougall's perspective of a society that can and should be divided between the haves and the have-nots, where the former are more deserving to reap societal benefits, should control the latter, and need to reproduce at greater levels than the latter. A close examination of *The Bell Curve* reveals that Herrnstein and Murray (1994) incorporated, in whole or in part, these pseudo-scientific, historically refuted ideas into their work. Interestingly, the fundamental question that Herrnstein and Murray posed was remarkably similar to what McDougall (1921) posed nearly 80 years ago in *Is America Safe for Democracy?*: "Does the social stratification of society correspond to, is it correlated with, a stratification of intellectual capacity?" (p. 62). In sum, if one is to reject 1920s hereditarianism as bad science and policy, then *The Bell Curve* must be rejected on the same basis.

Findings From Scientific Research Investigations

In the previous subsections, our focus was on the works of scholars (e.g., Shuey, Jensen) who offered opinions and conjectures (more so than scientific evidence) that between-group differences in measured intelligence (with an emphasis on African American-White differences) are mostly attributed to genetic factors. In this subsection, we turn to a brief coverage of those empirical studies that sought to examine the hypothesis that the persistent gap in measured intelligence in African Americans and Whites is largely genetic in basis. To address the scientific evidence for and against the genetic hypothesis, there are two forms of evidence: indirect and direct (Brody, 1992; Flynn, 1999). Indirect types of evidence are derived, for example, from (a) studies of heritability estimates of intelligence, (b) investigations in which there are attempts to control for environmental variables (e.g., SES),[52] and (c) analyses of racial differences in brain size.[53] Direct evidence, by contrast, is garnered in studies of (a) transracial adoptions and (b) racial admixture. The scholarly pursuit of understanding racial differences in measured intelligence regarding arguments for and against genetic bases is a complex enterprise. For explications of the logic, methodology, and evidence, we refer the interested reader to the works of, for example, Brody (1992), Flynn (1999), and Mackenzie (1984). Our intent here is merely to provide a brief summary and conclusions of studies that have focused on indirect or direct forms of evidence.

Let us first examine the indirect evidence. We confine our discussion to heritability of intelligence. As we have discussed, methodological and statistical issues have been raised concerning how heritability estimates of intelligence are derived. Notwithstanding these concerns, a strong consensus exists that intelligence in the White population is a significantly heritable trait. There is an enormous amount of data to support this assertion. The reader might be surprised, however, that very little data exist regarding heritability of intelligence in African American samples and that the data that do exist are inconclusive.

In *Race Differences in Intelligence*, one of the most comprehensive reviews of the nature-nurture controversy pertaining to African American-White differences in measured intelligence, Loehlin, Lindzey, and Spuhler (1975) identified just a handful of pertinent studies pertaining to heritability of intelligence in the African American population (e.g., Jensen, 1973a; Nichols, 1970; Osborne & Gregor, 1968; Osborne & Miele, 1969; Scarr-Salapatek, 1971; Vandenberg, 1969, 1970). Loehlin et al. (1975) addressed several problems with these studies including that (a) sample sizes were too small to allow accurate estimations of heritability; (b) there were strikingly different conclusions (e.g., heritability estimates were lower for Blacks than for Whites [Nichols, 1970; Vandenberg, 1970], heritability estimates for Blacks and Whites were comparable in magnitude [Osborne & Miele, 1969]); and (c) different measures of intelligence were used at different grade levels (Scarr-Salapatek, 1971).

An extremely important, but often overlooked, issue regarding the nature-nurture debate and racial differences in intelligence is the following. Let us suppose that the heritability estimates for the African American and White populations were equal and high in magnitude. If this were the case, then could genetics be hypothesized as the basis for largely explaining the 1 standard deviation difference in measured intelligence? Brody (1992) noted, "It is easy to answer this question in one word—no" (p. 299). As Brody and others (Lewontin, 1975; Mackenzie, 1984) have discussed, this resounding *no* is based on the logic derived from an understanding of the sources of between- and within-group differences. In explicating this logic, scholars frequently cite the brilliant classic example (from plant genetics) that was given by Lewontin (1970), who pointed out that the fundamental error of Jensen's (1969) hypothesis about between-group heritability was that he confused heritability of a trait within a population with heritability of the difference between two populations. That is, it is possible to have a situation in which genetic factors explain nearly all of the variance within two different populations, yet environmental factors could explain an average difference between the two. Given that Lewontin's (1970) famous example is quite long, here is an abbreviated version offered by Flynn (1999):

He [Lewontin] imagined a sack of seed corn with plenty of genetic variation. The corn is randomly divided into two batches, each of which will therefore be equal for overall genetic quality. Batch A is grown in a uniform and optimal environment, so within that group all height differences at maturity are due to genetic variation; Batch B is grown in a uniform environment that lacks enough nitrates, so within that group all height differences are also genetic. However, the difference in average height between the two groups will, of course, be due entirely to the unequal quality of their two environments. (p. 13)

We now shift to a short discussion of investigations of direct evidence tests of the genetic hypothesis of racial differences in intelligence. Our focus is on transracial adoptions and racial admixture. Based on a review by Brody (1992), there are two studies that have measured the intelligence of African American children raised in predominantly White environmental settings (Moore, 1986; Scarr & Weinberg, 1976).[54] We have located several others (Scarr & Weinberg, 1977; Scarr, Weinberg, & Waldman, 1993; Weinberg, Scarr, & Waldman, 1992). The design of transracial adoption studies lends itself to an interesting challenge. Scarr and Weinberg (1976) noted that under the genetic hypothesis, Black adopted children probably will fall in IQ performance "below that of other children reared in white upper middle-class homes. On the other hand, if black children have a range of reaction similar to other adoptees, their IQ scores should have a similar distribution" (p. 727).

Space limitations do not permit a thorough review of the transracial adoption literature, so we provide a few highlights of these findings. In their study, Scarr and Weinberg (1976) reported that the Black and interracial children (adopted by highly educated, above average in occupation, White parents in Minnesota) had a mean IQ lower than that of the natural children of the adopted parents but about 20 points above the mean IQ of Black children raised by their natural parents in the north central region.[55] Scarr and Weinberg interpreted the data to suggest that "the high IQ scores of the socially classified black adoptees indicate malleability for IQ under rearing conditions that are relevant to the tests and the schools" (p. 726). In addition, Scarr and Weinberg concluded that the social environment plays a paramount role in shaping the mean IQ of the participating Black children "and that both social and genetic variables contribute to individual variation among them" (p. 739).[56] In a 10-year follow-up of the children, Weinberg et al. (1992) found that there were no significant differences in IQ scores (from Time 1 to Time 2) between the transracial adoptees and the natural children of the adopting parents.

Moore (1986) undertook a study of traditionally adopted Black children (i.e., Black children adopted by middle-class Black parents) and transracially adopted Black children (i.e., Black children adopted by middle-class White parents). In all, the mean Wechsler Intelligence Scale for Children (WISC) IQ of both groups (tested between 7 and 10 years of age) was significantly higher than the mean IQ typically observed (in general) in Black children, thus providing support for the environmental interpretation proffered by Scarr and Weinberg (1976). Moore (1986) did find, however, that the mean IQ of the transracially adopted group ($M = 117.1$) was significantly higher than that of the traditionally adopted group ($M = 103.6$). Based on an analysis of styles of responding to test demands demonstrated by the two groups, Moore concluded that the transracial group scored higher on the WISC because of being raised "in environments that are [more] culturally relevant to the tests" (p. 321).

Another research design to examine the genetic versus environmental hypothesis of racial differences in measured intelligence is that of racial admixture, which can be assessed in three ways: genealogical records, biochemical assays and protein markers, and visible indicators of race such as skin color (Mackenzie, 1984).[57] Given that the African American population is racially heterogeneous (estimated 20% to 30% European [White] ancestry; Reed, 1969),[58] racial admixture studies can serve as potentially powerful tools for examining environmental and genetic conjectures about sources of racial differences in measured intelligence (Mackenzie, 1984; Scarr, Pakstis, Katz, & Barker, 1977).

A genetic hypothesis suggests that the amount of White ancestry in African Americans is positively correlated with measured intelligence. As we saw in Chapter 1 of this book, this is not a new hypothesis. The study of racial admixture was a prominent focus among race psychologists of the 1920s, who examined the relation between skin color and intelligence with African American, Mexican American, and American Indian children (Valencia, 1997b).

An earlier review of the racial admixture literature is provided by Loehlin et al. (1975). Based on their analysis of the available literature, Loehlin et al. noted that higher IQ correlated with lighter skin color among Blacks. The authors did qualify matters, however, stating that "these correlations have tended to be quite low and thus compatible with either the genetic or environmental explanations" (p. 132). Loehlin et al. concluded, "Recent studies of U.S. interracial matings and ability-blood group correlations suffer from methodological limitations and small samples but on the whole fail to offer positive support to hereditarian positions concerning between-group differences" (p. 132).

The investigation by Scarr et al. (1977) stands as one of the most comprehensive and ambitious racial admixture studies undertaken. More than 400 Black and White children (ages 7 to 10 years) attending schools in Philadelphia served as participants. Based on analyses of skin color reflectance, blood group markers, and ancestral phenotype frequencies, Scarr et al. concluded that the relation between degree of African ancestry "and intellectual skills failed to provide evidence for genetic differences in intelligence" (p. 84). Other scholars who have reviewed the racial admixture literature have rendered similar or unequivocal conclusions (see, e.g., Brody, 1992; Mackenzie, 1984; Nisbett, 1998).

Conclusions

What does the preceding discussion lead us to conclude about the possibility of genetic factors playing a significant role in explaining mean differences in measured intelligence between African American and White populations? Based on (a) the fact that there are extremely limited data on heritability estimates of intelligence for the African

American population, (b) the inability to draw conclusions about between-population heritability differences in intelligence from heritability estimates for the White population (which appear to be about .50), and (c) the nonsupportive indirect and direct evidence regarding the genetic hypothesis of racial differences in intelligence, we can conclude that *the hypothesis of genetic differences between African American and White populations in phenotypic differences in intelligence cannot be confirmed*. We certainly are not alone in drawing such a conclusion. Scarr and Carter-Saltzman (1982), drawing from Scarr and associates' research, noted, "The hypothesis of genetic differences between the races [Black and White] fails to account for the IQ performance differences" (p. 815). Similar conclusions, based on the available evidence, have been offered by a number of other scholars (see, e.g., Brody, 1992; Loehlin et al., 1975; Neisser et al., 1996; Nisbett, 1998; Taylor, 1980).[59] In *Intelligence*, which has been described by Plomin et al. (1997) as the "major textbook on intelligence" (p. 138), Brody (1992) concluded,

> While it may be difficult to definitely rule out a genetic hypothesis on the basis of the available evidence, I think it is also fair to say there is no convincing direct or indirect evidence in favor of a genetic hypothesis of racial differences in IQ. (p. 309)

Regarding environmental factors, we are in agreement with Nisbett (1998), who asserted that there is "strong evidence for a substantial environmental contribution to the IQ gap between blacks and whites" (p. 101). We also would extend this point to be germane in explaining the IQ gap that exists between the White population and other minority groups (e.g., Mexican Americans, Puerto Ricans). Our discussions in this book on SES (Chapter 3), home environment (Chapter 4), and test bias (Chapter 5) contained considerable evidence to support the position that environmental factors are strongly involved in explaining the average differences in measured intelligence between White and caste-like minority groups.

So, as we enter the new millennium, what is likely to transpire regarding research on the hypothesis that racial/ethnic differences in measured intelligence are significantly genetic in basis? To address this query, we are drawn to William H. Tucker's *The Science and Politics of Racial Research* (Tucker, 1994). The central concern that Tucker raised in his book—an issue, by the way, that needs to be discussed more frequently—is the "why" question of racial research on intellectual differences. Tucker asked why so much research has been done on this topic given that this area has generated little scientific value and that, although having the imprimatur of science, such research invariably has been linked (in actual and suggested ways) to policies of maleficence. And, indeed, there have been many such repressive policies, starting with slavery and unfolding through the decades as witnessed by anti-miscegenation laws, forced school segregation, immigration restrictions, curtailment of social welfare programs, opposition to school integration, calls for birth control, and pressures to abandon compensatory education programs. As we traverse the portal to the new millennium, we urge the reader to consider the words of Tucker:

For well over a hundred years, some of the finest scientific researchers with the best academic credentials have investigated racial differences in intelligence, yet this considerable investment of resources has produced little of scientific value. . . . Even if there were convincing proof of genetic differences between races, as opposed to flawed evidence that has been offered in the past, it would serve no purpose other than to satisfy curiosity about the matter. While the desire for knowledge, whether or not it has practical value, is not to be denigrated, a judicious use of our scientific resources would seem inconsistent with the pursuit of a goal that is probably scientifically chimerical and certainly lends itself to socially pernicious ends. (p. 8)

We do not want to leave the reader with the impression that we are advocating that genetic research on *individual differences* in measured intelligence should be abandoned. Such research should be encouraged. As Plomin and Petrill (1997) discussed, there are several exciting discoveries and ongoing research pertaining to genetics and intelligence. These advances involve research on genetic causes of mental retardation, developmental genetic analysis (genetic influence on intelligence through the life span), multivariate genetic analysis (genetic overlap among specific cognitive abilities), environmental genetics (developmental interface between nature and nurture), and molecular genetics (moving toward the identification of specific genes that are responsible for a significant influence on measured intelligence).

On a final note, we need to underscore the point raised by Angoff (1988), who argued that the controversy over whether measured intelligence is largely genetically or environmentally influenced is actually irrelevant when discussing racial/ethnic group differences. According to Angoff (and we strongly agree with his position), "The real issue, the one that should instead be the focus of debate in this context, is whether intelligence can be *changed*, and if so, how and under what circumstances" (p. 713). Referring to research demonstrating that intellectual performance and aptitude can be increased under appropriate conditions, Angoff commented, "These studies have important implications for future performance of minorities in this society" (p. 713). Angoff also noted that there would need to be efforts of substantial proportions to eliminate the mean differences between White and minority groups in measured intelligence. Given the robust connection between schooling attainment and measured intelligence (Ceci, 1991), we assert that a major starting point to eliminate the gap would be to provide equal educational opportunities for caste-like minority groups.

PART III

Assessment Issues

7

Race/Ethnicity, Intelligence, and Special Education

The public school system is the largest consumer of individually administered intelligence tests in the United States. Scores obtained on intelligence measures have been identified by many scholars as a primary factor in the determination of special education eligibility, classification, and placement. Numerous controversies have arisen in this educational arena with regard to charges of cultural test bias (see Chapter 5 of this book) and overrepresentation of particular racial/ethnic groups (e.g., African Americans) in specific special education categories (i.e., mild/educable mentally retarded and learning disabled). Acknowledgment of the overrepresentation problem was brought to the forefront several decades ago by Dunn (1968, cited in Artiles & Trent, 1994).

Recently, New York State was cited as failing to meet the requirements of the Americans with Disabilities Act (Hernández, 1999). Specifically noted was the disproportionately high number of Black and Latino male students in special education programs in New York City. In addition, concern was raised regarding the often restrictive placements (i.e., separate self-contained classrooms) and the limitations this placed on the possibility of these students graduating with regular high school diplomas. To place these concerns in perspective, the Special Education Program in the state of New York consists of 400,000 students (84,000 of them in New York City alone), with a current annual budget of $1.4 billion (Hernández, 1999).

The situation pertaining to minority students in special education programs in New York represents only a fraction of the problems that continue to confront the U.S.

educational system with regard to providing appropriate services to all children. Controversies involving appropriate referrals for special education services, as well as accurate classification and placement, have existed for decades in the case of minority students. The purpose of this chapter is to address the following areas pertaining to these concerns: (a) historical issues regarding the use of intelligence tests in educational placement, (b) the role of intelligence tests in the definitions of handicapping conditions, (c) defining mental retardation and LD, (d) race/ethnic incidence data regarding the mentally retarded and learning disabled, and (e) future directions for special education.

Historical Issues Regarding the Use of Intelligence Tests in Educational Placement

Public Law 94-142 and Related Amendments

Issues of test bias have been addressed in heated debates within the judicial system. Results have been conflicting, depending on the context of the particular case and those determining the final outcome. Public Law (P.L.) 94-142 ("Education of Handicapped Children," 1977) required that all children (i.e., school age) must receive free and appropriate education in the least restrictive environment. This included the right to appropriate psychological and educational assessments as well as a number of other provisions (e.g., mandate of nondiscriminatory assessment). A downward extension of P.L. 94-142 took the form of the Education of the Handicapped Act amendments of 1986 (P.L. 99-457), mandating services for children between 3 and 5 years of age. P.L. 99-457 also provided incentives to states for providing free and appropriate services for children from birth to 2 years of age. Stipulations of the Individuals With Disabilities Education Act (IDEA) of 1990 included that multidisciplinary teams conduct psychoeducational evaluations and that nondiscriminatory assessment methods be used including selecting and administering measures that are not culturally or racially discriminatory (López, 1997). In addition, measures must be validated appropriately for specific purposes. Furthermore, Part H of the IDEA specifically focuses on the assessment of infants and toddlers providing for early intervention programs. Part H "requires a comprehensive multidisciplinary evaluation in the areas of cognitive development, physical development, communication development, social or emotional development, and adaptive development" (cited in Bracken & Walker, 1997, p. 486).

Most recently, the Individuals with Disabilities Education Act amendments of 1997 (IDEA, 1997) highlight the educational climate prior to P.L. 94-142 and after its implementation. Given these changes, the IDEA amendments also emphasize the importance of the federal government becoming more responsive to the "growing needs of an increasingly more diverse society." The rapidly changing demographic racial/ethnic profile of the United States is noted:

> By the year 2000, this nation will have 275,000,000 people, nearly one of every three of whom will be either African-American, Hispanic, Asian-American, or American Indian.

Taken together as a group, minority children are comprising an ever larger percentage of public school students. Large-city school populations are overwhelmingly minority. For example, for fall 1993, the figure for Miami was 84 percent; Chicago, 89 percent; Philadelphia, 78 percent; Baltimore, 84 percent; Houston, 88 percent; and Los Angeles, 88 percent.

The IDEA amendments also noted that the numbers of school-age children who are limited English-proficient (LEP) are quickly growing in the United States. In addition, discrepancies in the levels of referral and placement of LEP children in special education have been reported in the IDEA amendments.

With regard to minority students, the following concerns were noted:

1. Greater efforts are needed to prevent the intensification of problems connected with mislabeling and high dropout rates among minority children with disabilities.
2. More minority children continue to be served in special education than would be expected from the percentage of minority students in the general school population.
3. Poor African American children are 2.3 times more likely to be identified by their teachers as having mental retardation than are their White counterparts.
4. Although African Americans represent 16% of elementary and secondary enrollments, they constitute 21% of total enrollments in special education.
5. The dropout rate is 68% higher for minorities than for Whites.
6. More than 50% of minority students in large cities drop out of school.

Court Cases Affecting Testing for Special Education and Placement

Numerous court cases have arisen due to problems regarding the classification and provision of services to special education students. As we introduced in Chapter 1 of this book, many of these cases pertain to the concerns of minority students and their families. In a review of the *Yearbook of School Law* (1950-1953, 1955-1997) and Sattler's (1990) extensive review, we identified approximately 21 seminal court cases that challenged the use of particular assessment procedures with minority students. Although many of these cases involve complaints of minority plaintiffs who claimed that they were inappropriately placed in classes for children with mild mental retardation (based on, in part, intelligence tests), other cases pertain to issues that involve other forms of testing (e.g., high-stakes tests used for high school diploma award or denial).

The use of various testing practices within the educational system has been cited in the literature as negatively affecting minority students from the early 1960s to the mid-1980s. It was interesting to find that no cases regarding testing of minority students were cited in the *Yearbook of School Law* editions published during the 1990s. It should be noted that our review specifically focused on the sections focusing on cases pertaining to "pupils."

Based on our review, the past legal issues surrounding testing appear to center on the following areas:

1. The use of tests as part of the admissions process
2. Educational tracking of students (e.g., the use of tests in placement decisions)
3. Testing of non-native English-speaking students
4. The use of tests as promotion and graduation requirements
5. Classification of educable mentally retarded students

With regard to the use of tests as admission standards, a number of early cases focused on issues pertaining to the desegregation of schools and application of admissions standards for Black students (e.g., *Jones v. School Board of the City of Alexandria*, 1961, cited in Garber, 1961). As noted, procedures for admission (i.e., testing practices) were not uniformly applied to Black and White students. In the cited cases, the court ruled in favor of fair test practices for all students regardless of race/ethnicity.

Educational tracking based on test scores has been an issue for decades (e.g., *Hobson v. Hansen*, 1967, cited in Sattler, 1990). Tracking based on test scores is of great concern for minority students because these special education settings often offer little in terms of future educational potential (e.g., fewer numbers of students obtain high school diplomas). Students in special education tend to experience lower expectations from their teachers, drop out of school at higher rates, and are more likely to be unemployed subsequent to leaving school (Glennon, 1995).

The testing of language-minority students also continues to be problematic (see, e.g., the historical case of *Diana v. State Board of Education*, 1970, cited in Sattler, 1990, and Overton, 1997). There are clear needs for better assessment and evaluation strategies as professionals struggle to determine whether low-achieving language-minority students are having difficulties due to LEP status or learning disabilities (LD; Gersten & Woodward, 1994). Studies have indicated that LEP students in special education make minimal gains (Wilkinson & Ortiz, 1986, cited in Gersten & Woodward, 1994). In addition, the effectiveness of programs designed to address the needs of language-minority students have been called into question (Goldenberg, 1996). Furthermore, concerns regarding the training of bilingual educators and bilingual special educators have been noted (Baca & Chinn, 1982).

One review article noted that many of the court cases focusing on the overrepresentation of Black, Latino, and American Indian students often were "decided in favor of the minority plaintiffs or were settled by consent decrees acceptable to the plaintiffs" during the early 1970s (Reschly, 1979, cited in Reschly, 1981, p. 1095). "These cases settled prior to 1975 generally involved a variety of poor and sometimes unethical practices in addition to the test fairness issue" (p. 1095; see also MacMillan, 1977). Several of the major problems summarized by Reschly (1981) include that (a) bilingual students were classified as mentally retarded using "verbal IQ scales that unfairly penalized them for lack of familiarity with English" (p. 1095); (b) short-form and group-administered verbal intelligence scales were used by "poorly trained personnel as the basis for classification of bilingual students" (p. 1095); (c) on occasion, parents were "not even informed, let alone accorded rights of consent, when their children were referred, evaluated, and placed in a special education program" (p. 1095); and, finally, (d) "The plaintiffs described deplorable conditions in special education programs. These included little academic emphasis, poor facilities, and inadequately trained teachers" (p. 1095).

These identified issues clearly compound concerns regarding the appropriate classification of minority students in special education programs. Unfortunately, many of these situations remain today with regard to special education placements and the overall effectiveness of our educational system. In particular, classification of language-minority students and poorly trained personnel with limited experience in urban special education programs continue to exist. Due to a failure to adequately address these concerns, the academic achievement of minority students in special education programs continues to be problematic.

The Role of Intelligence Tests in the
Definitions of Handicapping Conditions

Intelligence testing has been a focal point of controversy in the classification of students into the special education categories of mental retardation and LD. Many writers, however, have pointed out that intelligence testing is only one part of the overall process of referral and classification of students for special education services (Brady, Manni, & Winikur, 1983; Maheady, Towne, Algozzine, Mercer, & Ysseldyke, 1983; Reschly, 1988). In the best educational settings, numerous steps are taken to remediate concerns *prior* to an evaluation. Figure 7.1, based in part on a discussion by Rogers (1998), presents a flowchart of procedures beginning with the presentation of concerns regarding a student's functioning in regular education.

Concerns Identified by School Personnel

As many researchers have noted, most students are not tested with individually administered intelligence tests unless learning problems are identified in the regular education classroom and a referral is made for an evaluation (e.g., MacMillan & Meyers, 1980). As Lambert (1981) reported, "It is failure in school, rather than test scores, that initiates action for special education consideration" (p. 940). Low achievement scores on measures administered prior to the evaluation could define failure in school. This, however, does not rule out the possibility that the intelligence test administered to the minority student is culturally/linguistically biased.

In a seminal study of urban schools, it was reported that more than 60% of all referrals made by the classroom teachers were for children performing poorly in the academic setting including those with suspected speech and language problems (Gottlieb & Wishner, 1993, cited in Gottlieb, Alter, Gottlieb, & Wishner, 1994); fully 85% of these referred students were found eligible for special education services. The referring teachers often reported that they were unable to provide appropriate instruction in the regular education classroom. However, Gottlieb et al. (1994) also noted, "Many children are

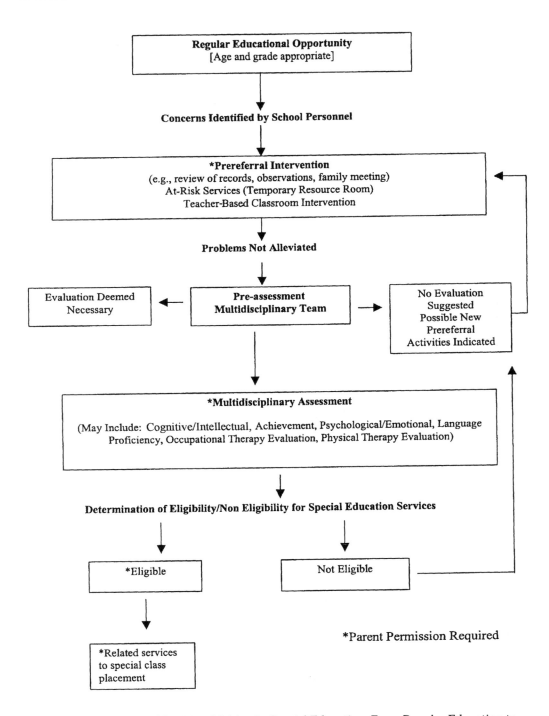

Figure 7.1. Flowchart of Decision Making in Special Education: From Regular Education to Placement

referred because teachers cannot control them" (p. 464). Problems in special education often are based on systemic problems within the general education setting. "It is unrealistic to expect that the severe needs of inner-city children can be met in large classes with inexperienced teachers, many of whom are poorly trained and are offered little if any support in the classroom" (p. 464).

Gottlieb et al. (1994) also provided data to support the important impact of poverty on the referral and placement of students in special education (see also Chapter 3 of the present book regarding socioeconomic status [SES] as a test performance factor). In their sample, 90% of the children in special education were receiving public assistance, and 95% of the students were members of minority groups. In addition, these students tended to come from single-parent homes; examination of mobility data indicated that many had experienced discontinuity in their educational experiences. Other authors also have noted the important connection between poverty and placement: "The degree of black student overrepresentation in programs for students with mild mental retardation reported over the past decade closely matches what might be expected from poverty statistics" (Reschly & Ward, 1991, p. 265). It must be underscored, however, that what goes on in the home intellectual environment of children is a stronger predictor of children's intellectual performance than is SES (see Chapter 4 of the present book).

Prereferral Interventions

This step of the referral process is vital and often overlooked in terms of overall effectiveness. Reschly (1988) reported that too often school personnel look to special education as the "only option for low-achieving students" (p. 317). During this stage, school personnel review records, conduct classroom observations, interview family members and teachers, and so forth to obtain more information about the students and their learning problems. Based on this information, decisions are made as to what form of intervention is needed. Interventions may include temporary placements in a resource room and/or teacher-based classroom interventions.

Maheady et al. (1983) identified five alternative instructional practices that were believed to improve the academic performance of "minority students at risk for special educational referral" (p. 450). These included a direct instructional approach, teacher-based instructional methods (i.e., the use of time, analysis of teaching techniques, individualized instruction plans incorporating positive reinforcement), classroom-wide peer tutoring, and a structured curriculum plan (including family involvement, multiage grouping, and staff development). Reschly (1988) cited National Academy of Sciences reforms including a focus on prereferral intervention, learning process assessment, assessment of biomedical factors, an emphasis on adaptive behavior, and instruction and curriculum. Maheady and colleagues concluded, "We believe that the pursuit of quality in testing has diverted our attention away from a much more pressing concern, that is, the provision of quality instructional services prior to referral for special education" (p. 454).

Preassessment Multidisciplinary Team

If problems are not alleviated, then a preassessment multidisciplinary team reviews the data and makes a determination as to whether the student should be evaluated. Considerations of other prereferral interventions may be discussed at this time. If an evaluation is deemed necessary, then the referral is processed along with the information obtained from the previous interventions.

Multidisciplinary Assessment

Students referred for learning problems often will be administered a variety of tests as part of the overall evaluation. This may include the following types of evaluation: intelligence, achievement, psychological/emotional, and language. In addition, it may include an occupational therapy evaluation and/or a physical therapy evaluation. Evaluations generally are conducted by a multidisciplinary team of professionals with expertise in a particular area (e.g., school psychologist, educational evaluator, school social worker). It is at this point that intelligence tests often are administered as part of the evaluation.

Concerns have arisen given the differential performance of particular minority groups on traditional measures, rendering classification and placement more likely (e.g., lower overall Full Scale IQ, higher percentage of significant Verbal Scale and Performance Scale discrepancies). In addition, there is evidence that minority students with LD might obtain different score profiles in comparison to the majority of students with LD (Zarske & Moore, 1982; for concerns that some predictive validity studies indicate that minority students might be overidentified as learning disabled, see also Chapter 5 of the present book).

Barona (1989) found that all of the Wechsler Intelligence Scale for Children–Revised (WISC-R) factors (i.e., Verbal Comprehension, Perceptual Organization, Freedom From Distractibility) contributed "differentially and significantly" to special education eligibility decisions for mentally retarded, learning disabled, and "not eligible" groups. Patterns of factor score profiles differed between Black and Mexican American students with LD and those who were deemed not eligible. The samples with mental retardation maintained the same factor profiles regardless of race/ethnicity.

Despite these profile differences, studies on the most popular scales (i.e., WISC, Stanford-Binet) seem to support the use of these scales with minority students in special education (Greene, Sapp, & Chissom, 1990; Knight, Baker, & Minder, 1990). However, as noted in Chapter 5 of the present book, more studies are needed to establish the reliability and validity of the more recent versions of these tests (e.g., WISC-III) with specific racial/ethnic groups.

Other studies have examined the role of examiner ratings of minority students in the evaluation process. Sattler and Kuncik (1976), for example, found that psychologists gave higher IQ estimates to Black and Mexican American students in comparison to those for White students with the same profiles of WISC subtest scores. The participating psychologists in this study indicated that the WISC was more valid for White students than for minorities and that ethnic background was an influential factor in forming judgments of an individual's IQ. Bilingualism was noted to be associated with lower IQ, as was the perceived cultural and educational "deprivation" experienced by Black and Mexican American students. It should be noted that SES was not identified as a significant variable. Subtest profiles with a great deal of "scatter" were seen as indicating higher intelligence than those with little scatter.

In a related study, Oakland and Glutting (1990) found that White examiners viewed Black and Mexican American students as more cooperative, attentive, and self-confident in comparison to their White peers with the same IQs (based on observations of their behavior following the testing). In addition, the authors found that the same positive attributions were given to low-SES students as to middle-SES students.

Determination of Eligibility

Based on all test data gathered, a determination is made regarding whether the student is eligible for special education services. A discussion of the criteria for special education classification is provided in the next section of this chapter. If the team decides that services are not warranted, then the student will be referred back to the regular education classroom. Assistance in the form of consultation services may be offered to the classroom teacher. For those students identified as eligible, issues of setting and instruction become of great importance. As noted in the introduction to this chapter, concerns have arisen regarding minority students (e.g., Blacks) being placed in more restrictive educational placements that hinder their chances for successful completion of high school (i.e., receiving a high school diploma).

Based on these steps in the referral process, it is clear that the use of intelligence tests constitutes only one part of the overall decision-making process to determine eligibility. Despite this fact, many critics continue to identify intelligence tests as the major culprit in the overrepresentation of minority students in special education.

Defining Mental Retardation and LD

The two special education categories most influenced by intelligence tests are mental retardation and LD. The following subsections cover the criteria for classification.

Mental Retardation

In a 1996 report of the U.S. Department of Education (cited in Henley, Ramsey, & Algozzine, 1999), 570,865 students had been identified as mentally retarded and were participating in special education programs nationwide in 1995; fully 90% of these students were classified in the mildly retarded range. Minorities (e.g., African Americans) were overrepresented relative to their proportions in the population. Later, we discuss this long-standing issue.

According to the latest report from the American Association on Mental Retardation (AAMR; 1999), "Mental retardation is a particular state of functioning that begins in childhood and is characterized by limitation in both intelligence and adaptive skills." More specifically, it is represented by (a) significantly subaverage intellectual functioning existing concurrently with (b) related limitations in two or more of the following applicable adaptive skill areas: communication, home living, community use, health and safety, leisure, self-care, social skills, self-direction, functional academics, and work.

In addition, mental retardation (c) manifests before 18 years of age. The AAMR 1999 also noted "four assumptions essential to the application of the definition":

1. Valid assessment considers cultural and linguistic diversity, as well as differences in communications and behavioral factors.

2. The existence of limitations in adaptive skills occurs within the context of community environments typical of the individual's age peers and is indexed to the person's individualized need for supports.

3. Specific adaptive limitations often coexist with strengths in other adaptive skills or other personal capabilities.

4. With appropriate supports over a sustained period, the life functioning of the person with mental retardation generally will improve.

Although the AAMR promotes differentiation of "cases of mental retardation based on needed levels of support across four domains" (MacMillan, Siperstein, & Gresham, 1996, p. 356), many professionals continue to adhere strictly to intelligence test score ranges to determine levels of retardation. The different levels of mental retardation determined by IQ ranges include the following (Bunch, 1994). Individuals with IQs below 25 are considered profoundly retarded and in need of custodial care at home or in an institutional setting. Those with IQs between 25 and 40 are considered severely retarded, also requiring custodial care. IQs between 40 and 55 indicate that the individuals are moderately retarded and trainable. They may attend public school in a special placement. IQs between 50 and 75 are designated to be in the mildly retarded range; such students are deemed educable in a public school special education classroom with partial mainstreaming.

The complexities of the classification process have been noted by research findings, indicating that students are labeled based on identification within the educational system, that is, behavior and IQ. Mercer (1970, cited in MacMillan et al., 1996) noted that some minority students who score within the mildly mentally retarded range never are referred for testing. Although Mercer's (1970) study is dated, her findings demonstrated that screening of students with the WISC revealed that there were many individuals with IQs below existing cutoff scores for mental retardation. The sample was comprised of Spanish-surname ($n = 509$), African American ($n = 289$), and White ($n = 500$) students who never had been referred for evaluations. Mercer reported that 78 Spanish-surname, 36 African American, and 6 White students scored below the IQ cutoff of 80 (criterion score for mild mental retardation at that time).

Further distinctions in the mentally retarded classification are based on etiology. For example, cultural-familial or familial retardation is indicated when there is no organic cause identified. Subgroups identified by Zigler and Hodapp (1986, cited in MacMillan et al., 1996) include the following:

1. Familial mental retardation, with at least one parent who also qualifies as having mental retardation

2. "Polygenic isolates," with parents of average intelligence and home environment considered nurturant and adequate for promoting intellectual development

3. "Socioculturally deprived" cases, where persons qualify as having mental retardation as a result of prolonged exposure to an impoverished environment (p. 359)

Cultural-familial retardation is defined by intelligence test scores below 70 (Hodapp, 1994). By contrast, intelligence test scores below 50 indicate organic mental retardation. There are more than 300 known causes of organic mental retardation including 225 prenatal, 49 perinatal, and 90 postnatal conditions (Belmont, 1994). Organic forms of mental retardation are estimated to account for 25% to 50% of all cases (Zigler & Hodapp, 1986, cited in Hodapp, 1994). Individuals with cultural-familial mental retardation tend to be overrepresented by minority group members and those from low-SES backgrounds.

Controversies have arisen regarding the use of adaptive behavior measures and overrepresentation of Black students in the mild mentally retarded category (Reschly & Ward, 1991). Reschly and Ward (1991) found that adaptive behavior measures were not used as frequently as intelligence tests with Black or White students despite legal mandates. In general, few discrepancies existed between measured intelligence and adaptive behavior, "suggesting that an equal treatment conception of fairness was achieved despite substantial overrepresentation of black students in programs for students with mild mental retardation" (p. 257). Of the students studied, 93% had been tested with an "appropriate" intelligence test. For 21% of the sample, adaptive behavior was not assessed, and for 31% of the sample, adaptive behavior was examined with a "non-normed" measure (p. 261). In sum, about one third of the students with mild mental retardation were placed without any indications of an adaptive behavior measure being administered. Reschly and Ward did note, however, that Black students were "slightly more" likely to have been administered an adaptive scale and less likely to have been placed without a documented adaptive behavior rating. Black students had higher overall adaptive behavior ratings than did White students in particular areas (e.g., communication skills), although special education teachers reported no observable differences.

To be sure, most controversial has been the classification of the educable mentally retarded (EMR) or mildly retarded student. "Mildly retarded persons are believed to constitute the vast majority of the mentally retarded," according to Reschly (1981, p. 1099). He continued,

Mild mental retardation, in contrast to the moderate, severe, or profound levels, typically is *not* associated with biological anomaly, is *not* permanent or lifelong, and is *not*
comprehensive in the sense of causing incompetence across most social roles and settings. Mild mental retardation typically is identified only in the public school context,
which establishes demands for abstract cognitive skills. (p. 1099)

Reschly went on to indicate that concerns regarding this classification are exacerbated
due to ignorance and confusion regarding what the mild mentally retarded label actually means and failure to distinguish it from the "other levels of mental retardation in
terms of comprehensiveness or the breadth of the handicap, etiology, and prognosis"
(p. 1099). He also noted the concerns regarding negative social consequences due to confusion with respect to the meaning of the label.

Others differ, however, in their assessments regarding the "curability" and "stability" of this diagnosis. According to MacMillan et al. (1996), for example, "Despite the
tendency of many people to embrace the belief in the curability of MMR [mild mental re-
tardation], the stability of intellectual behaviors associated with MMR are real" (p. 364).

It is important to note that many students previously classified in the mildly retarded range are now being identified as students with specific LD (Gottlieb et al., 1994).
Average IQs for students with LD have declined over the years (MacMillan et al., 1996).
There is a clear need to distinguish those students with generalized LD (with academic
problems in all areas) from those students with specific LD. A blurred distinction currently appears to exist.

LD: A Social Construction?

Prior to the passage of P.L. 94-142, the prevalence estimate of children with LD was
1% to 3%. In 1977, fewer than 800,000 children were diagnosed as learning disabled; in
1990, that number had risen to more than 2 million (Lerner, 1993). Lerner (1993) noted
that various figures cited in the literature indicate that 1% to 30% of the school population might have some form of learning disability. Reasons for this increase have
included

greater awareness, improvement in procedures for identifying and assessing . . . social
acceptance and preferences of LD classification . . . , cutbacks in other programs and lack
of general education alternatives for children who experienced problems in the regular
class . . . , court orders, [and] judicial decisions to reevaluate minority children. (p. 16)

Interestingly, this increase in numbers of children with LD has been in relative proportion to a decrease in the proportion of children and youths identified as mentally retarded. Lerner noted, that between 1976-1990 there was a major decline (i.e., over 30%)
of the number of students diagnosed with mental retardation. It is believed that some of
these students might have been identified as learning disabled. Brosman (1983), for example, reported an inverse relationship between the numbers of learning disabled and

mentally retarded individuals identified in California. "[The number of] students identified as LD has increased in California as the number of students identified as mentally retarded has decreased" (p. 523).

Sleeter (1986) noted that educational classifications were formulated, historically, in response to the school reform movement that took place after the launching of the Russian satellite Sputnik in 1957. This movement eventually led to both academic standards being raised and rigorous testing programs being introduced into the schools. Sleeter commented,

> Students unable to keep up with raised standards were placed into one of five categories. Four of the categories were used primarily to explain the failures of lower class and minority children; [the] learning disabilities [category] was created to explain the failures of white middle-class children in a way that gave them some protection from the stigma of failure. (p. 46)

The four categories for "lower class" and minority children were "slow learners," "mentally retarded," "emotionally disturbed," and "culturally deprived." Slow learners and mentally retarded students were identified based on IQ scores (i.e., students with IQs between 75 and 90 were identified as slow learners, and students with IQs below 75 were identified as retarded). Disproportionately high numbers of poor and minority children were classified in these categories as well as under the emotionally disturbed label.

The term *learning disabilities* was officially established with the formation of the Association for Children with Learning Disabilities in 1963 (Sleeter, 1986). According to the most recent federal definition noted in the IDEA amendments of 1997, a specific learning disability refers to "a disorder in one or more of the basic psychological processes involved in understanding or using language, spoken or written, which disorder may manifest itself in imperfect ability to listen, think, read, write, spell, or do mathematical calculations." These disabilities include conditions such as perceptual disabilities, brain injury, minimal brain dysfunction, dyslexia, and developmental aphasia. They do not include problems that result from "visual, hearing, or motor disabilities of mental retardation, of emotional disturbance, or of environmental, cultural, or economic disadvantage." It should be noted that the last phrase of the "rule out" clause (i.e., environmental, cultural, or economic disadvantage) is difficult to assess and often is not specifically addressed given current evaluation practices.

The role of intelligence tests in the determination of LD is noted in the criteria of a severe discrepancy between potential (based on intelligence tests) and achievement in one or more of seven areas (Lerner, 1993). The determination of a severe discrepancy varies with each state, school district, or evaluation team. Frankenberger and Fronzaglio (1991) conducted a review of the various state criteria currently being used to identify students with LD. Information from all states and the District of Columbia was obtained. Results indicate that 40% of those surveyed revised their criteria between 1988 and 1990. Although the specific reasons for the revisions were not noted, these changes attest, in some ways, to the fluidity of the defining criteria for LD. A total of 14 of the states and the

District of Columbia specified IQ cutoffs for LD as falling within the "average" range. Another 8 states did not define "average," and 6 states indicated that the IQ had to be higher than scores falling within the retarded ranges. A significant discrepancy between ability and achievement ranged from 1 standard deviation (6 states) to 2 standard deviations (3 states). Overall, 49% of the states use definitions varying from that in federal guidelines. The number of states specifying strict IQ cutoff scores, as well as ability and achievement discrepancies, has increased. Cutoff scores below which students would not be eligible for services for LD are more clearly defined.

Researchers, however, have challenged the role of intelligence tests in determining this discrepancy (Aaron, 1997; Siegel, 1989). In a comprehensive review, Aaron (1997) examined the use of the discrepancy formula with students with reading disabilities, the most common form of LD. He cited research indicating that IQ is "not a potent indicator of reading potential" (p. 462). He concluded that the discrepancy formula should "be abandoned and that a pragmatic approach which identifies the source of the reading problem for all children and focuses on remedial efforts [should be adopted]" (p. 462). Siegel (1989) cited empirical findings suggesting that "poor readers at a variety of IQ levels show similar reading, spelling, language, and memory deficits [and that therefore,] on logical and empirical grounds, IQ test scores are not necessary for the definition of learning disabilities" (p. 469).

Some researchers have advocated that little difference exists between learning disabled and non-learning disabled low achievers on tests used to classify students (Ysseldyke, Algozzine, Shinn, & McGue, 1982, cited in Gresham, MacMillan, & Bocian, 1996). The sample from the Ysseldyke et al. (1982) study, however, was noted to have been made up primarily of White students. More recently, a study conducted by Gresham et al. (1996), which included minorities, noted that one of the distinguishing features of those students classified as either learning disabled, low achieving, or mildly mentally retarded were scores obtained on measures of aptitude and achievement. The Gresham et al. sample consisted of 152 White, Black, and Hispanic students, Grades 2 to 4, from Southern California. Students had been classified as learning disabled, low achieving, or mildly mentally retarded and were compared on 41 different measures of ability, academic achievement, social skills, problem behavior, academic engaged time, perceptual motor skills, and school history. Findings indicated that the learning disabled group scored higher on cognitive ability measures (i.e., WISC-III, Raven Coloured Progressive Matrices) than did the low-achieving and mildly retarded groups. The learning disabled group obtained average intelligence test scores (Full Scale IQ = 92.46, Verbal Scale IQ = 91.67, Performance Scale IQ = 94.90), the low-achieving group had mean scores in the low average range (Full Scale IQ = 82.50, Verbal Scale IQ = 81.70, Performance Scale IQ = 86.32), and the mildly mentally retarded group obtained scores in the low average to borderline range (Full Scale IQ = 67.86, Verbal Scale IQ = 67.18, Performance Scale IQ = 73.39). The low-achieving group scored higher on academic achievement measures than did the learning disabled and mildly retarded groups. No significant differences were found on the other variables including social skills, problem behaviors, academic engaged time, and school history.

Racial/Ethnic Incidence Data Regarding Students Classified as Mentally Retarded and Learning Disabled

Tables 7.1 and 7.2 present the *most recent* data from the U.S. Department of Education, Office for Civil Rights (OCR; 1997, hereafter the 1997 OCR survey), biennial compliance report regarding the incidence of mental retardation and learning disability classification by racial/ethnic group; incidence rates are based on survey data collected in 1994 (for a full description of this database, see Chapter 8 of this book). We present data from the 1997 OCR survey with respect to overall national statistics as well as the top 10 states with the highest combined kindergarten through Grade 12 (K-12) minority enrollments (for the numerical counts of these students, see Note 38 in Chapter 8). The disparity analysis presented in Tables 7.1 and 7.2 is derived from our own calculations.

With regard to representation figures for mental retardation, findings indicate that Black students continue to be overrepresented in this overall category (Table 7.1). The 1997 OCR survey did not provide data regarding the various levels of mental retardation. Given earlier references in this chapter (see "Mental Retardation" subsection), however, it is assumed that the overwhelming majority of these students fall within the mildly mentally retarded category. The overall national percentage of Black students indicates a very high overrepresentation (+86.65%). Based on data in the OCR survey, this pattern of disproportionately high numbers of Black students in EMR classes has gone unabated over the past 20 years.[1] An unrelenting pattern of overrepresentation also is noted for the 10 states that we analyzed. That is, in the top 10 states with the highest combined K-12 minority enrollments, Table 7.1 shows that Black students are overrepresented in each state—ranging from a high of +94.24% (Florida) to a low of +31.50% (Illinois).

Regarding Asian/Pacific Islander (A/PI) students, they are underrepresented in the mentally retarded category (−63.71%) at the national level as well as in the 10 states we analyzed. For American Indian students, overrepresentation is seen at the national level (+13.46%) and in 4 of the 10 states. With respect to Hispanic students, they are underrepresented nationally (−35.09%). In comparison to overrepresentation of Hispanic students in the category of mental retardation, as seen during the early 1970s, it appears that the disproportionally high rates of Hispanics in classes (EMR) no longer is a pattern at the national level. Since 1978, the national pattern actually is the converse, that is, one of underrepresentation.[2] Of the top 10 states in combined minority enrollments, Table 7.1 indicates that Hispanics are underrepresented in 8 states and overrepresented in 2 states in the mental retardation category. Of these 2 states, a fairly high overrepresentation rate is seen in New Jersey (+62.25%), a state with a sizable Puerto Rican student population. Finally, White students are underrepresented at the national level (−12.04%), and the same underrepresentation pattern holds for all of the 10 states except Illinois (which shows a very small overrepresentation of +1.65%).

With regard to LD, a more mixed representation is apparent for all racial/ethnic groups (Table 7.2). Overall, national figures indicate an overrepresentation of American Indian, Hispanic, and Black students, but at much lower disparities than those seen in

TABLE 7.1 Disparity Analysis for Mental Retardation: For Nation and Top 10 States in Combined Minority Enrollment (K-12)

State	American Indians			Asian/Pacific Islanders			Hispanics			Blacks			Whites		
	Enroll-ment[a] (%)	MR[b] (%)	Dis-parity[c] (%)	Enroll-ment (%)	MR (%)	Dis-parity (%)	Enroll-ment (%)	MR (%)	Dis-parity (%)	Enroll-ment (%)	MR (%)	Dis-parity (%)	Enroll-ment (%)	MR (%)	Dis-parity (%)
Nation	1.04	1.18	+13.46	3.72	1.35	-63.71	12.71	8.25	-35.09	16.85	31.45	+86.65	65.68	57.77	-12.04
1. California	0.76	1.12	+47.37	11.55	7.47	-35.32	38.15	40.23	+5.45	8.40	11.79	+40.36	41.14	39.39	-4.25
2. Texas	0.22	0.26	+18.18	2.08	0.79	-62.02	35.62	31.65	-11.15	14.60	24.83	+70.07	47.48	42.48	-10.53
3. New York	0.36	0.46	+27.78	4.83	1.81	-62.53	16.68	16.23	-2.70	20.25	29.14	+43.90	57.88	52.35	-9.55
4. Florida	0.20	0.13	-35.00	1.72	0.79	-54.07	14.78	11.37	-23.07	24.56	47.95	+95.24	58.73	39.76	-32.30
5. Illinois	0.13	0.09	-30.78	2.94	1.07	-63.61	10.92	4.58	-58.06	22.89	30.10	+31.50	63.12	64.16	+1.65
6. Georgia	0.12	0.07	-41.67	1.63	0.44	-73.01	2.02	0.96	-52.48	38.89	61.16	+57.26	57.34	37.37	-34.83
7. New Jersey	0.32	0.31	-3.13	5.86	1.24	-78.84	11.47	18.61	+62.25	18.29	32.23	+76.22	64.06	47.06	-26.54
8. North Carolina	0.66	1.09	+65.15	1.18	0.27	-77.12	1.70	0.85	-50.00	32.03	60.02	+87.39	64.43	37.76	41.39
9. Virginia	0.16	0.05	-68.75	3.53	1.57	-55.52	3.32	2.60	-21.69	27.15	48.90	+80.11	65.84	46.90	-28.77
10. Louisiana	0.50	0.44	-12.00	1.21	0.43	-64.46	1.09	0.54	-50.46	44.06	67.67	+53.59	53.14	30.92	-41.81

SOURCE: U.S. Department of Education, Office for Civil Rights (1997).

a. % enrollment = percentage of racial/ethnic group in total K-12 enrollment.

b. % MR = percentage of racial/ethnic group in mental retardation category.

c. In the percentage disparity category, a plus sign (+) indicates overrepresentation and a minus sign (–) indicates underrepresentation.

TABLE 7.2 Disparity Analysis for Specific Learning Disabilities: For Nation and Top 10 States in Combined Minority Enrollment (K-12)

State	American Indians			Asian/Pacific Islanders			Hispanics			Blacks			Whites		
	Enroll-ment[a] (%)	LD[b] (%)	Dis-parity[c] (%)	Enroll-ment (%)	LD (%)	Dis-parity (%)	Enroll-ment (%)	LD (%)	Dis-parity (%)	Enroll-ment (%)	LD (%)	Dis-parity (%)	Enroll-ment (%)	LD (%)	Dis-parity (%)
Nation	1.04	1.37	+31.73	3.72	1.35	-63.71	12.71	13.01	+2.36	16.85	17.22	+2.20	65.68	67.05	+2.09
1. California	0.76	1.00	+31.58	11.55	3.98	-65.54	38.15	35.08	-8.05	8.40	12.66	+50.71	41.14	47.28	+14.92
2. Texas	0.22	0.23	+4.55	2.08	0.42	-79.81	35.62	36.54	+2.58	14.60	16.68	+14.25	47.48	46.14	-2.82
3. New York	0.36	0.36	0.00	4.83	1.43	-70.39	16.68	19.11	+14.57	20.25	23.25	+14.81	57.88	55.84	-3.52
4. Florida	0.20	0.19	-5.00	1.72	0.46	-73.26	14.78	12.74	-13.80	24.56	24.56	0.00	58.73	62.05	+5.65
5. Illinois	0.13	0.08	-38.46	2.94	0.77	-73.81	10.92	5.71	-47.71	22.89	15.49	-32.33	63.12	77.95	+23.49
6. Georgia	0.12	0.10	-16.67	1.63	0.47	-71.12	2.02	1.43	-29.21	38.89	27.52	-29.24	57.34	70.48	+22.72
7. New Jersey	0.32	0.50	+56.25	5.86	1.61	-72.53	11.47	10.71	-6.63	18.29	18.45	+0.87	64.06	68.72	+7.27
8. North Carolina	0.66	0.91	+37.88	1.18	0.25	-78.81	1.70	1.02	-40.00	32.03	31.32	-2.22	64.43	65.51	+1.68
9. Virginia	0.16	0.14	-12.50	3.53	1.43	-59.49	3.32	3.67	+10.54	27.15	26.31	-3.09	65.84	68.45	+3.96
10. Louisiana	0.50	0.64	+28.00	1.21	0.18	-85.12	1.09	0.70	-35.78	44.06	56.58	+28.42	53.14	46.90	-11.74

SOURCE: U.S. Department of Education, Office for Civil Rights (1997).
a. % enrollment = percentage of racial/ethnic group in total K-12 enrollment.
b. % LD = percentage of racial/ethnic group in learning disabilities category.
c. In the percentage disparity category, a plus sign (+) indicates overrepresentation and a minus sign (−) indicates underrepresentation.

203

the mental retardation category. However, a major observation that we glean from Table 7.2 is that the overrepresentation/underrepresentation pattern for American Indian, Hispanic, and Black students in the LD category is unclear for the top 10 states in combined minority enrollment. For example, note that in California, Blacks are overrepresented by +50.71%, whereas in Illinois, they are underrepresented by –32.33%. These fluctuations at the state level deserve scholarly attention. As can be seen in Table 7.2, American Indians were overrepresented in 6 of the 10 states cited, A/PI students were underrepresented across all 10 states, Black students were overrepresented in 6 states, Hispanic students were overrepresented in 4 states, and Whites were overrepresented in 8 states.

It should be noted that the issue of LEP status also might partially account for the underrepresentation in the LD category for particular groups (i.e., Hispanics, A/PIs). As reported earlier in this chapter, a significant problem for evaluators is ascertaining whether learning problems are primarily due to second-language acquisition or actual LD. This issue is of concern given that some LEP Hispanic and A/PI students might be rightly excluded from an LD classification due to language differences (not deficits); however, LEP status in itself may serve to deny services to those students who have genuine LD *and* are LEP.

It also should be noted that examining overrepresentation figures has yielded criticisms from various researchers. Artiles and Trent (1994), for example, noted that "overrepresentation figures for certain minority groups often vary depending on the authors or agency reporting or interpreting the data" (p. 413). One problem with comparing study results is that different calculation methods often are used in individual studies to determine proportions of representation. Nonetheless, patterns of racial/ethnic representation in special education categories need to be scrutinized. The persistent and pervasive overrepresentation of African American students in the mild mentally retarded category, for example, should send up red flags to researchers and practitioners alike. Questions should be asked as to the likely factors involved in such overrepresentation.

Examination of the percentages of minority students classified as mildly mentally retarded provides only a partial picture of the overrepresentation issue. It also is important to obtain information regarding the *actual* placements of these students into educational settings. In evaluating the *Larry P.* decision of the early 1970s (*Larry P. v. Riles*, 1972), Prasse and Reschly (1986) reported that Black students constituted 10% of the K-12 student population in California. However, they represented 25% of the students placed in mildly mentally retarded classes. Therefore, many assumed that 25% of Black students were in classes for the mildly retarded. In actuality, the percentage of Black students placed in special classes for the mildly retarded was 1%. Thus, "overrepresentation statistics need to be carefully analyzed to avoid exaggeration and distortion" (Reschly, 1988, p. 320).

In addition, Reschly (1981) reported, "Analyses of overrepresentation have largely ignored the variables of gender and poverty as well as the other steps in the referral-placement process" (p. 1095). A linkage between poverty and LD as well as mental retardation has been noted in the literature repeatedly (e.g., Brosman, 1983). In a study of special education programs in California, "low-SES school districts placed twice as many students in LD classes as did high-SES school districts . . ., [and] minority students from low-SES districts were overrepresented" (Brosman, 1983, p. 523).

Recently, Lester and Kelman (1997) examined interstate differences in the diagnosis of LD. Their findings in relation to minority students, based on the 1990 census, suggest that as the percentage of African Americans in the population increases, the proportion of students diagnosed with LD in restrictive settings also increases. In addition, growing numbers of African Americans were positively correlated with "increases in the proportion of the resident population that was nonmainstreamed LD or EMR" (p. 604). The overall outcome was that greater numbers of Black students were placed in restrictive settings far above their actual representation in the population. As noted earlier, levels of poverty also were positively related to the proportion of students with LD who were in restrictive settings.

Artiles and Trent (1994) noted the following with regard to overrepresentation:

(1) . . . The larger the minority student population is in the school district, the greater the representation of students in special education classes.
(2) The bigger the educational program, the larger the disproportion of minority students (Heller, Holtzman, & Merrick, 1982, cited in Artiles & Trent, 1994).
(3) . . . Variability in overrepresentation data has been found as a function of the specific disability condition and the ethnic group under scrutiny. (p. 414)

Future Directions for Special Education

Although intelligence tests have been the focus of many controversies surrounding the classification and placement of students in special education programs, numerous writers assert that the current disproportionate rates of placement for minority students are not the fault of the tests but rather a systemic problem (Lambert, 1981; Reschly, 1981). Reschly (1981) noted that bans on intelligence tests in some states (e.g., California) had "no appreciable effect on the degree of disproportionate placement of Black students in special education until the court ordered what amounted to rigid quotas in 1979" (p. 1100). In addition, both Reschly and Lambert hypothesized that the alternative to standardized testing procedures (i.e., subjective judgments) would not be favorable. On this, Reschly (1981) commented, "The basis for educational decisions in the posttesting era would undoubtedly be dominated by subjective impressions. There is every reason to believe that subjective procedures would be considerably less valid for everyone and more discriminatory toward minority students" (p. 1102). In a similar vein, Lambert (1981) asserted,

To eliminate IQ tests as a remedy to overrepresentation will solve nothing. Without aptitude measures, many failing children would be considered retarded, and expectations for them would be based solely on school performance rather than on a combination of their measured abilities and classroom assessment. IQ tests at least demonstrate that many failing children, including minority children, have abilities that are not being tapped by educational programs. (p. 941)

As noted in the preceding sections of this chapter, criticisms continue regarding current testing practices in special education. Based on his extensive writing in this area, Reschly (1981) recommended the following: (a) abandonment of the term "IQ" due to the negative connotations associated with this term; (b) multifactored assessment and consideration of "adaptive behavior, primary language, and sociocultural status" (p. 1101); and (c) the use of criterion referenced tests, behavioral assessment, and consultation that could provide more important methods in the assessment of children and the effectiveness of educational interventions. In addition, clearer definitions of handicapping conditions and criteria for classification are needed. Furthermore, school psychologists have identified areas of need including more extensive training to serve LEP and bilingual students (Ochoa, Powell, & Robles-Pina, 1996).

It appears that despite legal mandates supporting change, the special education system remains overburdened. The system appears unable to sustain new practices due to pressures to move more quickly through the evaluation process to meet federally mandated time lines. For example, adaptive measures continue to be used inconsistently.

Given current concerns regarding the achievement of students, many states have become more firmly entrenched in the testing enterprise in attempts to assess academic success more objectively. As a case in point, this chapter began with a discussion of the current problems facing the state of New York. Current reform efforts in New York highlight the increased emphasis being placed on accountability for student performance. Based on current plans, school districts and administrative personnel will be accountable for achievement results. School districts that are found to have "significant disproportionality in the identification of students as disabled or in the placement of students in the least restrictive settings among the racial/ethnic groups" will be scrutinized through a review process (State Education Department, 1999).

Despite the increased focus on academic standards, many believe that the most important issue facing special education is the quality of instruction provided to alleviate the learning problems of youths (e.g., Gersten & Woodward, 1994). In the *Larry P. v. Riles* (1972, 1979) trial, Judge Peckham "excoriated the efficacy of special classes no less than 27 times, classes which he characterized as dead end, inferior, and stigmatizing" (cited in Prasse & Reschly, 1986, p. 342). This impression often is shared by others as we see few minority students moving out of the special education system. Evidence regarding the effectiveness of special education intervention has remained inconclusive (Artiles & Trent, 1994). These scholars suggest that the reasons for this include the complexities of the field (e.g., differences based on setting, instructional methods, methodological flaws) as well as systemic problems regarding a lack of agreed-on definitions of basic constructs and faulty referral and assessment procedures.

Clearly, the use of intelligence tests in special education remains a controversial issue. However, of greater concern is the effectiveness of the resulting placements. As noted by Brady et al. (1983),

> The task that faces special education and evaluation team personnel is not simply to evaluate a child effectively. Rather, the task is to intervene, change behavior, and shape programs so as to make the child more functional within the educational domain. (p. 300)

Although the term "functional" implies only a level of basic skills, it is our opinion that the goals must be higher than just functional given the important impact that these programs potentially have on the future of special education students.

Important changes and reform throughout our educational system currently are being discussed and implemented. Standardized group tests, as well as individualized measures such as intelligence tests, will continue to play a crucial role in the evaluation process of all students so as to identify those having difficulties. It is imperative that our research, training, and practice in the process of identification for special education services be as culturally sensitive as possible. This will include recognizing the sociopolitical history, current underrepresentation and overrepresentation rates in certain categories, and systemic problems (e.g., poorly trained personnel, overcrowded classrooms, limited remedial services) that continue to plague our educational system.

8

Gifted Minority Students

As we have noted earlier in this book, since the advent of intelligence testing during the 1920s, a voluminous body of scholarly research has documented the consistent finding that minority students (e.g., African Americans, Latinos, American Indians) perform below the norm, on average, on most standardized intelligence tests. We also have underscored that minority students, like White students, exhibit considerable variability in measured intelligence. Our focus in this chapter is on those children and youths who perform substantially above the norm on tests of intelligence as well as on other measures (e.g., achievement tests, indexes of talent).

Regarding terminology, we use the omnibus concept of "gifted" in this chapter. When children perform at superior levels on scholastic-type skills (e.g., verbal abilities) as measured by intelligence tests, they typically are called "gifted." Children who demonstrate exceptional ability (e.g., in music, in the visual arts) are described as "talented." We are in agreement with Winner (1996), who commented, "Two different labels suggest two different classes of children. But there is no justification for such a distinction. Artistically or athletically gifted children are not so different from academically gifted children" (pp. 7-8). According to Winner, both types of children exhibit three characteristics of giftedness: *"precocity," "an insistence on marching to their own drummer,"* and *"a rage to master"* (p. 3).

The study of gifted students and their curricular emphasis, gifted education, is now a distinct scholarly domain in education and psychology. As evidenced by literature citations, the study of the gifted and gifted education clearly is an active area of scholarly pursuit. As of July 1999, there were 4,531 citations on "gifted," "children," "youth," and "students" listed in *Current Index to Journals of Education (CIJE)* as well as 2,544 such citations listed in *Psychological Abstracts (PA)*.[1] When our searches were delimited to minority children, youths, and students, the number of citations shrunk significantly. In *CIJE*, 384 (8%) of the 4,531 total citations dealt with minorities; in *PA*, 102 (4%) of the total 2,544 citations had references to minorities. These very low percentages confirm those of previous searches. For example, 10 years ago, Ford and Harris (1990) reported, "An examination of the relevant literature since 1924 reveals that of 4,109 articles found on the gifted and talented, less than two percent (75) addressed minority group members" (p. 27).[2] In sum, there is a large amount of information about gifted White students (particularly of middle- and upper-socioeconomic status [SES] backgrounds). Regarding gifted minority students, however, a similar claim cannot be made.

The scant attention paid to giftedness among minority students can best be understood when analyzed in the broadest context of the study of the gifted. Notwithstanding the strides that the study of giftedness has made over time, the field is beset with major problems. One major issue is what some scholars describe as society's "love-hate" relationship or ambivalence with giftedness and talent (Colangelo & Davis, 1997; Gallagher & Weiss, 1979). Gallagher (1993) cogently captured this broad-based sentiment:

> The attitude toward gifted students at a personal and societal level has often been one of ambivalence, in both the educational setting and in society at large. We may love the creative products of their mental processes but still feel the sting of envy when we observe some persons doing, with apparent ease, what is so difficult for others to accomplish. Such conflict between the public interest and personal feelings has been felt in societies and has been a barrier to the education of gifted and talented students. (p. 83)

Given that gifted students, in general, do not receive the attention they deserve, it is not surprising that gifted minority students are even more neglected. Not only does society's love-hate relationship hit minority students harder due to their subordinate status, but other issues that we discuss later (e.g., elitism) help to explain why gifted minority students (i.e., African Americans, Latinos, American Indians) are the "neglected of the neglected"—a dominant theme of this chapter.

This chapter is divided into three parts. The first section offers a historical sketch of giftedness in the United States with a focus on how minority students have fit in. The second section presents a number of issues—conceptions and measurement of giftedness, elitism, excellence versus equity, and underrepresentation of minority students in gifted programs. The sections in the third part contain discussions of select literature on gifted African American, Latino, Asian/Pacific Islander (A/PI), and American Indian students. A special focus is given to those conceptions and practices that show promise for increasing the number of gifted minority students.

Historical Sketch of Giftedness in the United States

Giftedness in the United States is predated by centuries of international interest in superior youths. The identification and education of gifted children have intrigued "virtually all societies in recorded history" (Colangelo & Davis, 1997, p. 5). For example, in ancient Sparta, a highly stratified society, the Spartiates (descendants of the Dorian conquerors) valued and built Spartan society around military and civic activities (Good, 1960). As such, superior abilities in military skills, physical strength, and leadership were highly valued and nourished. Little emphasis was placed on intellectual education.

In ancient Athens, all sons of Athenian upper-class citizens were given opportunities to learn to read in highly unregulated private schools whose fees were paid by parents.[3] In addition to reading, other areas such as music, literature, mathematics, and logic were valued curricular aspects (Good, 1960). Children from very wealthy backgrounds attended school for longer periods of time. Regarding higher education, Plato, the preeminent Greek philosopher, founded his academy circa 387 B.C. Plato was highly selective in who he admitted to his academy:

> In addition to its educational function, the school also had a selective function. Pupils were to be sent to school only as long as they received real benefit from the instruction, and Plato thought many pupils would be dismissed early in the course because they lacked capacity for advanced study. He considered that only a few were able to profit from the study of logic or advanced mathematics. His scheme of selection was fairly complex and was applied in a series of stages which were intended to select gradually but more and more closely the most able from the less gifted. His criteria were such as the following: love of knowledge, ability to learn, strength and skill, self-control, devotion to the public good, aptness to resist evil and deceit, and capacity for abstract thinking. Those who passed the successive tests and reached the highest levels of wisdom and devotion to the state were to rule the state. The government was, therefore, to be based upon knowledge of principle and truth. Not power or propaganda, but science and philosophy were to control. The philosopher was to be king. (Good, 1960, p. 34)

In the United States, the identification and education of gifted children were slow in coming (Newland, 1976). With the advent of the common school movement during the mid-1800s, the schooling of children became less of a church, city, or private matter and more of a state responsibility. An increasing heterogeneity of the student population and compulsory attendance laws led to a somewhat more diverse curriculum. In high schools, a college preparatory curriculum was considered by many to be the pathway for the most capable pupils. This curriculum emphasis did not, however, constitute a formal program of instruction designed specifically for meeting the needs of bright students (Newland, 1976). The little attention that was given to the gifted during the early years also was due to the idea that special classes should be devoted to "defective" children (e.g., the blind, the deaf, the "feeble-minded"; Race, 1918). Although the gifted were largely neglected in being identified and provided services, there were some note-worthy exceptions:

- In 1870, St. Louis initiated tracking, allowing some students to complete the first eight grades in less than eight years.
- In 1884, Woburn, Massachusetts, created the "Double Tillage Plan." After the first se-mester of the first grade, bright children were accelerated directly to the second semes-ter of the second grade.
- In 1886, Elizabeth, New Jersey, began a tracking system that permitted bright students to progress more rapidly than others.
- In 1891, schools in Cambridge, Massachusetts, developed a similar double-track plan. Students capable of even more accelerated work were taught by special tutors. About 1900, some "rapid progress" classes telescoped three years of school work into two.
- In 1901, Worcester, Massachusetts, opened the first special school for gifted children. (Colangelo & Davis, 1997, p. 5)

The Inception of Gifted Classes

At the turn of the 20th century, particularly during the second decade, attention began to turn from children with impairments to children with superior abilities (Race, 1918). For example, Edward L. Thorndike, a noted educational psychologist, stated in a 1914 address that millions of dollars were being spent—justifiably so—on people with low mental ability, criminals, and paupers, but nothing was being provided for the gifted (cited in Race, 1918). By 1916, however, reports on the establishment of specific classes for gifted students (other than tracking) began to appear in the educational liter-ature (Newland, 1976).[4] Race (1918) described a class for exceptional children that was organized in October 1916 under the auspices of the Louisville, Kentucky, city schools. Regarding selection, 111 children were administered the Stanford-Binet (the major in-strument used for selection).[5] Of these, 21 boys and girls were selected for the "opportu-nity class." The children's IQs ranged from 120 to 168 (median = 137). Students were pro-vided rigorous instruction and were able to perform schoolwork about 2 to 3 years in advance of the normal child.

Another example of a gifted program during this period was reported by Stedman (1924). This 5-year "experiment," as described by Stedman, was initiated in January 1918 in the training department of the Los Angeles State Normal School (later called the Southern Branch of the University of Southern California). Based on a cutoff Stanford-Binet IQ of 125 for program eligibility, 10 children (ages 7 to 11 years) were selected for enrollment in the "opportunity room." The children, whose IQs ranged from 125 to 167, formed what Stedman called her "charter class." In her book, *Education of Gifted Children*, she described in detail 16 cases of children enrolled in the senior opportunity room for the 1921-1922 and 1922-1923 school years. Interestingly, some of the intellectually supe-rior children also were described as possessing what we presently refer to as exceptional "talent." For example, R. W., a girl with an IQ of 147, was a musical prodigy. K. D., a girl with an IQ of 140, had unusual adroitness in manual dexterity (e.g., woodwork). Regarding racial/ethnic background of the children, Stedman's genealogical reporting described the 16 children as having European ancestry (e.g., Dutch, English, Irish, French). Notwithstanding some exceptions, the children's parents were college educated and had professional careers. It is noteworthy that Stedman's experiment took place in the Los Angeles area between 1918 and 1923, a time when Mexican American and African American children constituted a growing segment of the public school enrollment (González, 1990). Yet, none of these minority children were represented in Stedman's gifted program.

It appears that by the mid-1920s, the young and growing gifted movement had been launched. A particularly significant contribution to the literature was the National Society for the Study of Education's 23rd yearbook, *Report of the Society's Committee on the Education of Gifted Children* (Whipple, 1924). The society's report was a compendium of knowledge, opinions, and advocacy regarding the gifted. Topics included personal, social, and physical characteristics, as well as history and selection strategies, of the gifted; curriculum; and academic/career attainments. Illustrative of the growing literature on gifted children and their education was the annotated bibliography (in the yearbook) compiled by Henry (1924). This extraordinarily large listing contained 453 publications published between 1894 and 1923. It appears that the gifted children of interest in Henry's bibliography were White. As we discuss later, gifted minority children (i.e., African Americans) were not studied until the late 1920s.

Another major work that garnered attention to the young and growing gifted movement was Hollingsworth's (1929) book *Gifted Children: Their Nature and Nurture*. Her coverage of "frequency [of giftedness] as related to race" (p. 68) is particularly germane to our discussion. Although Hollingsworth acknowledged that so little data were available on "race" and prevalence of giftedness "that it is perhaps scarcely worthwhile to discuss the topic except to say that we are ignorant of the facts" (pp. 68-69), she nevertheless drew some conclusions:

1. Citing Terman (1925), which we discuss next, Hollingsworth (1929) noted that highly gifted children were disproportionately represented among the Scottish, English, and Jewish populations.[6]

2. Based on the mental testing literature to date, Hollingsworth (1929) commented, "Comparatively few . . . [Negro] children are found within the range which includes the best one percent of White children" (p. 69). To Hollingsworth, it appeared that gifted Negro children were highly anomalous in frequency. She asserted, "*It is possible by prolonged search to find an occasional negro or mulatto child testing above 130 IQ*" (p. 70, italics added). Later, when we discuss gifted African American students, we shall see that there is some evidence refuting Hollingsworth's claim. For example, in a study published 2 years before Hollingsworth's book appeared, Bond (1927) found that the gifted Negro child was not an anomaly.

The Terman (1925) Study of "Genius"

In 1921, Lewis Terman, a Stanford University professor, began one of the most ambitious and comprehensive studies ever undertaken in the history of educational psychology (Terman, 1926).[7] Funded with research grants totaling more than $50,000 (a huge sum in those days) and a large, highly trained research team, Terman's goal was to canvass parts of California to identify public school students (Grades 3 to 8) who had IQs that placed them in the top 1% of the population.

Based on test data from more than a quarter million cases, Terman (1926) identified 643 gifted students for inclusion in his main (Group I) study (based on a canvassing of schools in Los Angeles, San Francisco, and the East Bay [e.g., Oakland, Alameda, Berkeley]) and 356 gifted students in Group II.[8] Thus, his final sample consisted of 999 students. For Terman's main study, he used "a preliminary sifting method to determine which children should be tested" (p. 21). Step 1 in the search for participants involved teacher nominations of students (i.e., "most intelligent," "next most intelligent," "third most intelligent," "youngest pupil," and "brightest child from previous school year"[9]). In Step 2, nominated students were administered the National Intelligence Tests (NIT). Those students who ranked in the top 5% of the NIT were given an abbreviated Stanford-Binet (Step 3). Pupils who scored IQs of 130 or higher on the abbreviated Stanford-Binet were given the full Stanford-Binet (Step 4). For final inclusion in the main study, IQs of 132 to 140 were set (according to age levels) as provisional limits. The IQ scores of the 643 children ranged from 130 to 200 ($M = 151.0$, $SD = 10.2$). In addition to mental test data, Terman collected numerous other data (e.g., medical, anthropometric, character and personality, academic achievement, parental background). In all, about 100 pages of data were collected for each child.

Regarding the "racial" origins of the 643 gifted children in the main group, Terman (1926) listed 37 different "racial stocks."[10] Origin was based on the ancestry of the children's grandparents. In descending order, the percentages and numbers of the top 10 racial stocks were as follows:[11]

English:	30.7%, $n = 197$	
German:	15.7%, $n = 101$	
Scottish:	11.3%, $n = 73$	
Jewish:[12]	10.5%, $n = 68$	
Irish:	9.0%, $n = 58$	
French:	5.7%, $n = 37$	
Scottish Irish:	2.8%, $n = 18$	
Swedish:	2.5%, $n = 16$	
Italian:	1.4%, $n = 9$	
Welsh:	1.4%, $n = 9$	

Terman (1926) suggested that the top racial stocks in frequency probably were overrepresented in comparison to their presence in the California population. For example, he noted, "The proportion of Jews in the total population of the three main cities covered (Los Angeles, San Francisco, and Oakland) is approximately 5 percent. According to this estimate, the amount of Jewish blood in our group is about twice the expected [amount]" (p. 56). In sharp contrast to the high percentage of White gifted children in Terman's study, racial/ethnic minority students were greatly underrepresented. Note the following percentages and numbers:

Japanese:[13]	0.6%, $n = 4$	
Negro:	0.1%, $n = 1$	
Indian:	0.1%, $n = 1$	
Mexican:	0.1%, $n = 1$	

From a present-day perspective, it is incredulous that of 168,000 children canvassed statewide for the main study, Terman (1926) included only 4 gifted Japanese children, 1 gifted Negro child, 1 gifted Indian child, and 1 gifted Mexican child. When viewed in the context of the elitist Zeitgeist of the 1920s (see Chapter 1 of the present book), however, the invisibility of gifted minority children in Terman's study is not particularly unexpected. First, the use of teacher nominations to identify gifted children (the first step in the selection process) might have resulted in a selection bias. As Ceci (1990) voiced, "Teachers probably had a social class bias (as did Terman himself) that tilted the odds against nominating lower class children" (p. 56).[14] Given the collinearity of race/ ethnicity and SES, minority children were disproportionately of low-SES background. Hence, they might have been less likely to be nominated. Second, *no* minority children were part of the standardization samples of the Stanford-Binet or the NIT, the two mental tests that served as the intelligence tests in Terman's study. These tests were largely geared for White, English-speaking, middle-SES children. Furthermore, these highly verbally loaded tests tended to penalize children whose mother tongues were not English (see Klineberg's [1935b] research discussed in Chapter 1 of the present book). Third, in light of the widespread segregation of minority students in inferior schools in California (González, 1990; San Miguel & Valencia, 1998), many of these students did not have the opportunity to learn the school-related content for which the Stanford-Binet and NIT partially tested.

On a final note, it is worthy to point out what Terman (1926) proffered as a possible explanation for the extreme underrepresentation of gifted minority children in his investigation. Commenting on the "very low" percentage (we would argue the *virtual absence*) of gifted children of "Latin blood," Terman stated,

The total population of Latin extraction in the cities covered is not known, but it is certainly very large in comparison with the number of Latin children in our group. Intelligence tests on many Latin groups in America have yielded consistently low scores, with a median IQ usually between 75 and 85. Perhaps a median IQ of 80 for the Italian, Portuguese, and Mexican school children in the cities of California would be a liberal estimate. How much of this inferiority is due to the language handicap and to other environmental factors it is impossible to say, but the relatively good showing made by certain other immigrant groups similarly handicapped would *suggest that the true causes lie deeper than environment.* (p. 57, italics added)

It appears that in his reference to "deeper than environment," Terman was implying genetic factors as the "true causes" of why gifted Italian, Portuguese, and Mexican children were greatly underrepresented in his study. Such a conclusion by Terman, a hereditarian, was not unexpected. Keep in mind that 10 years earlier, in reference to the low intelligence test performance among Portuguese and Mexican children (as well as Spanish Indian and Negro children), he had asserted, "Their dullness seems to be racial or at least in the family stocks from which they come" (Terman, 1916, p. 91).[15]

In the final analysis, Terman's (1925) study of gifted children is considered to be one of the most remarkable research accomplishments in the annals of educational psychology.[16] His study also can be viewed as an indictment of the attitudinal and measurement bias toward minority gifted children during this era.

From the Post-Terman Era to the Sputnik Era

Perhaps the most significant conclusion that stemmed from Terman's (1925) *Genetic Studies of Genius* was his conceptualization of giftedness; that is, giftedness was conceived as general intellectual superiority as measured by an intelligence test. The tendency to equate giftedness with high IQ would be Terman's legacy for several decades to come. With the advent of group-based intelligence tests during the 1920s and their impact on sorting students for curricular assignments (Valencia, 1997b; see also Chapter 1 of the present book), schools began to narrowly define giftedness based on such test scores. As noted by Resnick and Goodman (1993),

The paradigm for the identification of gifted by intelligence tests was solidified between the two world wars, and a high test score remained the major or sole determinant of eligibility for participation in gifted programs in most states and districts into the early 1960s. (p. 114)

Although scholarly interest in gifted children rose during the post-Terman era, exiguous attention was paid to their actual education. For example, in a survey conducted by Heck (1930, cited in Tannenbaum, 1983), it was found that of 762 cities with populations of more than 10,000 people, only 30 cities (4%) had special classes or schools for gifted children. Given that gifted children in general received such meager consideration in their schooling, one can surmise that gifted minority children were doubly shortchanged.

At the federal level, governmental involvement began during the 1930s, as seen in the creation of the U.S. Office of Education, Office of Exceptional Children and Youth (DeLeon & Vandenbos, 1985). Notwithstanding this development, scant attention continued during the 1930s and 1940s toward gifted children and their education. Tannenbaum (1983) described matters as follows:

The 1930s and the war years saw some schools making valiant, if not always successful, efforts to provide enrichment [for gifted children] in the regular classroom. For the most part, the existence of special opportunities of any kind for the gifted has never been widespread even in years when interest grew considerably. In 1941, the National Education Association reported on the nature of education for the gifted in several hundred high schools throughout the country, but a subsequent check with many of these same schools thirteen years later indicated that most of the plans had long since been discarded for one reason or another. Interest probably reached its lowest ebb during World War II and immediately after. (p. 15)[17]

Regarding published literature on gifted children, the bibliographic search and content analysis by Albert (1969) are quite telling. Based on the *Cumulative Subject Index to Psychological Abstracts* (1927-1960) and the *Psychological Abstracts* (1960-1965), Albert identified all references using the following descriptions: Category A—"genius," "distinction," "eminence," "fame"; Category B—"creative," "gifted children," "giftedness." Here are some major findings we culled from his review:

1. Of the total number of identified references from the psychological literature base ($N =$ 264,453) from 1927 to 1965, only 1,318 (0.5%) pertained to Categories A and B combined.

2. Our own analysis of the data from Category B (which is most germane to our discussion) shows that there were 1,126 references (only 0.4% of the total N). In descending order of frequency of Category B citations, there were 618 references regarding creativity (55%), 306 regarding gifted children (27%), and 202 regarding giftedness (18%).

3. Of the 1,126 references in Category B, only 23% of the studies were published between 1927 and 1944, whereas 77% were published between 1945 and 1965.[18]

4. Given Albert's (1969) data in Category B, we calculated a trend analysis for the decades of the 1930s, 1940s, and 1950s. During this time frame, there were a total of 498 citations. Of these total studies, 117 (24%) and 123 (25%) were published during the 1930s and 1940s, respectively. By sharp contrast, 258 studies (52%, a slight majority) were published during the decade of the 1950s.

In sum, Albert's (1969) comprehensive bibliographic review informs us that the scholarly publications on gifted children and their education began slowly during the post-Terman era but escalated after World War II. It must be kept in mind, however, that the proportion of studies in this area was but a dearth in comparison to the general volume of psychological literature. On a yearly basis from 1927 to 1965, the percentage of studies in Category B accounted for less than 0.5% of the total yearly volume, on average. The primary emphases were on the identification and education of gifted children. Another conclusion that we draw from Albert's review is that publications on gifted minority children were scant in number. Albert noted that "interest . . . in racial and national origins is quite minimal" (p. 750). Thus, Albert's review of the extant literature on gifted children and their education supports our thesis that gifted minority children were indeed the "neglected of the neglected."

Historically, one of the most significant single events that spurred interest in the gifted and their education was the launching of Sputnik by the Russians in 1957. Sputnik, a satellite designed to explore outer space, initiated the "space race" between the United States and the Soviet Union. As we have seen, prior to this period, the United States gave little attention to developing its reservoir of talented youths. On this issue, Tannenbaum (1983) noted,

> There was no serious action in America's schools until Sputnik was launched in 1957. At that time, the rhetoric started to become more strident and the research more abundant, and together they either produced or accompanied radical changes in public education. We were convinced that the Russians slipped ahead of us in space technology because we had insufficient manpower to advance the sciences. Predictably, the schools were singled out as scapegoats. (p. 9)

With respect to Tannenbaum's (1983) reference to increased research, the previously discussed bibliographic review by Albert (1969) is very informative. In 1957, the year when Sputnik was launched, 15 studies were published in Albert's Category B (i.e., creative, gifted children, giftedness). The following year, the number of such publications more than doubled ($n = 36$). In 1959, the number increased to 86. From 1960 to 1965, the number of published studies ranged from 58 to 168, averaging about 100 per year during the Sputnik and post-Sputnik eras.

Notwithstanding the newfound research and schooling interests in the gifted and talented, such concerns were short-lived. According to Ross (1997), this interest came about by misdirected priorities. She commented,

> During this era, the rhetoric supporting programs for able students emerged from passionate concerns about our "race" with the Soviet Union and our ability to win the Cold War. Channeling able students into mathematics and science fields was the way to improve our national standing with the Soviet Union. The focus was on national need, not on self-fulfillment, and on the most able students, not the entire student body. With the success of the U.S. space program and the aversion of the Sputnik "crisis," support for able students waned. (pp. 553-554)

During the 1960s, the further decline of attention to the gifted was evident. One particular issue that influenced this diminished interest was a focus on the education of economically disadvantaged racial/ethnic minority students (Tannenbaum, 1983). In particular, allegations of cultural bias in intelligence tests and test abuse (see Chapters 1 and 5 of the present book) brought forth an attack on the Termanian legacy of the high IQ = giftedness conception. Critics argued that a reliance on this notion of giftedness underidentified gifted minority students and, as such, denied them the enrichment/ accelerated curriculum that, advocates argued, was their rightful access. Tannenbaum (1983) cogently captured this tension as follows:

Thus we see that American education was not able to reconcile its interest in the gifted with its concern about the disadvantaged, nor could it design a satisfactory methodology for locating and cultivating giftedness among these minority groups. The dilemma was easy to resolve insomuch as it reduced itself to a choice between battling for social justice as against pursuing excellence, and there was no doubt as to which of the two would better fit the mood of the 1960s. (p. 29)

The Contemporary Era

In our periodization scheme, the contemporary period gets under way during the early 1970s, signaling a renewed interest in the gifted child and his or her education. Drawing from the historical sketches written by Jackson (1979), Resnick and Goodman (1993), Ross (1997), Tannenbaum (1983), and Zettel (1979), the following represent some key events and developments:

1. During the early 1970s, Congress asked for a study on the status of gifted *and* talented children. The ensuing report, commonly referred as the "Marland Report" (U.S. Department of Health, Education, and Welfare, 1972), was coined after Sydney Marland, the commissioner of education. A major conclusion of this report was that gifted and talented children were largely undiscovered and that services for them were vastly underdeveloped. Furthermore, the Marland Report found that "minority and culturally different gifted and talented children were scarcely being reached" (Jackson, 1979, p. 48).

2. The 93rd Congress, in response to the Marland Report, in 1974 created an Office of Gifted and Talented in the U.S. Office of Education. Appropriations for the office were quite modest. For example, there was limited funding for training, research, and demonstration projects. Notwithstanding these federal efforts, state attention to gifted and talented children was slow in coming. In a nationwide survey of state agencies conducted in 1977 by the Council for Exceptional Children, it was noted that "there are fourteen states that make no reference at all to gifted and talented children in their state codes or statutory language" (Zettel, 1979, p. 65).[19] The survey also reported, "The responsibility for the development of specific . . . criteria . . . for the identification [of gifted and talented students] . . . is left largely to the discretion of the local school districts" (p. 66). Some state agencies elected to use the long-standing tradition of equating giftedness with high performance on an intelligence test. For example, in California during the early 1970s, "mentally gifted minors" were identified by superior performance on an intelligence test, that is, by demonstrating a "general intellectual capacity" within the top 2% on a measure of intelligence (Zettel, 1979).[20] Suffice it to say, as we discuss later, such a stringent criterion did not bode well for the identification of gifted minority students.

3. By the early 1980s, one can again see the vicissitudes of federal interest in the gifted. At this time, "the Office of Gifted and Talented had been closed, and funding for

the gifted programs had mainly been merged into block grants to be used at the discretion of individual states" (Resnick & Goodman, 1993, p. 111). Later during the 1980s, however there was a shift in sentiment, as the Office of Gifted and Talented was reinstated and increased funding ensued. A commissioned report prepared in 1990 noted that 47 states had legislation germane to gifted and talented children (Kleine, 1990, cited in Resnick & Goodman, 1993).

4. In 1988, the federal government passed Public Law (P.L.) 100-297, the Jacob K. Javits Gifted and Talented Students Education Act (1988). The passage of the Javits Act had great relevance for poor and working-class children as well as for the culturally/ linguistically diverse. "The legislation specified that the program place special emphasis on economically disadvantaged students, limited English-proficient students, and students with disabilities who are gifted and talented" (Ross, 1997, p. 555).

5. The most recent development in the field of gifted and talented education is seen in the context of broad educational reform. The report, *National Excellence: A Case for Developing America's Talent* (U.S. Department of Education, 1993, cited in Ross, 1997), is critical of a number of shortcomings in gifted and talented education (e.g., the need to establish challenging curriculum standards, the need to encourage appropriate teacher training). The report also zeroes in on the need to identify and nourish gifted and talented economically disadvantaged and minority students. Ross (1997), in reference to the report's recommendations, commented, "It is essential to remove barriers to participation of poor and minority [gifted and talented] children in advanced learning opportunities and to support research and demonstration projects that are developing ways to work with diverse populations" (p. 558).

The preceding historical sketch of gifted children and their education suggests a waxing and waning of interest since the turn of the 20th century to the present. Certainly, there have been major advances in the research and policymaking sectors with respect to gifted children. Yet, major issues continue. We now move to this discussion.

Issues

In this section, we examine four issues concerning gifted students that are particularly pertinent for minority children and youths: (a) conceptions and measurement of giftedness, (b) allegations of elitism, (c) excellence versus equity, and (d) underrepresentation of minority children in gifted programs.

Conceptions and Measurement of Giftedness

As we have emphasized, many scholars traditionally have "defined giftedness as high general intelligence as measured by a high global IQ score" (Winner, 1997, p. 1071).[21] Given the unidimensional and static nature of this conception of giftedness and its measurement, it is not surprising that during the past 10 to 15 years there has been a trend to conceptualize giftedness and how it is measured in multidimensional and dynamic ways. A common thread tying together the newer conceptions "is that we ought to define giftedness in a broader way that goes beyond what is measured by either the IQ or the achievement tests. . . . Giftedness cannot possibly be captured by a single number" (Sternberg, 1997b, p. 43).

Notwithstanding the shared feature of broadness that the newer conceptions of giftedness possess, they are quite different in how they express such multidimensionality. An excellent source that presents the rich variety of these new ideas of giftedness is Sternberg and Davidson's (1986a) edited book, *Conceptions of Giftedness*. Based on 16 different authors' theoretical approaches to understanding giftedness, Sternberg and Davidson (1986b) first divided the conceptual terrain into two major distinctions: implicit and explicit theories. The authors commented on the dissimilarity of these two theoretical frameworks as follows:

> Implicit theories are essentially dimensions that lie within the heads of the theorists, who may be experts or laypersons. Thus, implicit theories define what he or she means by giftedness and goes on to illustrate the implications of this definition. Because implicit theories are definitional, they cannot be empirically tested, other than in the sense of showing that the proposed implicit theory is in accord with other people's implicit theories.
>
> Explicit theories presuppose definitions and seek to interrelate such definitions to a network of psychological or educational theory and data. Such theories are testable by the usual empirical means and thus may be falsified. But the definitions upon which they are based cannot be falsified, so it is important in evaluating the explicit theories to be sensitive to the underlying conception of giftedness that has generated the theory and data and to evaluate whether this conception is a useful one. (p. 3)

Following their implicit/explicit division of theoretical approaches, Sternberg and Davidson (1986a) proceeded to further partition matters into four sections. The four parts (with their respective numbers of chapters) are as follows:

Implicit theoretical approaches $(n = 5)$

Explicit theoretical approaches: Cognitive theory $(n = 4)$

Explicit theoretical approaches: Developmental theory $(n = 2)$

Explicit theoretical approaches: Domain-specific theory $(n = 2)$

Space limitations do not allow us to discuss these four approaches and their constituent authors' models in detail. We can, however, provide a sampling of these conceptions of giftedness. Our focus is on those theoretical approaches that we believe offer the most useful insights into understanding giftedness among minority students and their identification for inclusion in gifted programs.

The work by Renzulli (1986) falls within the implicit theoretical approach conceived by Sternberg and Davidson (1986a). At the heart of Renzulli's (1986) theorizing is his "three-ring" conception of giftedness. Focusing on displays of "gifted behaviors" and not "gifted people," his model contains three clusters: (a) "above-average abilities" in

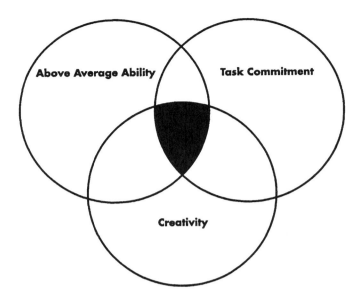

Figure 8.1.   Renzulli's Three-Ring Conception of What Makes Giftedness
SOURCE: From "The Three-Ring Conception of Giftedness: A Developmental Model for Creative Productivity," by J. S. Renzulli, in R. J. Sternberg & J. E. Davidson (Eds.), *Conceptions of Giftedness*, 1986, p. 66. Copyright 1986 Cambridge University Press. Adapted with permission of author.

either general ability (e.g., as measured by general aptitude and intelligence tests) or specific abilities (e.g., chemistry, mathematics, ballet); (b) "task commitment," also commonly referred to as "perseverance, endurance, hard work, dedicated practice, self-confidence, and a belief in one's ability to carry out important work" (p. 69);[22] and (c) "creativity," as evidenced by, for example, originality of thinking, curiosity, and divergent thinking. It is important to underscore that in Renzulli's three-ring conception, "no single cluster 'makes giftedness.' Rather, it is the interaction among the three clusters that research has shown to be the necessary ingredient for creative-productive accomplishment" (p. 65).[23] This intersection is illustrated in Figure 8.1.

In sum, Renzulli's (1986) three-ring conception of giftedness holds great promise for increasing the percentage of minority students in gifted programs. His model is based on the principle of inclusion. Traditional means of identification, intelligence tests, typically select 2% to 3% of the student population. In Renzulli's model, his "talent pool" "consists of the top 15-20% of the school population in general ability or any and all specific performance areas that might be considered high priorities in a given school's programming efforts" (p. 76). Referring to his Schoolwide Enrichment Model, which is a procedure for identifying and nurturing talent,[24] Renzulli and Reis (1997) commented that one particular

> reason for the growing popularity of the Schoolwide Enrichment Model is our concern for providing special enrichment opportunities for students from low socioeconomic backgrounds and for students who show potential for superior performance areas that are not easily assessed by traditional ability measures. (p. 150)

Another theoretical approach discussed in Sternberg and Davidson's (1986a) work is Sternberg's (1986) triarchic theory of human intelligence, an explicit theoretical approach: cognitive theory.[25] Given that we already introduced Sternberg's theory in Chapter 2 of this book, our focus here is on how the triarchic theory can serve as a basis for understanding what Sternberg called "extraordinary intelligence" or giftedness. Let us first review the three subtheories subsumed in the triarchic theory:

> The triarchic theory comprises three subtheories. The first subtheory [componential] relates intelligence to the internal world of the individual, specifying the mental mechanisms that lead to more or less intelligent behavior. The subtheory specifies three kinds of information-processing components that are instrumental in (a) learning how to do things, (b) planning what things to do and how to do them, and (c) actually doing the things. The second subtheory [experiential] specifies those points along the continuum of one's experience with tasks or situations that most critically involve the use of intelligence. In particular, the account emphasizes the roles of novelty and of automatization in exceptional intelligence. The third subtheory [contextualist] relates intelligence to the external world of the individual, specifying three classes of acts—environmental adaptation, selection, and shaping—that characterize intelligent behavior in the everyday world. (p. 223)

Regarding the componential subtheory, giftedness can be expressed as "analytic giftedness," a term coined by Sternberg (1997b) in more recent writings. Such giftedness "involves being able to dissect a problem and understand its parts. People who are strong in this area of intellectual functioning tend to do well on conventional tests of intelligence, which place a premium on analytical reasoning" (p. 43).

With respect to the experiential subtheory of intellectual giftedness, Sternberg (1997b) referred to "synthetic giftedness," which "is seen in people who are insightful, intuitive, creative, or just adept at coping with relatively novel situations. People who are synthetically gifted do not necessarily do well on conventional measures of intelligence" (p. 43). Synthetic giftedness, as conceptualized within the experiential subtheory, has considerable promise for understanding cross-racial/ethnic giftedness. Sternberg (1986) suggested that an experiential perspective is helpful in comprehending why it often is so difficult to compare intellectual levels of individuals across different sociocultural groups. That is, even if a particular intelligence test requires the same performance components across groups, the chances are very unlikely that the test will "be equivalent for the groups in terms of its novelty and the degree to which performance has been automatized prior to the examinee's taking the test" (p. 234).

Finally, there is the contextualist subtheory of intellectual giftedness, which has to do with what Sternberg (1997b) described as "practical giftedness." This "involves applying whatever analytic or synthetic ability you have to everyday pragmatic situations" (p. 44). The key point here is that Sternberg (1986) was conceptualizing intelligence contextualized in behavior in "real-world" environments that are deemed relevant to a person's life. As such, his conceptions of intelligence and giftedness are truly culture bound. On this point, Sternberg (1986) asserted,

An implication of this view is that exceptional intelligence cannot be fully understood outside a sociocultural context, and it may in fact differ for a given individual from one culture to the next. Our most intelligent individuals might come out less intelligent in another culture, and some of our less intelligent individuals might come out more intelligent. (p. 235)

Another model of giftedness presented in Sternberg and Davidson (1986a) is Csikszentmihalyi and Robinson's (1986) sociocultural constitution of giftedness, an explicit theoretical approach: developmental theory. Csikszentmihalyi and Robinson pointed out that their review of the literature on giftedness leaves an impression that giftedness is a stable trait of a person, that is, something one either possesses or does not possess. Contrary to this "objective fact" perspective, Csikszentmihalyi and Robinson have a very different view of giftedness. They summarized their position as follows:

1. Talent cannot be observed except against the background of well-specified cultural expectations. Hence it cannot be a personal trait or attribute but rather is a relationship between culturally defined opportunities for action and personal skills or capacities to act.
2. Talent cannot be a stable trait because individual capacity for action changes over the life-span, and cultural demands for performance change both over the life-span and over time within each domain of performance. (p. 264)

What we find particularly valuable about Csikszentmihalyi and Robinson's (1986) sociocultural constitution of giftedness approach is their emphasis that giftedness is a culture-bound social construction. In this perspective, giftedness becomes what society wants it to be. Given the reification of IQ during the 1920s and the subsequent use of IQ scores to measure giftedness, it is not surprising that the conceptualization of high IQ as giftedness has become such an entrenched social construction.

In sum, the newer conceptions of giftedness proffered by Sternberg (1986), Renzulli (1986), and Csikszentmihalyi and Robinson (1986) are inventions that offer considerable promise as inclusionary models. The implications of these conceptions for measuring and identifying giftedness among minority students are profound. Particularly significant are the dynamic connections among how society views the intellectual possibilities of minority children, the cultural rules about how giftedness among minority children can be expressed, and what these children may perceive as the rules. Speaking in a general context, but nonetheless having relevance for gifted minorities, Csikszentmihalyi and Robinson noted,

These considerations suggest that to reify talent as some kind of performed gift that exists within the child is a mistake. The homuncular view of giftedness does not fit the facts. It is not that the child's talent reveals itself and is recognized by society. It is closer to the truth that the possibility to reveal talent is provided by the cultural environment and that it is this possibility that the "talented" child recognizes. (p. 270)

Allegations of Elitism

Of the criticisms leveled against giftedness, none has been so trenchant as the charge of elitism (see, e.g., Goodlad & Oakes, 1988; Margolin, 1994; Newland, 1976; Richert, 1985b, 1987, 1997; Sapon-Shevin, 1994). According to Richert (1997), the three most frequently voiced complaints by influential scholars about elitism in programs for the gifted are the following:

1. Elitist identification practices and definitions of giftedness create school segregation by economic class and cultural groups.
2. The most motivating and challenging curriculum is found in programs for the gifted, while curriculum for other programs is often monotonous and devoid of interest.
3. The best trained and most effective teachers work with the gifted, denying the benefits of these teachers to students of other abilities. (p. 75)

In a related vein, Newland (1976) noted that some laypeople, school personnel, and scholars have commented that special provisions for the gifted "will contribute to the establishment of an intellectual elitism and, thereby, the creation of a nonelite of second-class citizens" (pp. 34-35). Several arguments have been advanced on the grounds that such special provisions are inimical to education in a democratic society:

1. Special education of the gifted would contribute to intellectual snobbishness.
2. The gifted themselves would be deprived of benefitting socially from association with the nongifted.
3. Nongifted children would be deprived of the experience of learning from the gifted—the stimulus to learn more and the picking up of the gifted's learning styles. (p. 35)

One of the most comprehensive, as well as provocative, discussions on the social construction of giftedness as an elitist notion is seen in Margolin's (1994) book, *Goodness Personified: The Emergence of Gifted Children*. Margolin described the purpose of his book as follows:

> Conceptualizing gifted child experts' findings as accounts . . . , that is, as an effort to shape people's beliefs in giftedness, I deliberately avoid the question of whether there "really" are gifted children or whether the methods used to describe them are empirically valid. Rather, I am interested in the way the experts see, describe, and explain gifted children. The purpose of this methodological stance is to specify the language and imagery used to make the concept *gifted children* appear representative of something real, obdurate, and objective. Put somewhat differently, this book explores the methods used to display gifted children as objects of nature rather than of human imagination, as something discovered rather than something created. This book, then, is not about the "emergence" of a new vocabulary or a new class of people. Nor is it primarily a description of claims made in conjunction with the gifted child label. Instead, this book examines these

methods by which and through which claims about the intended object—the gifted—were (and are) taken as true. (p. xix)

What we provide here is a synopsis of Margolin's (1994) analysis that is particularly germane to the minority gifted:

1. Drawing from the works on giftedness by, for example, Goddard (1928), Hollingsworth (1929, 1935, 1942), Stedman (1924), and Terman (1925), Margolin (1994) asserted that the gifted child was "assembled," historically, in an idealized manner, that is, as being highly intelligent and talented in many areas (such gifts were believed to be inborn, not resulting from SES advantages), morally superior, paragons of comportment in the classroom, strong, very healthy, beautiful, stable, and White.

2. Notwithstanding the renewed attention to the issue of underrepresentation of ethnic minority children in gifted programs, Margolin (1994) commented that racial discrimination continues. Much of this, he noted, is due to what we refer to as "deficit thinking" (Valencia, 1997a; see also the Introduction to the present book). His point is that "responsibility for the underrepresentation of minorities is attributed to factors located outside of, or peripheral to, gifted education itself" (p. 34). Margolin continued, "Most commonly, it is attributed to the excluded themselves, to their 'real learning deficits, not just hidden talents' (Tannenbaum, 1990, p. 85)" (p. 34).[26] Margolin (1994) asserted that notwithstanding all the statements about the need for inclusion and cultural pluralism in the gifted child discourse, the elitist nature of giftedness continues to exclude minority children: "What is conspicuously absent . . . is the suggestion that our very understanding of giftedness—the ways we study, validate, recognize, and describe this phenomenon—reflects and supports discrimination" (p. 34).

Margolin's (1994) treatise on the elitist nature of giftedness is a profound account of the epistemological problem of how our society has come to believe in and value the reality of children who are deemed gifted. Margolin's stirring discourse raises significant concerns about why gifted minority children continue to be underidentified and underserved. He asserted that much of this neglect has to do with power relations:

A "pedagogy of the oppressed"[[27]] directly implies a pedagogy of oppressors, that there is no power relation involving the poor without a correlative power relation involving the affluent (Bourdieu & Passeron, 1977). Moreover, I argue that gifted child education constitutes a mechanism specifically geared toward articulating this power relation on the affluent, that it is a strategy to develop a class of people who lead, direct, and originate. (p. 77)

In the most fundamental sense, Margolin's thesis of elitism in giftedness is that it rests on a zero-sum philosophy: "For the gifted to be elevated, other children have to be downgraded" (p. xxii). Stemming from this claim is the issue of excellence versus equity, which we turn to next.

Excellence Versus Equity[28]

A few public values have dominated educational policy in the United States—efficiency, equity, and excellence (Murphy, 1990). These values, nonetheless, have shifted in importance as American educators reform public schools. For example, Callahan (1962) indicated that educational policies of the 1920s and 1930s were concerned with the efficiency in which schools were conducted. Consolidation and reduction in expenditures were the two most cited reforms. As evidenced by the development of Operation Head Start during the mid-1960s and the passage of the Education for All Handicapped Children Act of 1975 (P.L. 94-142, "Education of Handicapped Children," 1977), there was a shift to a focus on equity as the dominant value. During the 1980s and 1990s, the reform movement redirected its efforts to excellence as the dominant value in educational policy (Bacharach, 1991).

Regarding the excellence versus equity debate, Gardner (1961) cogently captured the issue in his book, *Excellence: Can We Be Equal and Excellent Too?*, nearly four decades ago. Gallagher (1986) noted that Gardner attempted to wrestle with a conflict in societal values that continues to this day—"a deep and abiding desire to have equality of opportunity and treatment for all citizens while, at the same time, preserving the American tradition of high achievement and production as a national characteristic" (pp. 233-234). In a later publication, Gallagher (1993) commented on the American educational system's struggle reconciling these two significant societal values:

> The first of these [values] is *equity:* the promise that all children shall receive an equal opportunity for education. The second value is *excellence:* that full attention and stimulation will be given to the very best students—those who demonstrate their ability and superiority in the educational domain. (p. 96)

A good example of the nation's ambivalence in attempting to achieve these two apparently competing educational objectives is seen in the proposed national goals in education (Gallagher, 1991, 1993). In 1989, six major national education goals (four of which deal with equity and excellence), targeted to be met by the year 2000, were accepted by the 50 governors who made up the Governors' Task Force on Education (U.S. Department of Education, 1991a).[29] Goals 1 and 2, which focus on equity and have clear implications for improving education for low-SES minority students, are that "all children in America will start school ready to learn" and that "the percentage of students graduating from high school [will increase] to at least 90 percent," respectively. By contrast, Goals 3 and 4, which emphasize excellence and have implications for gifted students, are that students demonstrate "competency over challenging subject matter" (e.g., English, geography) and that "U.S. students will be first in the world in science and mathematics achievement," respectively.

The excellence versus equity issue, which is a national debate in the United States,[30] clearly has penetrated the gifted education movement. Renzulli and Reis (1991) referred to the impact of the growing debate as the "quiet crisis" in gifted education. A great deal of the tension stems from two trends in public elementary and secondary education as a whole. First, there is a movement to eliminate most forms of homogeneous grouping, commonly referred to as "detracking" at the secondary level. Colangelo and Davis (1997), editors of the *Handbook of Gifted Education,* referred to the detracking trend as a "damaging reform movement" (p. 4). On this claim of harm to gifted education, Colangelo and Davis asserted,

> Jeannie Oakes (1985) and others have argued that tracking plans are racist and discriminatory, deprive slow-track students of educational opportunities, and damage their self-concepts. Such arrangements certainly touch the heartstrings of democratic-minded persons. Unfortunately, this "national hysteria" (Reis et al., 1992) has led some districts to abolish not only accelerated classes, but [also] special classes for the gifted and gifted programs themselves. Says Renzulli (1991), with heterogeneous grouping, "bright kids learn nothing new until January." (pp. 3-4)[31]

Another "damaging reform movement" that harms gifted education, according to Colangelo and Davis (1997), is the cooperative learning trend. The idea behind cooperative learning is that learning is best promoted in heterogeneous groups. Citing the work of Slavin, Madden, and Stevens (1990), Colangelo and Davis (1997) commented that proponents of cooperative learning cite research that shows students' gains in academic achievement, cognition, motivation, social skills, and self-confidence. Colangelo and Davis conceded that cooperative learning is an effective teaching strategy—but not for gifted students. Such students, they asserted (in the absence of a research citation), prefer to work alone (because they can learn faster), frequently get stuck doing the work of slower students, and miss opportunities to engage in accelerated or enriched learning activities that are commensurate with their abilities. "In short, cooperative learning diverts attention from more valid educational needs of gifted students" (p. 4). It is important to note, however, that such assertions have not gone unchallenged. For example, Slavin (1991), a leading researcher and advocate of cooperative learning, strongly disagreed, that ability grouping is beneficial to high-achieving students.

In sum, the excellence versus equity drama continues to linger. Unfortunately, the noble goal (excellence) of providing the best possible education for our nation's brightest and most talented students and the noble goal (equity) of improving the educational lot for low-achieving students (particularly low-SES, inner-city minorities) have largely been cast as *competing* goals. Need this be the case? Some scholars think not. For example, Renzulli and Reis (1991) took the position that to solve the "quiet crisis" and achieve a resolution between the two noble goals, "what should be done is to extend the technology that has been developed in gifted and talented programs to a broader spectrum of general education" (p. 32). That is, through the use of innovations already used in gifted education (e.g., thinking skills applications, independent study, acceleration of subject content, more challenging textbooks, learning centers), there will be a greater "inclusion of at-risk and underachieving students in our [gifted] programs" (p. 34).

Underrepresentation of
Minority Students in Gifted Programs

The underrepresentation of African American, Latino, and American Indian students in gifted programs has been a persistent and pervasive finding from Terman's (1925) study to the present.[32] In a brief review of reported evidence of giftedness among U.S. racial/ethnic groups for the period 1925 to 1964, Adler (1967) reported that the highest incidences were seen among people of Jewish, German, English, and Scottish ancestry. By sharp contrast, "at the other end of the scale is the Negro group. This group tends to be represented in far lesser numbers than would be expected" (p. 105). Adler also reported that American Indians and people of Mexican origin had low incidences. Since Adler's review, numerous studies and reports have documented or discussed the underrepresentation of gifted African American, Latino, and American Indian students.[33] Likewise, the underrepresentation of these minority groups in gifted programs is a commonplace discussion in major texts on the gifted and their education.[34]

The most comprehensive data sources on the underrepresentation of minority students in gifted programs are the Office for Civil Rights (OCR) biennial compliance reports. Since 1968, the OCR has surveyed elementary and secondary schools nationally regarding student placements in special education categories (e.g., mental retardation, specific learning disabilities), gifted and talented programs, and other categories.[35] National, state, district, and local school data are provided. With respect to the most recent OCR data on student enrollment in the various categories by race/ethnicity, the OCR collected information in 1994 and made it available to the public in 1997 (U.S. Department of Education, OCR, 1997, hereafter the 1997 OCR survey). In the 1994 data set, 5,173 districts were surveyed. The districts were drawn from a universe of 14,814 school districts using a stratified random sampling approach. Of the 5,173 districts sampled, 4,671 are on the final data file.[36] Although the selection process used random sampling procedures, there were some school districts that were sampled with certainty (personal communication, Peter McCabe, July 26, 1999). School districts sampled with certainty were (a) those with 25,000 or more students, (b) those with 25 or more schools, (c) those with existing court orders that are being monitored by the Department of Justice, and (d) for the period 1990 to 1998, those with former settlement agreements with the OCR.

In Table 8.1, we present the 1994 OCR data regarding racial/ethnic group disparities (over- and underrepresentation) for the gifted and talented category. The disparity analysis is based on our calculations of the OCR data.[37] For the nation as a whole, two racial/ethnic groups show an overrepresentation: A/PIs (58.60%) and Whites (19.84%). That is, A/PI and White students are enrolled at disproportionately higher rates in the gifted and talented category than expected based on their percentages in the national school-age population. By contrast, American Indian, Hispanic, and Black students show underrepresentation. They are enrolled at disproportionately lower rates in the gifted and talented category than expected based on their percentages in the national school-age population. American Indian students are underrepresented by 23.08%, and Hispanic and Black students are underrepresented at nearly identical rates—50.83% and 50.56%, respectively.

TABLE 8.1 Disparity Analysis for Gifted and Talented, 1994: For Nation and Top 10 States in Combined Minority Enrollment (K-12)

State	American Indians			Asian/Pacific Islanders			Hispanics			Blacks			Whites		
	Enroll-ment[a] (%)	Gifted/ Talented[b] (%)	Dis-parity[c] (%)	Enroll-ment (%)	Gifted/ Talented (%)	Dis-parity (%)	Enroll-ment (%)	Gifted/ Talented (%)	Dis-parity (%)	Enroll-ment (%)	Gifted/ Talented (%)	Dis-parity (%)	Enroll-ment (%)	Gifted/ Talented (%)	Dis-parity (%)
Nation	1.04	0.80	−23.08	3.72	5.90	+58.60	12.71	6.25	−50.83	16.85	8.33	−50.56	65.68	78.71	+19.84
1. California	0.76	0.58	−23.68	11.55	18.94	+63.98	38.15	18.10	−52.56	8.40	4.51	−46.31	41.14	57.88	+40.69
2. Texas	0.22	0.16	−27.27	2.08	3.82	+83.65	35.62	21.76	−38.91	14.60	10.28	−29.59	47.48	63.98	+34.75
3. New York	0.36	0.31	−13.89	4.83	3.48	−27.95	16.68	2.77	−83.39	20.25	5.88	−70.96	57.88	87.56	+51.28
4. Florida	0.20	0.18	−10.00	1.72	3.43	+99.42	14.78	7.68	−48.04	24.56	7.67	−68.77	58.73	81.04	+37.99
5. Illinois	0.13	0.09	−30.07	2.94	5.07	+72.45	10.92	3.33	−69.51	22.89	13.37	−41.59	63.12	78.15	+23.81
6. Georgia	0.12	0.08	−33.33	1.63	3.09	+89.57	2.02	0.54	−73.27	38.89	12.46	−67.96	57.34	83.84	+46.22
7. New Jersey	0.32	0.45	+40.63	5.86	11.32	+93.17	11.47	4.19	−63.47	18.29	9.03	−50.63	64.06	75.00	+17.08
8. North Carolina	0.66	0.30	−54.55	1.18	1.95	+65.25	1.70	0.48	−71.77	32.03	9.68	−69.78	64.43	87.58	+35.93
9. Virginia	0.16	0.12	−25.00	3.53	5.74	+62.61	3.32	1.64	−50.60	27.15	11.16	−58.90	65.84	81.34	+23.54
10. Louisiana	0.50	0.20	−60.00	1.21	3.76	+210.74	1.09	0.94	−13.76	44.06	17.30	−60.74	53.14	77.79	+46.39

SOURCE: U.S. Department of Education, Office for Civil Rights (1997).
a. Percentage enrollment = percentage of racial/ethnic group in total K-12 enrollment.
b. Percentage gifted/talented = percentage of racial/ethnic group in gifted/talented category.
c. In the percentage disparity category, a plus sign (+) indicates overrepresentation percentage and a minus sign (−) indicates underrepresentation percentage.

229

TABLE 8.2   Incidence Rates for Gifted and Talented, by Race/Ethnicity, 1994:
Nation (K-12)

Racial/Ethnic Group	Total Enrollment	Total Gifted/Talented Enrollment	Gifted/Talented Incidence Rate (%)
Asian/Pacific Islanders	1,588,124	151,335	9.53
Whites	28,017,376	2,017,376	7.20
American Indians	445,105	20,521	4.61
Blacks	7,193,038	213,551	2.97
Hispanics	5,425,976	160,210	2.95

SOURCE: U.S. Department of Education, Office for Civil Rights (1997).

Another way of analyzing the OCR data shown in Table 8.1 is by comparing the *incidence rates* of gifted and talented identification by race/ethnicity. For example, what percentage of the White student group is identified as gifted and talented in comparison to the Hispanic group? To address this question, and to make other racial/ethnic comparisons, we calculated estimates of incidence rates for the five racial/ethnic groups. These data are presented in Table 8.2. As can be observed, the incidence rates for the gifted/talented classification vary across racial/ethnic groups. A/PI students have the highest incidence rate (9.53%), followed by Whites (7.20%), American Indians (4.61%), Blacks (2.97%), and Hispanics (2.95%). In comparison to the White student population, A/PIs are 1.3 times more likely to be identified as gifted/talented. By contrast, in comparison to Whites, American Indians are 1.6 times less likely to be identified, and Hispanic and Black students both are about 2.4 times less likely to be identified as gifted/talented. If parity in incidence rates existed among the racial/ethnic groups, then many more American Indian, Hispanic, and Black gifted/talented students would be identified. That is, if the incidence rate for these three groups were identical to the White incidence rate of 7.20%, then the following *increases* in the number of gifted/talented minority students would be observed:

Racial/Ethnic Group	Current Number of Gifted/Talented Students	Projected Number of Gifted/Talented Students Using White Incidence Rate of 7.20%	Projected Net Increase of Gifted/Talented Students
American Indians	20,521	32,048	11,527
Blacks	213,551	517,899	304,348
Hispanics	160,210	390,670	230,460
Total	394,282	940,617	546,335

In sum, if American Indian, Hispanic, and Black students were being identified as gifted/talented at the same rate as White students (7.20%), then the total number of gifted/talented students from these three minority groups would increase sharply from the current total of nearly 400,000 to close to 1 million students—a net growth of more than a half million gifted/talented students. When one looks at the OCR data in this manner, it becomes clear why some scholars (e.g., Russo, Ford, & Harris, 1993) assert that the issue of underrepresentation of minority students in gifted/talented programs is so grave that it affects these students' legal and educational rights.

Thus far, our discussion of the 1997 OCR survey has focused on the national level. Given that any analysis at the aggregate level can, and often does, obscure variability at sublevels, we have calculated gifted/talented disparity analyses for the top 10 states in combined minority total (kindergarten through Grade 12 [K-12]) enrollments.[38] As the data in Table 8.1 show, the state-by-state disparities reflect, with some expectations, the national pattern for the respective racial/ethnic groups. For the White student population, there is overrepresentation in gifted/talented programs in each of the 10 states with the highest combined minority enrollments. With the exception of New Jersey, the overrepresentation rates are higher than the national percentage of 19.84%. New York, Louisiana, Georgia, and California (in that order) have the highest rates of overrepresentation among Whites.

Regarding American Indian students, the national pattern of underrepresentation is observed in 9 of the 10 states (New Jersey is the exception). These disparities range from a low of 10.00% in Florida to a high of 60.00% in Louisiana.

With respect to gifted/talented A/PIs, the analysis by states reflects, to a large degree, the national pattern of overrepresentation. There is, however, 1 outlier among the 10 states. In New York, A/PIs are *underrepresented* (27.95%). Why this is so is not clear. The overrepresentation percentages of A/PIs in the other 9 states exceed the national rate of 58.60%. In some cases, these overrepresentations are quite large (e.g., 210.74% in Louisiana, 99.42% in Florida).

For Hispanic gifted/talented students, the data presented in Table 8.1 show that the disparities in the top 10 states in combined minority enrollments reflect the national pattern of underrepresentation, as 6 of the 10 states have disparities that surpass the national rate of 50.83%. The lowest underrepresentation rates are seen in Louisiana, Florida, and Texas, suggesting that these states are the most successful in identifying gifted/talented Hispanic students. Although the 1997 OCR survey does not disaggregate the Hispanic student group by national origin, we can infer how Hispanic gifted/talented students fare by subgroups by examining the disparities in several states. Underrepresentations are observed in California and Texas (predominantly Mexican-origin subgroup), New York and New Jersey (predominantly Puerto Rican-origin subgroup), and Florida (predominantly Cuban-origin subgroup). Thus, it appears that the underrepresentation of gifted/talented students cuts across the major Hispanic subgroups.

Regarding Black gifted/talented students, they too can be characterized as showing disparities in the top 10 states in combined minority enrollments that mirror the national

TABLE 8.3   Underrepresentation/Overrepresentation for Gifted and Talented, 1994: 50 States (K-12)

Racial/Ethnic Group	Number of States Underrepresented	Number of States Overrepresented
American Indians	46	4
Asian/Pacific Islanders	7	43
Hispanics	49	1
Blacks	49	1
Whites	2	48

SOURCE: U.S. Department of Education, Office for Civil Rights (1997).

pattern of underrepresentation. With the exceptions of Texas, Illinois, and California, all states show disparities that exceed the national rate of 50.56% for Blacks. Some of the highest disparities for Black students are seen in New York, North Carolina, Florida, and Georgia.

In addition to examining racial/ethnic disparities for the top 10 states in combined minority enrollments, we sought to identify disparities at the state level for the nation as a whole, based on 1994 OCR data (U.S. Department of Education, OCR, 1997). These data for the 50 states are presented in Table 8.3. As can be seen, the data in the table underscore the pervasive underrepresentation of American Indian, Hispanic, and Black students in the gifted/talented category. Hispanic and Black students experience the greatest degree of underrepresentation. With the exception of 1 state (Ohio), these students are *underrepresented in every state* in the nation. For White students, their situation is the antithesis of what exists for Hispanics and Blacks. With the exception of 2 states (North Dakota and Ohio), White students are *overrepresented in every state* in the nation.

Another type of gifted/talented disparity analysis that can be derived from the 1994 OCR data (U.S. Department of Education, OCR, 1997) is at the school-by-school level within a school district. As a case in point, we examined the Austin Independent School District (AISD) in Austin, Texas. Our focus is on the elementary schools. As of 1994, the AISD had 66 elementary schools, with a total K-5/6 population of 37,217 students. The respective percentages by race/ethnicity are White (40.6%), Hispanic (38.9%), Black (17.6%), A/PI (2.3%), and American Indian (0.6%). For our analysis, we compared the top 10 schools to the bottom 10 schools regarding the percentage of identified gifted/talented students. These data are shown in Table 8.4. The pattern of percentage of gifted/talented students is very discernible: Students who attend high-enrollment minority schools are less likely to be identified as gifted/talented than are students who attend high-enrollment White schools. For the top 10 schools (7 of which are predominantly White in enrollment), 530 of 5,933 students were identified as gifted/talented, an incidence rate of 8.93%.[39] By sharp contrast, for the bottom 10 schools (9 of which are predominantly minority in enrollment), only 18 of 5,039 students were identified as gifted/

TABLE 8.4 Disparity Analysis for Gifted and Talented: Austin Independent School District Elementary Schools, 1994

Elementary School	Number of Students	Number of Gifted/Talented Students	Percentage of Gifted/Talented Students	Percentage Minority Enrollment in School
1. Barton Hills	347	50	14.41	18.7
2. Bryker Woods	367	50	13.62	23.7
3. Patton	1,026	99	9.65	20.8
4. Maplewood	413	37	8.96	73.6
5. Zilker	496	44	8.87	44.4
6. Menchaca	814	69	8.48	19.0
7. Casis	820	63	7.68	21.3
8. Oak Hill	798	58	7.27	13.4
9. Pease	297	21	7.07	62.3
10. Doss	555	39	7.03	16.9
Total	5,933	530	8.93	
57. Norman	305	3	0.98	95.4
58. Houston	717	4	0.56	90.0
59. Galindo	659	3	0.46	33.3
60. Harris	732	3	0.41	100.0
61. Pecan Springs	503	2	0.40	94.6
62. Zavala	373	1	0.27	98.4
63. Jordan	398	1	0.25	97.5
64. Allan	450	1	0.22	98.0
65. Brooke	346	0	0.00	97.1
66. Govalle	556	0	0.00	98.2
Total	5,039	18	0.36	

SOURCE: U.S. Department of Education, Office for Civil Rights (1997).

talented, an incidence rate of 0.36%. Such disparities raise two important questions. First, why are so few students in predominantly minority elementary schools in the AISD identified as gifted/talented? Second, why are some predominantly minority schools in the AISD (e.g., Maplewood) more successful in identifying gifted/talented students than are other very high-minority enrollment schools (e.g., Brooke)? Intensive research is needed to see whether the pattern of disparities in the AISD also holds for other multiracial/ethnic districts in Texas.

A final analysis we present of the OCR data on gifted/talented students is a look at disparities over time for the various racial/ethnic groups. In Table 8.5, we show disparities that we calculated from OCR data surveys done from 1978 to 1994.[40] Although the over- and underrepresentation rates fluctuate for each racial/ethnic group, the results of this trend analysis over the 16-year period nonetheless are quite clear. First, White and A/PI students are consistently overrepresented in the gifted/talented category. Second, American Indian, Hispanic, and Black students are consistently underrepresented at

TABLE 8.5 Disparity Comparisons for Gifted and Talented, 1978-1994: Nation (K-12)

Racial/Ethnic Group	1978[a] Percentage Disparity	1980[a] Percentage Disparity	1982[a] Percentage Disparity	1984[a] Percentage Disparity	1988[b] Percentage Disparity	1994[c] Percentage Disparity	Average Disparity (1978-1994)
American Indians	-60.76	-51.43	-49.02	-54.66	-50.00	-23.08	-48.16
Asian/Pacific Islanders	+137.32	+94.20	+84.38	+85.75	+65.71	+58.60	+87.66
Hispanics	-24.15	-39.87	-53.70	-45.76	-45.67	-50.83	-43.33
Blacks	-34.48	-44.69	-57.38	-47.31	-47.20	-50.56	-46.94
Whites	+7.41	+15.91	+28.06	+25.79	+23.17	+19.84	+20.03

NOTE: A minus sign (–) for percentage disparity indicates underrepresentation in gifted/talented category; a plus sign (+) for percentage disparity indicates overrepresentation in gifted/talented category.
a. Disparities calculated from data presented in Chinn and Hughes (1987).
b. Disparities calculated from data presented in 1988 Office for Civil Rights survey (U.S. Department of Education, Office for Civil Rights, 1991).
c. Disparities calculated from data presented in 1994 Office for Civil Rights survey (U.S. Department of Education, Office for Civil Rights, 1997).

each of the six data points from 1978 to 1994. Despite the attention that this under-representation issue has been receiving in the literature over the years, the OCR data leave little doubt that educators and testing personnel have not been successful in identifying these minority students as gifted/talented at rates that are commensurate with their proportions in the public school population.

Gifted African American Students

In this section on gifted and talented African American students, as well as in the following sections on other minority groups, we focus on literature that speaks to concerns that might be particularly relevant to the minority group (e.g., cultural considerations). In addition, we cover research findings or discussions about innovative ways in which a greater number of gifted minority students can be identified. Due to space limitations, we need to be brief in our discussions of germane literature. The reader who prefers more comprehensive coverage of gifted minority students might want to inspect the available books or reports on the topic.[41]

Of the four minority groups covered in this chapter, African Americans have the longest history of being subjects of interest and research in the scholarly literature on giftedness. At the turn of the 20th century, Du Bois (1903, cited in Cooke & Baldwin, 1979) posited that among Black Americans there was a "Talented Tenth"—a subgroup of talented Blacks whose giftedness should be identified and nurtured. Cooke and Baldwin (1979) noted that Du Bois's interest was to bring public attention to the Talented Tenth so as to educate these Black leaders "with the intellectual tools to set political, economic, and social goals of black liberation from poverty, social malaise, and educational deficit" (p. 388).

It appears that the first publication on gifted African Americans was by Bond (1927). In this brief article, Bond reported that of 30 Negro children tested on the 1916 Stanford-Binet, 8 (27%) scored IQs higher than 130.[42] Referring to this highly exceptional group, Bond commented, "This is an extremely rare group, as only 1% of white children (to use Terman's data . . .) reach or exceed this mark" (p. 259).[43] The study by Proctor (1929, cited in Jenkins, 1936) is considered to be the "first thorough study of Negro children of superior intelligence" (Jenkins, 1936, p. 177). Proctor's investigation, an unpublished master's thesis, presented case studies of Negro elementary school students in Washington, D.C. Based on administrations of the Stanford-Binet, Proctor identified 30 children whose IQs ranged between 129 and 175. These two studies by Bond (1927) and Proctor (1929) were not exceptions during this early era. From 1920 to 1935, 15 different investigations recorded African American children with IQs higher than 120 (Jenkins, 1936).[44]

Historically, educational psychologist Martin D. Jenkins was the most active and prominent scholar of giftedness among African American children and youths. After earning his doctorate in 1935 (dissertation titled *A Socio-Psychological Study of Negro Children of Superior Intelligence*), Jenkins went on to publish a number of articles in this area (see, e.g., Jenkins, 1936, 1943, 1948, 1950; Jenkins & Randall, 1948; Witty & Jenkins, 1934, 1935).[45] For his 1935 doctoral dissertation (later published as an article; Jenkins,

1936), Jenkins systematically sought "superior Negro children" (p. 175) in Grades 3 to 8 in seven public schools in the city of Chicago. Using a method of selection similar to that used by Terman (1925) in his study of gifted children in California, Jenkins identified 103 Negro children (from a population of 8,145 students) with Stanford-Binet IQs from 120 to 200 (M = 134.2). Jenkins's dissertation is considered a "classic" study (Guthrie, 1976) and "a truly significant contribution to the literature of its field" (Wilkerson, 1936, p. 130).[46] In a review of Jenkins's dissertation, Wilkerson (1936) commented that his study has significant implications for dealing "a severe blow . . . to the hypothesis of racial differences" (p. 130). Wilkerson juxtaposed, in parallel columns (p. 130), the tenets of the racial difference hypothesis and Jenkins's findings as follows:

Tenets of the Racial Difference Hypothesis	*Findings of the Present Investigation [Jenkins]*
1. The "gifted" Negro child is an anomaly.	1. The incidence and characteristics of "gifted" Negro children are approximately the same as for American children in general.
2. Superior Negro children [excel] by virtue of a [predominantly] white ancestry.	2. Superior Negro children come from predominantly Negroid stock, and as regards of racial admixture, are "strikingly in line with . . . the general Negro population."
3. Very bright Negro children are found chiefly in the primary grades and at younger years.	3. Brilliant Negro children are distributed fairly evenly among different age and grade levels.
4. The IQ of the superior Negro child retrogresses during later elementary school years.	4. The incidence of superior Negro children in the later elementary years reflects no retrogression of high IQ's.
5. Racial inequalities in education do not significantly influence racial differences in "test" intelligence.	5. "No one of these [Negro] children of [superior intelligence] has ever attended a school in a Southern state."
6. Negro ancestry is more potent than inferior socio-economic status as a determiner of low "test" intelligence.	6. Superior Negro children, like superior white children, come from a superior socio-economic [background].
7. Because of limitations in racial heredity, Negro children reveal relatively lower scholastic achievement in the linguistic and highly "verbal" school subjects.	7. Superior Negro children show greatest scholastic achievement in language and reading, highly "verbal" subject matter.

In sum, the pioneering work of Jenkins went a long way in dispelling the myth that gifted African American children did not exist. Guthrie (1976), in a series of biographical sketches of renowned African American psychologists, offered this tribute to Jenkins:

> While many psychologists were debating the question of equality of black-white intelligence test scores during this period, Jenkins' [1935] investigation discovered that intelligence levels for blacks were as high as those recognized for the white population. Furthermore, he identified a representative sample of "superior" black students, one of whom had the highest IQ then on record. (p. 144)[47]

To be sure, during the 1930s, 1940s, and 1950s, Jenkins and other scholars made tremendous progress in demonstrating that gifted African American students could be identified using conventional assessment tools. As we have seen, however, the under-identification of gifted African American children and youths persists to the present. Numerous scholars have discussed possible explanations for such underrepresentation. In our view, Ford and Webb (1994) cogently captured the many concerns of scholars who have advanced reasons to explain the relative absence of African Americans in gifted programs. The reasons offered by Ford and Webb included

> exclusionary definitions and theories of giftedness, culturally biased identification practices, a lack of understanding among educators regarding the effect of cultural differences on learning, inadequate training of teachers to recognize gifted students from diverse cultural backgrounds, a lack of encouragement for African American parental involvement in educational placement decision making, and inadequate definitions of underachievement among the gifted. (p. 359)[48]

In light of these many obstacles described by Ford and Webb (1994) in identifying gifted African American students, it is not at all surprising that some scholars have been creative in addressing these issues. In the remainder of this section, we focus on two select studies that attempted to use innovative, multimodal, multidimensional, pluralistic assessment strategies to increase the percentage of African American students in gifted programs.[49]

Woods and Achey (1990)

This study took place in the Greensboro, North Carolina, public schools. Target participants were 2nd- to 5th-grade students. Reflective of the national pattern, minority students (predominantly African American) in the Greensboro schools, who made up 55% of the district's 20,000 pupils, constituted only 13% of the district's gifted elementary student population. Thus, an innovative project, the Academically Gifted Project (AGP), was designed. The AGP was "based on enhancing existing referral and evaluation procedures and maximizing the efforts of school personnel rather than lowering or changing requirements for identification" (Woods & Achey, 1990, p. 21). The AGP procedures for identification of gifted children broke considerably from the traditional procedures that had been set in place. The new procedures involved (a) the identification of a "target group" by systematically reviewing *all* students' existing group-based achievement scores,[50] (b) informative group meetings at the schools for parents who had eligible students (at these meetings, parents signed permission slips to allow their children to be further evaluated),[51] (c) the automatic evaluation of target students on group intelligence and achievement tests (done twice) and the automatic evaluation of target students on individual intelligence and achievement tests,[52] and (d) the employment of two full-time educational diagnosticians who were assigned strictly to the AGP. These four innovative procedures were in contrast to the district's traditional procedures of identifying gifted students that relied heavily on nominations (particularly teachers), had limited parental involvement, and had no automatic evaluation of eligible students.

Woods and Achey (1990) reported results from the AGP for the first 3 years of operation (1986 to 1989). Regarding minority students who became part of the target group during this period, there were 705 2nd- to 5th-graders (95% African Americans, 5% other minorities). Of these 705 minority students, 200 (35%) eventually qualified for AGP classification. The net result of the AGP over the 3-year period was a 181% increase ($n = 99$ to $n = 278$) in gifted minority students.[53] Within the first year of operation alone, minority gifted students more than doubled in their incidence rate, from 13% to 28%. The most frequently occurring intelligence/achievement test combination in identifying gifted minorities was the Otis-Lennon School Abilities Test and the California Achievement Test.[54]

In summary, the study by Woods and Achey (1990) demonstrated that dramatic increases of gifted African American students can be realized by incorporating expanded conceptions of giftedness, early parental involvement, automatic evaluations, schoolwide cooperation, and the use of trained educational diagnosticians specifically assigned to identify gifted students.

Matthew, Golin, Moore, and Baker (1992)

This study sought to increase the percentage of gifted African American elementary students by using the System of Multicultural Pluralistic Assessment (SOMPA; Mercer & Lewis, 1979). Mercer (1989), the key architect of this innovative and pluralistic assessment battery, commented that the SOMPA

> is based on a modified version of the IQ paradigm. It is a battery of tests that measures the student's adaptive behavior, collects data on the student's health history, includes scales for measuring the sociocultural background of the student's family, and retains the Wechsler Intelligence Scale for Children–Revised (WISC-R). However, it treats the WISC-R as an individually administered achievement test, not as a measure of intelligence. (p. 296)

The SOMPA was standardized on 2,100 elementary school students in California (equally divided among White, African American, and Latino students). A child's conventional WISC-R Full Scale IQ is termed the "school functioning level," and the adjusted WISC-R Full Scale IQ (based on separate sociocultural norms for each racial/ethnic group) is called the "estimated learning potential" (Mercer & Lewis, 1979).

The participants in Matthew et al.'s (1992) study were 270 African American students in the 2nd to 5th grades attending public schools in a Pennsylvania school district (N = over 40,000 students). Of the 270 students, 215 were identified through conventional means (i.e., grade point average, group- and individually administered achievement tests [WISC-R or Stanford-Binet IQs of 125 or higher]). In the investigation by Matthew et al., these students were referred to as "non-SOMPA" pupils. The remaining gifted students (n = 55) also were identified through the traditional methods, but their WISC-R Full Scale IQs were adjusted via the SOMPA. This group was referred to as "SOMPA" students.

The major findings reported by Matthew et al. (1992) were as follows:

1. The IQs for the non-SOMPA and SOMPA students were extremely similar.[55]

2. No significant differences were found between the non-SOMPA and SOMPA groups in any of the academic measures (i.e., grade point average, group and individual achievement test scores).

3. On the Ross Test of Higher Cognitive Processes, 34 of 36 comparisons across Grades 2 to 5 showed no significant differences between the non-SOMPA and SOMPA groups.[56]

Although the SOMPA has not been free of criticism,[57] its use clearly demonstrates that gifted minority students can be identified at higher proportions than those seen when only traditional measures are used.[58] Matthew et al. (1992) concluded,

> The SOMPA students in this study were apparently penalized by the original WISC-R scores which did not accurately predict their ability to perform in the gifted program. By adjusting their scores to reflect background factors that might have affected their performance on a standardized intelligence test, the school district was able to include them in the educational program offered for gifted elementary students. (pp. 353-354) . . . The SOMPA procedures used to identify these gifted students may provide an alternative method to increase the proportion of minority students in gifted programs, particularly in states that use IQ cut-off scores for placement decisions. (p. 344)

Gifted Latino Students

In comparison to gifted African American students, gifted Latino students do not have a long history of being subjects of interest in the research literature.[59] Such attention did not arise until the early 1970s. This is not to say, however, that exceptionally bright Latino students did not exist in decades past. For example, in a 1931 master's thesis, Davenport administered the Goodenough Draw-a-Man Test (Goodenough, 1926a) to 124 White and 420 Mexican American students (Grades 1 to 3) attending public schools

TABLE 8.6   High-Performing Mexican American and White Students:
A Historical Example

IQ Interval	Mexican Americans (N = 420)		Whites (N = 124)	
	n	Percentage Total	n	Percentage Total
115 or higher	27	6.4	14	11.3
120 or higher	18	4.3	9	7.3
125 or higher	11	2.6	7	5.7
130 or higher	3	0.7	5	4.0
135 or higher	2	0.5	3	2.4
140 or higher	1	0.2	2	1.6

SOURCE: Tabular data calculated from raw data in Davenport (1931).
NOTE: IQ test is the Goodenough Draw-a-Man Test (Goodenough, 1926a). N refers to total number of students tested.

in San Antonio, Texas. Based on the author's original data for all 544 children, we calculated frequency of cases for IQ intervals as seen in Table 8.6. As the data indicate, Mexican American children certainly were represented at high levels of intellectual performance (e.g., 6.4% scored IQs of 115 or higher). Although the White children had greater incidence rates at these higher levels of intellectual performance, Davenport's data show that Mexican American children in the past clearly were present in the ranks of superior performance on intelligence tests.[60]

During the 1970s, interest in gifted Latino students (particularly Mexican Americans) began to surface. This attention can be traced, in part, to Moreno (1973), who reported that in the fall of 1971, Mexican American K-12 public school students in California made up 16% of the state's total K-12 population, but they constituted only 3.8% of the total K-12 gifted population, a disparity of 76.3%.[61] This growing interest in gifted Latino students during the 1970s also was influenced by the writings of Bernal (1974, 1978; see also Bernal & Reyna, 1975), a pioneer in the area of gifted Mexican Americans. There also was the development of the SOMPA (Mercer & Lewis, 1979) and its use in identifying gifted Mexican American children (Mercer, 1977).

Based on our review of the literature on gifted Latino students, we offer several observations. First, as is the case of research on gifted African American students, scholarly work on giftedness among Latino students primarily focuses on problems that thwart the identification of gifted Latino students as well as on principles and strategies that could lead to an increase in the number of identified gifted Latinos.[62] There are few empirical studies that actually have been designed to demonstrate the effectiveness of such principles and strategies (we discuss two of these later). Second, most of what is known about gifted Latino students pertains to Mexican Americans. Very little of this literature focuses on Puerto Rican, Cuban American, and other Latino subgroups. Third, there is a growing interest in identifying and providing education for gifted Latino students who are limited English-proficient (LEP) or bilingual (see, e.g., Barkan & Bernal, 1991; Evans de Bernard, 1985; U.S. Department of Education, Office of Educational Research and Improvement, 1998). Some of the latest thinking on how to increase the number of LEP and bilingual students in gifted education is to merge the fields of bilingual and gifted education in concerted ways (U.S. Department of Education, Office of Educational Research and Improvement, 1998). For example, the San Diego City School System currently has about 100 teachers who hold district certification in *both* bilingual and gifted/talented education (GATE) programs. Such merging of expertise clearly has been effective. In 1986, there were only 673 (8%) Latino (mostly LEP) students in the total GATE program enrollment of 8,205. In 1997, Latino students numbered 3,924 (19%) of the total 20,879 GATE program students (U.S. Department of Education, Office of Educational Research and Improvement, 1998). In sum, gifted Latino LEP students (as well as other language-minority groups [e.g., Chinese Americans]) present a major challenge in gifted education. Barkan and Bernal (1991) cogently captured this issue:

> The most perplexing populations for traditional educators of the gifted to select and educate are the gifted children from language-minority groups. These children, if they are "identified" at all, are typically admitted only after they have mastered English and can receive instruction in an all-English classroom. An obvious point—one that many educators fail to recognize, however—needs to be emphasized: One does not have to be fluent in English to be intelligent. There is no inherent need to delay the education of limited English-proficient . . . gifted children if bilingually competent teachers of the gifted are available. (p. 144)

We now turn to brief discussions of two select empirical studies that have sought to increase the percentage of Latino students through the use of innovative strategies.

Elliott, Argulewicz, and Turco (1986)

The purpose of this investigation was to examine how well the Scales for Rating the Behavioral Characteristics of Superior Students (SRBCSS; Renzulli, Smith, White, Callahan, & Hartman, 1976) predicted intellectual performance and academic achievement for gifted Mexican American ($n = 25$) and White ($n = 379$) students in Grades 3 to 6 in a southwestern school district. The SRBCSS, a rating scale completed by teachers, consists of 10 subscales, with the 4 major ones being Learning, Motivation, Creativity, and Leadership.

In this study by Elliott et al. (1986), all of the gifted students had been placed in a gifted program based on teacher recommendations and by scoring 2 standard deviations above the mean on either the WISC-R or the Stanford-Binet. Following placement, all students were administered the Stanford Achievement Test (SAT; Madden, Gardner, Rudman, Karlsen, & Merwin, 1974) by their teachers. About a week later, all teachers completed the SRBCSS on each student, with SAT scores being unknown by the teachers. The major finding was as follows:

The SRBCSS has minimal predictive validity when one's purpose is to predict Anglo students' performance on IQ or achievement tests. However, based on the data from this study, there may be value in its use in identification of gifted Hispanic students. (Elliott et al., 1986, p. 32)[63]

In the past, teachers' accuracy in identifying gifted students has been disappointing. In a brief review of the literature, Gear (1976) concluded that teachers' "effectiveness" (i.e., total number of gifted students identified divided by total number of confirmed gifted) ranged between 10% and 48% (median of 45%). Argulewicz and Kush (1984) noted that these poor effectiveness estimates have "been largely due to a lack of systematic and objective methods for rating behaviors thought to be indicative of giftedness" (p. 82). As such, rating scales (e.g., the SRBCSS) are ways in which to improve the identification of gifted students by operationalizing the behaviors that teachers observe on a daily basis. Nonetheless, there is some evidence that even when "objective" rating scales such as the SRBCSS are used in identifying potentially gifted students, Mexican Americans are referred at lower rates than are their White peers (High & Udall, 1983). "Whether this problem is related to negative teacher attitudes toward minority students as indicated by . . . studies,[64] or whether it is merely unawareness of the manifestation of giftedness in minority cultures, is a question which needs exploring" (pp. 163-164).

Tallent-Runnels and Martin (1992)

In this investigation, the authors sought to examine the Screening Assessment for Gifted Elementary Students (SAGES; Johnsen & Corn, 1987), which at that time was a relatively new measure of general intelligence for gifted students. The SAGES, which has a mean of 100 and a standard deviation of 15,

focuses on three areas of giftedness: reasoning, school-acquired information, and divergent production. Subtest I, reasoning, uses classification and analogies as measures of reasoning. Subtest II, school-acquired learning, contains items in mathematics, reading, science, and social studies. The divergent production subtest is composed of items using pictures and figures to elicit responses. (Tallent-Runnels & Martin, 1992, p. 940)

The participants in the Tallent-Runnels and Martin (1992) study were 122 White and 41 Mexican American students (3rd to 5th grades) attending public schools in a western Texas school district. All participants had been nominated for inclusion in the district's program for gifted students.[65] The authors' major objective in this investigation was to examine whether SAGES subscale scores were "unbiased," that is, "relatively neutral with respect to ethnicity" (p. 940). The investigators hypothesized that if the SAGES was unbiased, "then it should not be possible to classify subjects with respect to ethnicity, using scaled scores as predictors" (p. 940). A direct discriminant function analysis confirmed the authors' hypothesis. That is, the results showed that the SAGES did not predict racial/ethnic group membership. These results were validated by the use of a second sample of 25 White and 14 Mexican American gifted students (3rd to 5th grades).

In sum, Tallent-Runnels and Martin concluded, "The vector of scores used in the discriminant analysis seemed to treat individuals similarly, i.e., as from a single group" (p. 941). The authors also performed a multivariate profile analysis and found small score differences between the White and Mexican American groups, thus reflecting that similar cognitive processes were being assessed. The authors concluded that the SAGES appears to function fairly with both racial/ethnic groups.

Gifted Asian/Pacific Islander Students

Compared to the number of scholarly publications on gifted African American and Latino students, there is scant literature on gifted Asian/Pacific Islander (A/PI) students.[66] As Plucker (1996) asks, "Why so little attention?" (p. 318). According to Plucker, there are two likely explanations for this limited attention paid to gifted A/PI students: (a) demographic patterns and (b) the widely held—and false—perception that the A/PI population is very homogeneous.[67]

Regarding demographic patterns, Plucker (1996, drawing from Tsai, 1992) noted that Chinese Americans and Japanese Americans, for example, are not uniformly distributed across the nation. As a case in point, the vast majority of Chinese Americans live on either the West Coast (52%) or the East Coast (27%). Such patterns of intense localization "result in most school districts [across the country] having small or nonexistent Asian-American populations" (Plucker, 1996, p. 318). Given their very small presence in many school districts, it was not unexpected that in a nationwide needs assessment in gifted education, research on gifted Asian American students was judged to be a low priority (Renzulli, Reid, & Gubbins, 1991, cited in Plucker, 1996).

With respect to the erroneous perception of A/PI homogeneity, there is abundant evidence that this population is extremely heterogeneous. Numerous ethnic groups make up the A/PI people who reside in the United States (e.g., Chinese, Japanese, Asian Indian, Korean, Vietnamese, Filipino, Hawaiian, Samoan). Contrary to popular belief, there is tremendous diversity between and within the various A/PI ethnic groupings (e.g., as seen in cultural expressions, language, level of acculturation, SES, immigration experience, and schooling attainment).[68] One particular issue that stems from the false perception of homogeneity of the A/PI population is what has become known as the "model minority" belief. As noted by Kim and Hurh (1983),

> During the mid-1960s, both the major mass media and scholarly works promoted the image of Asian Americans as successful minorities. The media first cited Japanese and Chinese Americans as minorities that had attained success in American society. In the 1970s, Koreans were included in this success story, as were recent refugees from Indochina. Currently the success theme is applied to most Asian ethnic groups. (p. 3)[69]

To some extent, the model minority myth has played a role in creating a climate of little attention being paid to gifted A/PI students (see, e.g., Chen, 1989; Gallagher, 1989; Hasegawa; 1989; Maker & Schiever, 1989b). The implications of this meager attention are serious. It is important to acknowledge that although A/PI students as a group are overrepresented in gifted programs, there are a number of divergences from this pattern. For example, precocious LEP A/PI students face problems similar to those experienced by bright LEP Latino students in being underidentified as gifted. Thus, language concerns should be considered during identification (see, e.g., Chen, 1989; Gallagher, 1989; Plucker, 1996). Another example of a break from the pattern of overrepresentation of gifted A/PI students has to do with SES. Not all A/PI families can be described as middle or upper SES.[70] Given the positive correlation between SES and being identified as gifted, it is likely that A/PI students from low-SES backgrounds are underrepresented in gifted programs.[71]

Notwithstanding these divergences from the norm, it is important to underscore that A/PI students as a group have been relatively successful in being identified as gifted. How might one explain this? One major conjecture expressed in the available literature has to do with cultural values and socialization practices of A/PI families. For example, Kitano (1986) suggested,

> Although APA [Asian and Pacific American] cultures differ in many ways from the majority culture, the Asian values of educational attainment and obedience to authority clearly support achievement in American schools. Hence, assessment procedures designed to identify high achievers are consistent with APA values and, in fact, may be biased in favor of Asian minority students. (p. 54)[72]

Kitano's assertion about familial values and practices appears to be a dominant theme in the general literature on the A/PI people (Chan, 1986; Hartman & Askounis, 1989). Hartman and Askounis (1989) commented,

> Many researchers believe the major influence in Asian culture is the Confucian ideal that stresses family values and emphasizes education. The Confucian ethic drives people to work, excel, and repay the debt they owe to their parents. This strong influence is present in Japan, Korea, China, and Vietnam. . . . The pressure to succeed academically for Asian students is extremely strong. Outstanding achievement is emphasized because it is a source of pride for the entire family. (p. 110)

In conclusion, school personnel have been quite successful in identifying A/PI students in general. One must be aware, however, that the model minority myth has helped to create a stereotype that *all* A/PI students are high achievers and academically successful and, therefore, are shoo-ins for being identified as gifted. As we have discussed, this false perception has led to meager scholarly attention to giftedness among these students as a group. A greater focus is needed in identifying A/PI students who are underrepresented among the ranks of the gifted, particularly LEP and low-SES students.

Gifted American Indian Students

Similar to gifted Latino and A/PI students, American Indian students do not have a long history of being subjects of research interest in the literature on giftedness. There is some evidence, however, that as far back as the 1920s some intellectually bright American Indian children existed. For example, Goodenough (1926b) administered the Goodenough Draw-a-Man Test to 79 American Indian children attending the Hoopa Valley Indian School in California. Although the mean IQ for the sample was only 85.8 (based on a standardization mean of 100), there were 2 children (2.5% of the total group) who had IQs of 120.[73]

Our assessment of available literature on gifted American Indian students leads us to the conclusions that (a) there is a dearth of studies, (b) a number of references cited by various authors of the published literature are unpublished studies, and (c) most of the literature focuses on problems in identification of gifted American Indian students and assessment principles/strategies that may lead to an increase in the identification of these students, with very little empirical research available that has been designed to demonstrate actual assessment practices that result in increased numbers of gifted American Indian students. In short, the knowledge base on gifted American Indian students is scanty.[74]

Of the four racial/ethnic minority populations we cover in this chapter, it appears that American Indians present the most arduous challenge in being identified as gifted. This assertion stems from the observation that there is immense diversity within the American Indian people. As of 1998, there were 555 *different* tribes that the U.S. government recognized and that were eligible for services and funding by the Bureau of Indian Affairs ("Indian Entities Recognized," 1998). Such diversity is appreciated when one examines five ways in which such differences are likely to manifest, that is, in the contexts of (a) geographic locations (e.g., urban, rural), (b) languages and cultures of the numerous tribal affiliations, (c) the types of schools attended by American Indian students (i.e., Bureau of Indian Affairs funded, private or mission, public),[75] (d) SES conditions, and (e) individual differences among the students (Callahan & McIntire, 1994). Notwithstanding this tremendous diversity among American Indians, one still sees stereotypes in the literature (Garrison, 1989; Montgomery, 1989; Sisk, 1989). For example, in "comparative continua of differential characteristics of American Indian and dominant American groups," Montgomery (1989) referred to the American Indian perspective as one of being "right brain" and the dominant American perspective as one of being "left brain" (p. 82).[76]

Regarding the identification of gifted American Indian students, the assessment framework developed by Brittan and Tonemah (1985, described in Tonemah, 1987) might have some promise. The American Indian Gifted and Talented Assessment Model (AIGTAM) uses a Tribal-Cultural checklist that has four derived categories: (a) Acquired Skills (e.g., problem-solving skills, language/communication skills), (b) Personal/Human Qualities (e.g., intelligence, inquisitiveness), (c) Aesthetic Abilities (e.g., artistic and dancing abilities), and (d) Tribal/Cultural Understanding (e.g., knowledge of tribal traditions, understanding of tribal culture and history). We are unaware, however, of any research that has demonstrated the effectiveness of the AIGTAM. Another limitation of the AIGTAM is that it might not be useful for urban American Indian students who have limited or no exposure to their tribal traditions and history.

A fairly common theme in the literature on giftedness among American Indian students is that, due to cultural differences, the mainstream or conventional concept of giftedness (e.g., intellectually superior, highly motivated, excellent student) might not be appropriate in identifying gifted American Indian students, particularly those who have strong tribal affiliations. Unfortunately, little research is available to support this assertion. There is, however, one study we can point out. Romero (1994), of the Center for Planning and Research, Santa Fe Indian School, New Mexico, undertook a comprehensive ethnographic study to investigate the concept of giftedness among the New Mexico Keresan Pueblo Native communities.[77] Based on 22 open-ended interviews with Keresan tribal members, Romero sought to address the following: "What characteristics do the respective Keresan Pueblo communities identify as indicative of giftedness from a traditional community perspective?" (p. 38). Included among her numerous findings were the following:

1. Among the Keresan Pueblos, no term exists for "giftedness." There are, however, "descriptive terms in the Keresan language referring to the possession of unique and special cultural abilities, traits, and talents in specific areas which retain their significance only in the Pueblo value system" (p. 40). For example, *Kaam "asruni"* refers to creativity entailing psychomotor skills (i.e., creating with the hands), and *Dzii guutuni* refers to one's knowledge domain (e.g., cultural knowledge of Native customs).

2. The concept of having unique and special abilities or talents in certain areas "is meaningful only as they are applied and utilized in *a way which benefits others*" (p. 40, italics added).

The study by Romero (1994) is very valuable in that it demonstrates, through qualitative research, that the concept of giftedness, as perceived among the Keresan Pueblos, contrasts significantly with the conventional mainstream notion. The implications for the identification of gifted Pueblo children are clear. Romero noted,

Particularly for gifted education, meeting the educational needs of Pueblo children and addressing the underrepresentation of American Indian children in gifted programs first requires an acceptance of the importance of Native principles, values, and traditions which may not always coincide with those of the mainstream society but play a vital role in the lives and educational performance of Pueblo children. Through the case histories shared by the Keresan Pueblo people involved in this study, a view of the Keresan Pueblo culture, lifestyle, values, and needs, including a special look at the learning processes employed by these people, was provided. This information, combined with Keresan perceptions of giftedness, can assist in the nationwide effort to advance gifted education theory and practice in relation to culturally different populations. But, more importantly, it will provide a different perspective of learning and culture, language, and community. Both Native and mainstream gifted concepts can contribute to the conceptual framework needed to design a challenging education program that provides opportunities for all children—*our gifts*. (p. 56)

Conclusions

As we have discussed, scholarly writings and research investigations concerned with gifted minority populations are quite limited in quantity. Suffice it to say, our knowledge base in this area is underdeveloped. Furthermore, this body of knowledge tends to focus on barriers to identification of underrepresented gifted minority students as well as innovative strategies to increase the number of such students. There are few empirically based studies designed to demonstrate the effectiveness of these novel identification procedures. To be sure, problems abound. School districts across the country continue to rely, although not exclusively, on an entrenched identification paradigm (e.g., teacher nominations, standardized measures of intelligence). Gifted language-minority students are vastly underidentified. The considerable underrepresentation of gifted American Indian, African American, and Latino students remains constant.

Notwithstanding such concerns, we have seen some bright spots in the existing literature. A number of studies have identified, through discourse and empirical findings, specific practices that lead to an increase in the number of gifted minority students. Frasier's (1989) discussion on "best practices" is a good example. Although her focus was on African American students, we believe that her points can be generalized to other students of color. Based on her extensive review of the literature on identification procedures, Frasier offered the following principles of identification that are derived from research and practice:

1. The focus should be on the diversity within gifted populations.
2. The goal should be inclusion, rather than exclusion, of students.
3. Data should be gathered from multiple sources; a single criterion of giftedness should be avoided.
4. Both objective and subjective data should be collected.
5. Professionals and nonprofessionals who represent various areas of expertise and who are knowledgeable about behavioral indicators of giftedness should be involved.
6. Identification of giftedness should occur as early as possible, should consist of a series of steps, and should be continuous.
7. Special attention should be given to the different ways in which children from different cultures manifest behavioral indicators of giftedness.
8. Decisionmaking should be delayed until all pertinent data on a student have been reviewed.
9. Data collected during the identification process should be used in determining curriculum. (p. 214)

With regard to Frasier's (1989) list of best practices, we suggest that the role of parents of gifted minority students needs to be considered as an additional principle in the identification process. The importance of minority parents in the identification of gifted children appears to be a common theme in much of the literature on gifted minorities. The investigation by Scott, Perou, Urbano, Hogan, and Gold (1992) is a case in point. Scott et al. surveyed White, African American, and Hispanic parents whose children attended schools in a large urban school district in Dade County, Florida.[78] The children (Grades 3 to 5) were enrolled in the district's gifted/talented program. As their two major findings, the researchers reported that (a) there were fewer African American and Hispanic parents, as compared to White parents, who actually had referred their children for possible inclusion in the gifted/talented program; and (b) parents from all three racial/ethnic groups were very similar in generating characteristics that originally led them to believe that their children were gifted. In short, all parents shared very common perceptions of giftedness, but fewer minority parents made requests that their children be evaluated for inclusion in the gifted and talented program. Scott et al. commented that a major practical implication emerged from their study:

> These data . . . suggest that the identification of minority students might be enhanced through a public education program which alerted parents to those characteristics which might indicate giftedness and which informed them about the availability and function of educational programs for gifted students. Awareness of superior abilities appears to be a necessary but not sufficient condition for an active role in the referral process. (p. 139)

On a final note, we mention that more than 20 years ago, Bernal (1979) commented, "The juxtaposition of minority and gifted still produces dissonance in the minds of many educators" (p. 395). We contend that this discordant frame of mind is likely the root problem of the nagging underrepresentation rates of minority students in gifted programs. It appears that to begin the rectification of this pervasive underrepresentation, there needs to be a collective and resounding affirmation by educators that gifted minority students do exist and need to be identified. This is the first step in the remediation of a long-standing problem.

9

A Multicultural Review of Cognitive Ability Tests

Despite criticisms and controversy, the field of intelligence testing has continued to grow, with greater numbers of measures on the market assessing a multitude of areas. The demand for cross-cultural validation of ability measures has led many to support the use of nonverbal "culturally reduced" measures and the modification of procedures for interpreting traditional intelligence tests. For example, many of these instruments do not specifically use the term "intelligence" in their titles (e.g., Cognitive Assessment System; Naglieri & Das, 1997). These may be identified as aptitude or cognitive measures that appear to assess areas parallel to those indicated within the intelligence domain. Thus, in this chapter, reference is made to cognitive ability tests as an overall encompassing area inclusive of intelligence tests. This chapter provides information about some of the most popular and most current editions of various cognitive ability and intelligence tests with regard to areas assessed, test development strategies, and general findings related to diverse populations (i.e., racial/ethnic ability profiles and overall score discrepancies).

Given the number of instruments currently available, it is beyond the scope of this chapter to provide a comprehensive examination of all measures in the cognitive area or of all research conducted using these tests with diverse populations. In the preparation of this chapter, we used the computerized version of the *Buro's Mental Measurements Yearbook* (WinSPIRS, 1989-present) to obtain current information about "intelligence" tests. The specific term used in selecting the grouping of measures to be included was the search term "intelligence," and 348 records were identified. Closer scrutiny of the test descriptions indicated that this subset of tests included measures beyond the parameters of this chapter. To obtain a more focused selection of cognitive ability tests, measures in the following areas were not included in the review: adaptive behavior scales, school readiness instruments, achievement tests (e.g., those solely assessing reading, mathematics, music, spelling, and writing), language scales, personality tests, behavior scales, specific population scales (e.g., attention deficit disorder), and school environment scales. In addition, tests in which the reviewers did not provide information regarding race/ethnicity were not included in the present chapter because this information is the focus of our review. Test manuals that were available to us were obtained, whenever possible, to gain more direct access to information. If some newer popular measures were not included in the databases (e.g., NEPSY; Korkman, Kirk, & Kemp, 1998), then we added these whenever possible.

This chapter focuses on the following topics: (a) areas assessed by current cognitive tests, (b) current test development strategies, (c) racial and ethnic profiles of abilities and discrepancies yielded by traditional cognitive tests, (d) current procedural modifications of traditional cognitive tests, and (e) conclusions.

Areas Assessed by Current Cognitive Instruments

Based on the selection procedures outlined in the preceding section, 59 test reviews were examined. The names of the tests reviewed, authors of the measures, years of publication, and areas assessed by the particular measures are presented in Table 9.1. As noted earlier, whenever possible, information obtained from test manuals was added to the present review. Even with the limitations of the methodology selected for this chapter, the purposes assessed by current cognitive instruments are very broad. In clustering the areas, we observed the following major measurement categories: developmental measures, nonverbal measures, multidimensional batteries (e.g., including verbal, nonverbal, *and* memory tests), memory and learning scales, qualitative measures (e.g., those based on interviews, self-reports, and rating scales), and neuropsychological test batteries.

Table 9.1 also highlights the numerous abilities and skills tapped by cognitive measures. There exists a great deal of overlap in the types of abilities assessed across measures. These include crystallized abilities, auditory and visual memory, quantitative reasoning, conceptual and abstract reasoning (e.g., matrices), perceptual and motor processing (e.g., form perception, mazes), sequential reasoning, attention and focus, communication skills, emotional coping and social skills, and learning. Many tests assess multiple abilities; for example, the developmental measures often assess memory, auditory and visual reasoning tasks, and motor abilities. Most tests include verbal and nonverbal reasoning abilities, with the exception of the purely nonverbal measures.

Some of the more unique instruments include the Multiple Intelligence Developmental Assessment System (MIDAS; Shearer & Branton, 1994-1996, cited in Packard, 1986-present, and Trevisan, 1986-present) and the Gifted Evaluation Scale (McCarney & Henage, 1987-1990, cited in Callahan, 1989-present, and Traub, 1989-present), two more

(Text continued on page 257)

TABLE 9.1 Years of Publication, Senior Authors, and Areas Assessed by "Intelligence" Measures Based on WinSPIRS *Buro's* Reviews

Test Name and Authors	Year(s)	Reviewers	Areas Assessed
Battelle Developmental Inventory (Newborg, Stock, Wnek, Guidubaldi, Svinicki)	1984	Stinnett, Paget	Adaptive, motor, communication, and cognitive abilities (e.g., perceptual discrimination, memory, reasoning, academic skills)
Bayley Scales of Infant Development (Bayley, Rhodes, Yow, Rush)	1969-1993	Rozkowski, Rysberg	Intellectual and psychomotor functioning (e.g., sensory and perceptual development, auditory and visual skills, memory)
Bilingual Verbal Ability Test[a] (Muñoz-Sandoval, Cummins, Alvarado, Reuf)	1998		Verbal abilities (i.e., picture vocabulary, oral vocabulary, verbal analogies); available in 16 languages
Canadian Test of Cognitive Skills (Canadian Test Centre Educational Assessment Services)	1992-1996	Hebert, Maller	Sequential reasoning, memory, analogies, verbal abilities
Children's Category Test (Boll)	1993	Shriver, Vacc	Nonverbal learning and memory, concept formation, problem-solving abilities
CID Preschool Performance Scale (hearing-impaired, language-impaired, and normal children) (Geers, Lane, Central Institute for the Deaf)	1984	Bracken, Oosterhof	Intelligence (e.g., manual planning and dexterity, part-whole relations, form perception, perceptual motor skills)
Cognitive Abilities Scale (Bradley)	1987	Hightower, Robertson	Cognitive development (e.g., language, reading, math, handwriting, enabling behavior)
Cognitive Assessment System[b] (Das, Naglieri)	1997		Planning, attention, simultaneous, and successive processes
Cognitive Control Battery (Santostefano)	1987-1988	Brown, Hartman	Nonverbal cognitive activity (e.g., focal attention, selective attention) and emotion
Comprehensive Test of Nonverbal Intelligence (Hammill, Pearson, Wiederholt)	1996	Aylward, van Lingen	Nonverbal intellectual abilities (e.g., analogical reasoning, categorical classifying, sequential reasoning)

(Continued)

TABLE 9.1 Continued

Test Name and Authors	Year(s)	Reviewers	Areas Assessed
Detroit Tests of Learning Aptitude (third edition) (Hammill)	1935-1991	Mehrens, Poteat	Verbal, nonverbal, attention, and motor abilities; simultaneous and successive processing
Detroit Tests of Learning Aptitude–Adult (Hammill, Bryant)	1991	Dinero, Druva-Roush	Linguistic (verbal, nonverbal), attentional, and motoric domains
Detroit Tests of Learning Aptitude-Primary (second edition) (Hammill, Bryant)	1986-1991	Ackerman, Williams	Linguistic, attention, motoric, and general mental abilities
Differential Ability Scales (Elliott)	1990	Aylward, Reinehr	Verbal, nonverbal, spatial, memory, and conceptual abilities
General Ability Measure for Adults (Naglieri, Bardos)	1997	Fitzpatrick, Goldman	Analogies, sequencing, construction, problem solving
Gesell Preschool Test (Haines, Ames, Gillespie)	1980	Kaufman, Naglieri	Motor, adaptive, language, personal-social (adaptive similar to mental ability)
Gifted and Talented Scale (Dallas Educational Services)	n.d.	Brody, Colangelo	Numerical reasoning, vocabulary, synonyms-antonyms, similarities, analogies
Gifted Evaluation Scale (Teacher Rating Scale) (McCarney, Henage)	1987-1990	Callahan, Traub	Intellectual, creative, academic, and leadership ability; performing and visual arts
Halstead-Reitan Neuropsychological Test Battery (Reitan)	1979-1993	Dean, Meier	Executive functioning, cognitive abilities, language, memory, motor and perceptual abilities
Halstead Russell Neuropsychological Evaluation System (Russell, Starkey)	1993	Mahurin, Retzlaff	Executive functioning, cognitive abilities, language, memory, motor and perceptual abilities (same core set of measures from original battery by Halstead plus other clinical instruments)
Infant Mullen Scales of Early Learning (Mullen)	1984-1989	Hart	Cognitive and gross motor abilities (e.g., visual organization, visual expressive organization, receptive and expressive language)

Test	Date	Description
Hammill Multiability Intelligence Test[a] (Hammill)	1998	Verbal (understanding, integrating, and using spoken language) and nonverbal abilities (i.e., perceptual, logical, and abstract reasoning); eight subtests of the Detroit Tests of Learning Aptitude (fourth edition)
Kaufman Adolescent and Adult Intelligence Scale (Kaufman, Kaufman)	1993 Flanagan, Keith	Fluid and crystallized abilities, composite intelligence (e.g., memory, visual and auditory reasoning, deductive reasoning)
Kaufman Assessment Battery for Children[a] (Kaufman, Kaufman)	1983	Mental processing (e.g., spatial and perceptual tasks, numerical reasoning) and achievement (e.g., vocabulary, reading, arithmetic, information)
Kaufman Brief Intelligence Test[a] (Kaufman)	1990	Verbal and nonverbal abilities (e.g., expressive vocabulary and definitions, matrices)
Learning and Memory Battery (Schmidt, Tombaugh)	1995 Long, Ries, van Gorp	Auditory and visual memory (verbal-nonverbal, short and long term)
Leiter-R[a] (Roid, Miller)	1997	Nonverbal intelligence (e.g., fluid reasoning and visualization, visuospatial memory, attention)
Luria Nebraska Neuropsychological Battery–Children's Revision (Golden)	1987 Hooper	Motor functions, tactile functions, visual reasoning, receptive speech, arithmetic, memory (includes intellectual processes)
Luria Nebraska Neuropsychological Battery–Forms I and II (Golden, Purisch, Hammeke)	1980-1985 Snow, van Gorp	Motor functions, tactile functions, visual reasoning, receptive speech, arithmetic, memory (includes intellectual processes)
Matrix Analogies Test (Naglieri)	1985 McMorris, Rule, Steinberg	Nonverbal reasoning (e.g., visual matrices)
McCarron-Dial System (McCarron, Dial)	1973-1986 Garbin, Solly	Verbal-spatial-cognitive, sensory, motor, emotional, integration-coping
McCarthy Scales of Children's Abilities (McCarthy)	1970-1972 McCarthy	Verbal, perceptual performance, quantitative, memory, motor skills

(Continued)

TABLE 9.1 Continued

Test Name and Authors	Year(s)	Reviewers	Areas Assessed
Multiple Intelligence Developmental Assessment System (questionnaire) (Shearer)	1994-1996	Packard, Trevisan	Multiple intelligences—musical, kinesthetic, logical-mathematical, spatial, linguistic, interpersonal, intrapersonal, leadership, logic
NEPSY: Developmental Neuropsychological Assessment[a] (Korkman, Kirk, Kemp)	1998		Attention/executive functions, language/sensorimotor functions, visuospatial processing, memory and learning
Non-Reading Intelligence Tests–Levels I to III (Young)	1989-1992	Krauthamer, Diamond	Reasoning abilities (e.g., object identification, analogies)
Otis-Lennon School Ability Test (sixth edition) (Otis, Lennon)	1977-1990	Anastasi, Swerdlik	Abstract thinking and reasoning
Primary Test of Cognitive Skills (Huttenlocher, Levine)	1990	Bain, Barnes, McIntosh	Verbal, spatial, memory, and conceptual abilities
Psychoeducational Profile–Revised (autistic and communication handicapped) (Schopler, Reichler, Bashford, Lansing, Marcus)	1979-1990	Mirenda, Tindal	Behavioral and developmental areas (e.g., cognitive performance, cognitive verbal)
Raven's Coloured Progressive Matrices[a] (Raven, Court, Raven)	1990 edition		Eductive (nonverbal abilities)—making sense of complex situations, deriving meaning from events, perception
Raven's Standard Progressive Matrices[a] (Raven, Court, Raven)	1992 edition		Eductive (nonverbal abilities)—making sense of complex situations, deriving meaning from events, perception
Raven's Advanced Progressive Matrices[a] (Raven, Court, Raven)	1994 edition		Eductive (nonverbal abilities)—making sense of complex situations, deriving meaning from events, perception
Scholastic Abilities Test for Adults (Bryant, Patton, Dunn)	1991	Raju, Smith	Verbal reasoning, nonverbal reasoning, quantitative reasoning, general aptitude and achievement (e.g., reading, math, writing)
Scholastic Aptitude Scale (Bryant, Newcomer)	1991	Ayers, Hopkins	Quantitative, verbal, nonverbal, and general reasoning

Test	Dates	Author	Description
Shipley Institute of Living Scale (Shipley, Zachary)	1939-1986	Deaton	General intellectual functioning (vocabulary, conceptual reasoning, abstraction abilities)
Slosson Full-Range Intelligence Test (Algozzine, Eaves, Mann, Vance, Slosson)	1988-1994	Hanna, Tindal	Cognition, verbal, abstract, quantitative, memory, and performance abilities
Slosson Intelligence Test-Revised (Slosson, Nicholson, Hibpshman)	1961-1991	Kamphaus, Watson	Verbal cognitive ability
Southern Ordinal Scales of Development (Ashurst, Barnberg, Barrett, Bisno, Burke)	1977-1985	Camp, Rosenthal	Cognition, communication, social, affective, practical, fine/gross motor
Stanford-Binet Intelligence Scale (fourth edition)[a] (Thorndike, Hagen, Sattler)	1916-1986		Verbal reasoning, abstract visual reasoning, quantitative reasoning, short-term memory
Structure of Intellect Learning Abilities Test (Meeker, Meeker, Roid)	1975-1985	Cummings, Newman	Cognition, memory, evaluation, convergent production, divergent production
Swanson-Cognitive Processing Test (Swanson)	1996	Callahan, Darr	Mental processing ability, working memory, potential (e.g., verbal, auditory, and visual abilities)
Test of Memory and Learning (Reynolds, Bigler)	1994	Geller, Maller	Verbal, visual, and auditory memory
Test of Nonverbal Intelligence (third edition)[a] (Brown, Sherbenou, Johnson)	1982-1997		Nonverbal abilities
Universal Nonverbal Intelligence Test[a] (Bracken, McCallum)	1998		Memory and reasoning abilities, symbolic and nonsymbolic mediation
Wechsler Adult Intelligence Scale (third edition) (Wechsler)	1995-1997	Wechsler	Verbal (e.g., general fund of knowledge, vocabulary) and performance abilities (e.g., visual analysis, visual synthesis)
Wechsler Intelligence Scale for Children (third edition)[a] (Wechsler)	1949-1991	Wechsler	Verbal (e.g., general fund of knowledge, vocabulary) and performance abilities (e.g., visual analysis, visual synthesis)

(Continued)

TABLE 9.1 Continued

Test Name and Authors	Year(s)	Reviewers	Areas Assessed
Wechsler Memory Scale[a] (Wechsler)	1997	Wechsler	Auditory and visual memory, working memory
Wechsler Preschool and Primary Scale of Intelligence–Revised (Wechsler)	1967-1989	Bracken, Braden	Verbal (e.g., general fund of knowledge, vocabulary) and performance abilities (e.g., visual analysis, visual synthesis)
Wide Range Assessment of Memory and Learning (Sheslow, Adams)	1990	Clark, Medway	Verbal/visual memory and learning
Woodcock-Johnson Psychoeducational Battery–Revised[a] (tests of cognitive ability) (Woodcock, Johnson)	1977-1991		Cognitive abilities (e.g., memory, processing speed, auditory and visual comprehension, fluid reasoning, scholastic aptitude, oral language)

SOURCE: All information for this table was obtained from the WinSPIRS (1989-present) database for the *Buro's Mental Measurements Yearbook*.
a. Test manual used.

qualitative tools that appear to assess multidimensional aspects of intelligence. The MI-DAS includes reference to the multiple intelligences put forth by Gardner's (1983) multiple intelligences theory. The Gifted Evaluation Scale includes the assessment of an individual's leadership and performing arts abilities.

Many of the other instruments appear to be somewhat "traditional" in focus, with little to differentiate the type of scales or the abilities assessed from earlier versions of the various instruments. Although overall formats and item content might have been altered, it seems that the instruments remain fairly "mainstream." By this, we mean that many instruments assess the same ability areas using similar formats.

In addition, many test developers continue to rely on observed correlations with other long-standing measures to establish a newer test's validity (e.g., concurrent, predictive). This increases the probability that there will be little change in the measurement scales or abilities assessed. For example, of the 59 measures cited in Table 9.1, 39 (66%) of the reviewers and/or test manuals cite validity studies using one or more of the Wechsler scales as the criterion measure on which newer tests have been compared. The next most frequently used scales for establishing concurrent validity were the Stanford-Binet Intelligence Scale (cited 14 times, 24%), the Kaufman Assessment Battery for Children (K-ABC, cited 11 times, 19%), the Woodcock-Johnson Psychoeducational Battery (WJ-PB, cited 9 times, 15%), and the McCarthy Scales of Children's Abilities (MSCA, cited 6 times, 10%). Other scales noted were the Peabody Picture Vocabulary Test (PPVT), the Differential Ability Scale (DAS), the Bayley Scales of Infant Development, the Test of Nonverbal Intelligence, and the Wide Range Achievement Test (WRAT). Please note that there is overlap in these figures given that some tests were validated on multiple measures.

Current Test Development Strategies

Information regarding the age range of examinees for which the particular test is intended, size of norming sample, and stratification variables is presented in Table 9.2. Norms are critical to examine given that most test users assume that the norming sample is representative of the population on which the test will be used. The norms of cognitive ability tests enable the examiner to infer "meaning" with regard to an individual's performance (Woodcock, 1994). Decisions often are made based on raw scores that are converted into standard scores. These standard scores are developed as a result of the norming process for a given test. Woodcock (1994) noted that norming samples for most individually administered intelligence tests range from 2,000 to 6,000 participants. The large and representative sizes of the norming samples represent, to some extent, the largest and most extensively configured databases (i.e., with regard to stratification variables) often obtained in research on a particular test or test battery. Therefore, numerous studies often are conducted based on the standardization sample. For example, the Wechsler standardization samples (e.g., Wechsler Intelligence Scale for Children–Revised [WISC-R; Wechsler, 1974]) have been used in a number of studies examining racial and ethnic group differences (e.g., Suzuki & Gutkin, 1993).

(Text continued on page 268)

TABLE 9.2 Cognitive Tests and Characteristics of Their Standardization Samples as Noted by WinSPIRS *Buro's* Reviews

Test Name	Ages	Norming	Racial/ Ethnic Represen- tation	Socio- economic Status	Sex	Region of Country	Urban/ Rural Represen- tation	Edu- cation	Special Edu- cation	Bias Exam
Batería Woodcock-Muñoz Revisada	2 years or over	4,000 Limited information[a]								X
Battelle Developmental Inventory	Birth to 8 years	(1981) 800	X? 84% White	X?	X	X	X? 75% urban		—	X?
Bayley Scales of Infant Development–II	1 to 42 months	(1988) 1,700	X		X	X		X		X
Bilingual Verbal Ability Test[b]	5 years to adult	5,602 Over- sampling	X	X	X	X	X	X	X	X
Canadian Test of Cognitive Skills	Grades 2 to 12	36,000 Canadian	—	?		X	X		X	X?
Children's Category Test	5 years 0 months to 16 years 11 months	(1988) 920	X		X	X		X	X	

Test	Ages	N								
CID Preschool Performance Scale	2 years 0 months to 5 years 5 months	978	—	—	×	—	×		×	
Cognitive Control Battery	4 to 12 years	1,100	X?c	×	×	×	—		×	
Comprehensive Test of Nonverbal Intelligence	6 years 0 months to 18 years 11 months	2,129	×	×	×	×	×	×	×	×
Detroit Tests of Learning Aptitude (third edition)	6 years 0 months to 17 years 11 months	(Census?) 2,587 (Detroit Tests of Learning Aptitude [second edition]) 1,832 Norming not specified					×			
Detroit Tests of Learning Aptitude–Adult	16 years or over	1,254	×	×	×	×	×		×	?
Detroit Tests of Learning Aptitude Primary (second edition)	3 to 9 years	(1985/1990?) 2,095	×	×	×	×	×		×	×

(Continued)

TABLE 9.2 Continued

Test Name	Ages	Norming	Racial/ Ethnic Represen- tation	Socio- economic Status	Sex	Region of Country	Urban/ Rural Represen- tation	Edu- cation	Special Edu- cation	Bias Exam
Differential Ability Scales	2 years 6 months to 7 years 11 months 5 years 0 months to 17 years 11 months	(1988) 3,475	X Over- sampling		X	X		X	X	X
General Ability Measure for Adults	18 to 96 years	More than 2,000	X		X	X		X		
Gesell Preschool Test	2.5 to 6.0 years	(1960) 640	—	X	X					
Gifted and Talented Scale	Grades 4 to 6	Limited in- formation	—	—	—					
Gifted Evaluation Scale (Teacher Rating Scale)	4.5 to 19.0 years	2,276		?		X?				?
Halstead-Reitan Neuropsychological Test Battery for Adults	15 years or over	Limited in- formation								

Test	Population / Age	Sample							
Halstead-Russell Neuropsychological Evaluation System	Neuropsychological patients	576 brain-damaged 200 non-brain-damaged All Veterans Administration hospital patients			—	?			
Hammill Multiability Intelligence Test[b]	6 years 0 months to 7 years 11 months	(1996) 1,350	X	X	X	X	X	X	X
Infant Mullen Scales of Early Learning	Birth to 36 months	1,231	X		X	X	X	X	X
Kaufman Adolescent and Adult Intelligence Scale	11 to 85 years or over	(1988) 2,000	X	X	X	X?	X		X
Kaufman Assessment Battery for Children[b]	2.6 to 12.5 years	2,000 Oversampling	X	X	X	X?	X	X	X
Kaufman Brief Intelligence Test[b]	4 to 90 years	(1985/1990) 2,022	X	X	X			?	
Learning and Memory Battery	20 to 79 years	(Canada) 480	?	X	X	?	X		X

(Continued)

TABLE 9.2 Continued

Test Name	Ages	Norming	Racial/ Ethnic Represen- tation	Socio- economic Status	Sex	Region of Country	Urban/ Rural Represen- tation	Edu- cation	Special Edu- cation	Bias Exam
Leiter-R[b]	2 years 0 months to 20 years 11 months	(1993) Visual Reasoning 1,719 Atten- tion and Memory 763	X		X	X	X	X	X	
Luria-Nebraska Neuropsychology Battery–Children's Revision	8 to 12 years	125	—	—	X					
Luria-Nebraska Neuropsychology Battery–Forms I and II	15 years or over	(Form I) 50 (Form II) 73	—	—	Limited in- formation	—	—	Limited in- formation	—	
Matrix Analogies Test	5 to 17 years		X	X	X	X				X
McCarron-Dial System	3 years to adult	More than 2,000 Limited in- formation		X		X				
McCarthy Scales of Children's Abilities	2 years 4 months to 8 years 7 months	(1970) 1,032	X	X		X	X		X	?

Test	Age range	Standardization sample								
Multiple Intelligence Developmental Assessment Scales (questionnaire)	14 years or over	Not specified	?							
NEPSY: A Developmental Neuropsychological Assessment[b]	3 to 12 years	(1995) 1,000		X	X	X	X	X		X
Non-Reading Intelligence Tests–Levels 1 to 3	6 years 4 months to 8 years 3 months / 7 years 4 months to 9 years 3 months / 8 years 4 months to 10 years 11 months	Not specified		X				X		
Otis-Lennon School Ability Test (sixth edition)	Grades K-12	(1980) 356,000	X			X	X	X	X	X
Primary Test of Cognitive Abilities	Grades K-1	16,066	X				X		X	X
Psychoeducational Profile–Revised (autistic and communication handicapped)	6 months to 7 years Functioning	(North Carolina) (1985) 420 (normal)						—	?	

263

TABLE 9.2 Continued

Test Name	Ages	Norming	Racial/ Ethnic Represen- tation	Socio- economic Status	Sex	Region of Country	Urban/ Rural Represen- tation	Edu- cation	Special Edu- cation	Bias Exam
Raven's Advanced Progressive Matrices[b]	11 years or over	3,953[d] U.S. naval recruits data collected 1986 to 1990 Limited in- formation	?							
Raven's Coloured Progressive Matrices[b]	5 to 11 years Elderly	598 (Dum- fries)[e] Limited in- formation							X	X
Raven's Standard Progressive Matrices[b]	6 to 80 years 6 to 16 years in standard- ization sample	(1984/1986 U.S.)[f] more than 22,000 Limited in- formation	?		X					X
Scholastic Abilities Test for Adults	16 years or over	(1985) 1,005	X		X	X				

Test	Age range	Sample							
Scholastic Aptitude Scale	6 years 0 months to 17 years 11 months	1,448		—			X		
Shipley Institute of Living Scale	14 years or over	1,016 Grade 4 to college 290 16 to 45 years revised psychiatric		X	X				
Slosson Full-Range Intelligence Scale	5 to 21 years	(1980) 1,509[g]	—?	?	?	?	?	X?	
Slosson Intelligence Test	—?	1,854	—?	X	X	X	—	X	X?
Southern Ordinal Scales of Development (e.g., disability)	"Children"	508	X 68% White	X 35% professional	X 65% male	—	—	—	X 0.6% regular education
Stanford-Binet Intelligence Scale (fourth edition)[b]	2 years to adult	(1980) more than 5,000	X	X	X	X	X	X	X
Structure of Intellect Learning Abilities Test	Preschool-age to adult	Not specified	—	—	X	—			
Swanson Cognitive Processing Test	5 years to adult	1,611	—		X	—		X	

(Continued)

TABLE 9.2 Continued

Test Name	Ages	Norming	Racial/ Ethnic Represen- tation	Socio- economic Status	Sex	Region of Country	Urban/ Rural Represen- tation	Edu- cation	Special Edu- cation	Bias Exam
Test of Memory and Learning	5 to 19 years	(1990/1992) 1,342	X	X	X	X	X			X
Test of Nonverbal Intelligence (third edition)[b]	5 years 0 months to 85 years 11 months	3,451	X	X	X	X	X		X	X
Universal Nonverbal Intelligence Test[b]	5 to 17 years	(1995) 2,100	X		X	X	X	X	X	
Wechsler Adult Intelligence Scale (third edition)[b]	16 to 89 years	(1995) 2,450	X Over-sampling		X	X		X	X	X
Wechsler Intelligence Scale for Children–III[b]	6 years 0 months to 16 years 11 months	(1988) 2,200	X Over-sampling			X	X?	X		X
Wechsler Preschool and Primary Scale of Intelligence–Revised	2 years 11 months to 7 years 3 months	(1986) 1,700	X	X	X	X	X	X		X
Wechsler Memory Scale[b]	16 to 89 years	(1995) 1,250	X Over-sampling	X	X	X	X	X	X	X

Test	Age range	(Year) N	Norming					
Wide Range Assessment of Memory and Learning	5 to 17 years	(1980) 2,363	? 78% White		X	X	X	—
Woodcock-Johnson Psychoeducational Battery–Revised	2 to 90 years	(1980) 6,359	X Over-sampling	X	X	X	X	X

SOURCE: All information for this table was obtained from the WinSPIRS (1989–present) database for the *Buro's Mental Measurements Yearbook*. All test information (e.g., authors and years of tests) is the same as in Table 9.1.

NOTE: Education = parent/adult educational level; special education = special education students included (i.e., learning disabled, gifted and talented); bias exam = special efforts made to examine measure with respect to race/ethnicity. In norming column, year in parentheses = time of census representation; blank = no information provided by reviewer; X = stratification variable; — = reviewer's concerns regarding this variable; ? = mixed reviews; X? = reviewer notes general concerns related to psychometric issue; oversampling = oversampling of minority groups.

a. Consistent with U.S. norms of Woodcock-Johnson.

b. Test manual was examined in preparation of this table.

c. No Hispanic or Asian.

d. Information taken from Advanced Progressive Matrices manual (Raven, Court, & Raven, 1994, p. 42). The authors noted extensive data for U.S. norms collected in 1984 and 1990 in many parts of the United States. Norms are provided for school districts and ethnic groups. The Standard Progressive Matrices norms were converted to Advanced Progressive Matrices norms. Between 1986 and 1990, a study was conducted on 3,953 U.S. male naval recruits. Findings indicate that the mean score of Blacks was "markedly" lower, and the mean score of Hispanics was significantly lower, than the mean scores for Whites and Asians.

e. Information taken from Coloured Progressive Matrices manual (Raven, Court, & Raven, 1990). Dumfries standardization is the most recently cited in this section of the manual. Reference is made to the 1990 edition with U.S. norms; however, only little specific information is provided. Numerous studies are cited regarding reliability and validity with various racial/ethnic groups.

f. Information taken from Standard Progressive Matrices manual (Raven, Court, & Raven, 1992). The authors noted a series of standardization procedures. Of particular interest is the series of local norming studies that was carried out in U.S. school districts between 1984 and 1986. More than 22,000 students were tested. Norms varied with ethnic and socioeconomic composition of the school districts and geographic locations. Within districts, norms varied independently with ethnicity and socioeconomic status. Item analyses showed that the test scales performed the same way in each of the ethnic groups and had similar predictive validity. The authors noted that, based on these findings, the use of ethnic norms rather than national norms seems to be indicated. Norms are presented by age and percentile rank.

g. Caution must be used given that the normative sample was tested 10 years ago and comparisons were made to census data 20 years ago (Tindal, 1986–present).

Given the limitations of examining only information in the various reviews without the benefit of the test manuals, the blank spaces in Table 9.2 indicate that variables were not specifically mentioned. It should be noted that most census-based measures include stratification by sex even if this is not stated by reviewers. In addition, parental educational level, educational level of adult examinees, and parental occupation often were used as indicators of socioeconomic status (SES) by many test developers.

Based on the findings of Table 9.2, it appears that many of the cognitive measures cited include census-based sampling procedures to create the norms. Although the overall numbers of the norming samples might appear large, when broken down proportionally by census data, the representation of particular minority groups becomes very small. For example, the standardization data for the Leiter-R (Roid & Miller, 1997) includes 1,138 Caucasians, 286 African Americans, 217 Hispanics, 55 Asians, and 23 Native Americans spanning the ages of 2 to 20 years. Similarly, the standardization sample of the Kaufman Brief Intelligence Test (Kaufman & Kaufman, 1990) includes 1,456 White, 298 Black, 190 Hispanic, and 78 other minorities spanning the ages of 4 to 90 years.

Note that only a small number of Asian and Native American participants are included at many of the age ranges (e.g., only 1 to 5 Native Americans and 1 to 6 Asians at each age level of the Leiter-R). In addition, on the Kaufman Brief Intelligence Test, "other" is used as the category for those who are not White, Black, or Hispanic (including Native Americans, Alaskan Natives, Asians, and Pacific Islanders).

Some test developers, however, have begun to include oversampling procedures of various minority groups (e.g., DAS [Elliott, 1990, cited in WinSPIRS, 1989-present], K-ABC [Kaufman & Kaufman, 1983a], SB-FE [Thorndike et al., 1986b], WISC-III [Wechsler, 1991], WJ-PB revised [Woodcock & Johnson, 1989]), moving beyond the census-based norming samples. More is discussed later in this chapter regarding this practice in terms of racial/ethnic bias examinations.

Most cognitive measures are based on norming samples stratified based on gender and region of the country. As noted earlier, socioeconomic indicators may include parental/self-occupation and/or parental/self-education. The urban/rural or community size variable also is indicated as a stratification variable on many scales given early findings that children living in rural areas generally have scored lower than those residing in urban areas (Vernon, 1979). Recent studies, however, have indicated that this discrepancy no longer is practically significant (Kaufman, 1990). Many tests also include members of special populations including individuals with learning problems (e.g., learning disabled, mentally retarded) and the gifted and talented. This is especially important given that many of these tests are used to identify children, adolescents, and adults with specific learning problems or disabilities.

Few tests were identified to address linguistic differences in relation to cognitive abilities. This might be due to the method used to identify the majority of tests cited in Tables 9.1 and 9.2. It should be noted that the introduction to the WinSPIRS *Mental Measurements Yearbook* database clearly stipulates that reviews are done primarily on English-language standardized measures. Thus, measures not based on English are not prioritized in the database. It also should be noted that there exist few non-English-based cognitive ability tests. Given changes in the demographic features of the United States (i.e., growing numbers of language-minority populations), it appears that the need for bilingual measures will continue to grow. The use of translators and interpreters in administering English-based measures remains a questionable practice given issues of equivalence noted in Chapter 2 of this book.

Examiners must be careful in using published measures in various languages that are based on English versions. For example, one of the more popular measures available in Spanish was the Batería Woodcock Psico-Educativa en Español (Woodcock, 1982, cited in WinSPIRS, 1989-present). This represents an earlier version of the current Batería Woodcock-Muñoz–Revisada (Woodcock & Muñoz, 1996, cited in Frary, 1989-present). The early version was used frequently given that there were very few Spanish-language-based tests available to clinicians and educators. The norms for the Batería Woodcock Psico-Educativa en Español (Woodcock & Muñoz-Sandoval, 1982-1996, cited in WinSPIRS, 1989-present) were based on data obtained from urban areas of Costa Rica, Mexico, Peru, Puerto Rico, and Spain. In the test manual, no mention is made of Spanish-speaking communities within the United States as serving in the norming process. Prendes-Lintel (1989-present) questioned the appropriateness of administering to U.S. populations a test that has been normed abroad. Currently, the revised edition incorporates specific examination of the item content regarding translation procedures to test the Spanish item bank in various countries and provides information regarding application to U.S. samples.

More recently, measures such as the Bilingual Verbal Ability Test (BVAT; Muñoz-Sandoval, Cummins, Alvarado, & Ruef, 1998) have been developed in the area of bilingual assessment. The BVAT is based on three subtests adapted from the language proficiency battery of the Woodcock-Johnson Tests of Cognitive Ability–Revised, namely Picture Vocabulary, Oral Vocabulary, and Verbal Analogies. If items on a given subtest are scored as incorrect in English, then this area subsequently is assessed in the examinee's first language. The BVAT "can provide a more accurate estimate of academic potential than assessments administered nonverbally, only in English, only in the student's first language, or through separate measures of the student's abilities in English and first language" (Muñoz-Sandoval et al., 1998, p. 12). The BVAT includes 16 different linguistic versions: Arabic, Chinese, English, French, German, Haitian-Creole, Hindi, Italian, Japanese, Korean, Polish, Portuguese, Russian, Spanish, Turkish, and Vietnamese. A standardized translation procedure was followed in developing the tests using 66 translators, editors, and professionals who were native speakers of the target language as well as fluent in English. Extensive procedures were followed in the test development process (Muñoz-Sandoval et al., 1998). These included back-translation procedures, field testing with native-speaking examiners and children, and age level reviews of test materials by assessment professionals. The test procedures followed in the development of the BVAT highlight the complexity of developing state-of-the-art bilingual measures.

In Chapter 5 of this book, we presented a comprehensive review of cultural bias investigations that scholars undertook after intelligence tests were developed and became available to researchers and clinicians. It is important to note, however, that bias examination procedures also are included as part of a number of test development strategies. Table 9.3 presents some of the measures from the *Buro's Mental Measurements Yearbook* database (WinSPIRS, 1989-present) in which reviewers specifically identified how "racial/ethnic bias"[1] was addressed in the test development process. In general, the bias examination strategies clustered as follows:

1. Racial/ethnic oversampling to examine potential item bias (e.g., expert panels to review item content, item analysis)
2. Specific reliability and validity studies conducted with diverse populations
3. The development of sociocultural norms

Helms (1997) specifically noted the limitations of using "expert panels" in that many times no criteria are provided by the test developer to clarify what constitutes bias with regard to the content of particular items. The issue often is "differential familiarity of test content" (p. 527) and the evaluation of item content appropriateness. Unfortunately, this process is not standardized; thus, "experts" often are left to their own devices to determine appropriateness.

Most contemporary intelligence test batteries use Rasch modeling procedures for item analysis (Harrison, Flanagan, & Genshaft, 1997). Rasch modeling examines issues of item difficulty based on item response theory. Rasch modeling also is referred to as a one-parameter latent trait analysis given its singular focus on item difficulty. The K-ABC, Wechsler scales, and DAS were specifically noted as having conducted bias analyses with regard to racial/ethnic minority samples (Harrison et al., 1997).

Helms (1997), in a provocative treatise, challenged the use of this technique given that "potential hindrances [in test performance] may be elicited by the internalized culture, race, and socioeconomic socialization under which the test taker performs" (p. 530), not just the individual's ability or difficulties within a specific area. In addition, performance on a particular item may reflect more than one ability. Helms noted,

> Thus, item response models do not resolve the issue of the psychometric cultural equivalence of CATs [cognitive ability tests]. Rather, they create the illusion that the problem has been solved when, in fact, it has not been truly addressed. As is the case for other forms of equivalence, models of culture external to the test are needed to establish person [characteristics] as distinguished from item characteristics. In the absence of such models, neither cultural equivalence nor fairness of CATs can be "proven." (p. 530)

Numerous studies have been conducted to examine the validity and reliability of particular measures by racial/ethnic group. The test manuals often include references to support the use of their reliable and valid measures. For example, the WISC-III and the SB-FE report a number of validity studies in the context of the test development process. Inclusion of specific racial/ethnic group members in the studies often is reported. However, separate analyses for particular racial/ethnic minority groups frequently are *not* provided.

TABLE 9.3 Methods Used to Examine Racial/Ethnic Bias in Tests Reviewed in WinSPIRS *Buro's* and Available Test Manuals

Test Name	Method of Racial/Ethnic Examination
Canadian Test of Cognitive Skills	Items evaluated for bias by expert panels; statistical analyses cited, but limited information provided
Detroit Tests of Learning Aptitude (primary second edition)	Delta values for non-Whites and Whites correlated .98
Hammill Multiability Intelligence Test	Specific reliability and validity information presented by race/ethnic group; item bias analysis using delta score approach
Kaufman Assessment Battery for Children	Special sociocultural norms for race and parental education; item and subtest analysis by ethnic groups indicating similar validities for Blacks, Hispanics, and Whites
Kaufman Brief Intelligence Test	Rasch analysis and item bias analysis completed; biased items eliminated from tryout edition; Mantel-Haenszel statistical procedure also used.
Leiter-R	Expert review panel; item analyses conducted using Rasch analysis on tryout and final edition; examination of area scores by racial/ethnic group conducted on tryout edition; examination of reliability and validity studies by race/ethnicity
McCarthy Scales of Children's Abilities	Analysis of standardization data for five age groups conducted separately for Blacks and Whites; examination of General Cognitive Indexes by race/ethnicity not significantly different; differences similar to those for conventional IQ tests
Otis-Lennon School Ability Test (sixth edition)	Examination of differential item functioning using Rasch model and Angoff delta analyses; examination of difficulty levels and discriminant values (biserial r values); specific attention paid to eliminating item bias related to gender, socioeconomic status, ethnicity, culture, and region part of test development process; expert panels also employed
Primary Test of Cognitive Skills	Statistical evaluation of tryout items based on item response theory; limited information provided
Raven's Progressive Matrices	Extensive information provided regarding studies conducted on validity and reliability of different test versions; information regarding race/ethnicity noted

(Continued)

TABLE 9.3   Continued

Test Name	Method of Racial/Ethnic Examination
Slosson Intelligence Test–Revised	Absence of mean raw score differences for ethnicity and gender noted by authors; question regarding analysis of raw scores rather than standard scores; latent trait analysis also conducted with regard to slope and intercept bias; no slope or intercept bias for any pair of ethnic groups yielded by pairwise comparisons and regression analyses on raw scores; question raised regarding representation of norming sample
Test of Memory and Learning	Item analysis conducted by different groups; item functioning consistent across ethnic groups; caution in using measure with minority populations cited by reviewer
Test of Nonverbal Intelligence (third edition)	Reliability and validity studies noted for African Americans; item bias examined using norming sample and item response theory and delta score approach; standard indexes of bias presented in manual; replication of validity studies by race/ethnic group
Universal Nonverbal Intelligence Test	Expert review panels; studies of item fit conducted by race/ethnicity; reliability and validity studies conducted by group; item characteristic curves, item performance, and item demographic correlations conducted; evidence of bias not supported by Mantel-Haenszel method
Wechsler Adult Intelligence Scale–III	Item bias analyses conducted with oversample of 200 African American and Hispanic participants; items removed based on content review and empirical analyses
Wechsler Primary Preschool Scale of Intelligence–Revised	Oversampling and examination by group
Wechsler Intelligence Scale for Children–III	Expert review panels; item analysis to examine differential item functioning; predictive validity contrasts (i.e., achievement scores, classroom grades); little predictive bias noted by gender or race/ethnicity; examination or slope and intercept
Wide Range Assessment of Memory and Learning	Person and item separation statistics; construct validity using Rasch measurement
Woodcock-Johnson Tests of Cognitive Ability	Expert review panels; Rasch model used at item development stage

Only a few measures include specific reference to special norms developed aside from the standardization sample. One such measure is the K-ABC, which provides sociocultural norms by race and parental education.

In summary, it appears that many qualitative techniques (i.e., expert review panels) and statistical techniques have been developed to address issues of psychometric bias. However, the challenges and complexities of examining racial/ethnic group and individual differences on cognitive measures remain. Helms (1997) suggested the need to move beyond "categorical operational definitions of cultural (racial and socioeconomic) factors at the test development phase" (p. 530). This includes (a) providing more specifics regarding what variables related to "racial, cultural, and/or socioeconomic socialization are controlled or reduced" (p. 531); (b) noting "how factors were measured at both the person and environmental levels" (p. 531); and (c) providing the "hypothesized psychological implications of these characteristics for CAT performance" (p. 531). It is clear that issues of racial/ethnic bias must be part of the test construction and conceptualization phase. This is particularly pressing given that cultural bias investigations by independent researchers of individually administered intelligence tests have declined greatly during recent years (see Chapter 5 of the present book).

Racial and Ethnic Profiles of Abilities and Discrepancies Yielded by Traditional Cognitive Tests

In this section, we specifically cover the most popular intelligence instruments—the WISC-III, the SB: FE, and the K-ABC—for closer scrutiny regarding overall scaled score "discrepancies" (i.e., differences in Full Scale IQ) and racial/ethnic ability "profiles" (i.e., differences in subtest scores). In addition, the Universal Nonverbal Intelligence Test (UNIT; Bracken & McCallum, 1998) is included because it represents one of the newest measures in the area of nonverbal assessment.

Wechsler Intelligence Scale for Children–III

Despite the growing number of instruments currently available to clinicians, the overwhelming favorite appears to continue to be the Wechsler scales. The WISC-III is considered to be the most frequently used intelligence test in North America. In a survey of school psychologists, one of the largest consumer groups of intelligence measures, Wilson and Reschly (1996) reported that each school psychologist respondent administered the WISC-III approximately 10 times per month. Of the other individually administered intelligence tests cited, average test administration approached 2 times per month only for the Stanford-Binet (SB and L-M scales combined). Results indicated that other scales were administered fewer than 1.5 times per month.

In the initial conceptualization of the Wechsler scales, David Wechsler was aware of the issues regarding intelligence testing with minorities. However, he "viewed the differences in mean scores not as indicators of lower intelligence among certain groups but as indicators of differences in our society and how variations in social, economic, political, and medical opportunities have an impact on intellectual abilities" (cited in Prifitera, Weiss, & Saklofske, 1998, p. 11).

With regard to the WISC-III, racial and ethnic group differences continue to be noted. However, predictive validity studies indicate that the WISC-III "predicts achievement scores equally well for African-American, Hispanic, and White children" (Prifitera et al., 1998, p. 10). For example, a study by Weiss, Prifitera, and Roid (1993) examined how well the WISC-III Full Scale IQ predicted standardized achievement test scores for Black, Hispanic, and White children. Results supported the absence of bias in predicting achievement based on IQ scores (but for a caveat about interpreting these results, see Note 4 in Chapter 5 of this book).

Average WISC-III scores (Full Scale IQ, Verbal Scale IQ, Performance Scale IQ) for African Americans, Hispanics, and Whites from the standardization sample were reported as follows: African Americans (Full Scale IQ = 88.6, Verbal Scale IQ = 90.8, Performance Scale IQ = 88.5), Hispanics (Full Scale IQ = 94.1, Verbal Scale IQ = 92.1, Performance Scale IQ = 97.7), and Whites (Full Scale IQ = 103.5, Verbal Scale IQ = 103.6, Performance Scale IQ = 102.9). It should be noted that no information was reported regarding Hispanic subgroups (e.g., Puerto Ricans, Mexican Americans). These results appear similar to scores obtained over the years on other traditional measures of intelligence (i.e., 15-point mean difference between African Americans and Whites on Full Scale IQ, with Hispanic and White average differences falling within this range).

When matched samples (i.e., equated on SES, parental education, sex, age, region of the country, and number of parents living at home) from the standardization sample were compared, average group score discrepancies decreased but still were present. For example, the difference between Whites and African Americans on Full Scale IQ was 14.9. When matched samples of White and African American children are compared, the difference decreases to 11.0. Prifitera et al. (1998) also presented difference scores on matched African American samples by age. These data indicate that the differences between groups are much lower for younger groups (6- to 11-year-olds) than for older groups (12- to 16-year-olds). Differences appear to be approximately 4 to 5 points higher for the older group.

Differences on WISC-III index scores provide information regarding the profile of ability areas. Based on the standardization sample, Prifitera et al. (1998) reported index scores by racial/ethnic group. They interpreted these index profiles as follows:

There is considerable variation in the differences among African Americans, Hispanics, and Whites among the other IQ index scores. For example, African Americans score only 6.1 and 7.4 points, respectively, below Whites on PSI [Processing Speed Index] and FDI [Freedom from Distractibility Index] scores. Hispanics continue to show a relatively higher PIQ [Performance Scale IQ] and POI [Perceptual Organization Index] score compared to their VIQ [Verbal Scale IQ] and Verbal Comprehension Index (VCI) scores, which is consistent with the previous literature. The difference in the PIQ between Hispanics and Whites is only 5.2 points. In addition, the Hispanic group's PSI score is virtually identical to that of Whites, and there is only a 7.7-point difference between the groups on the FDI score. These results strongly suggest that simply looking at the FSIQ

[Full Scale IQ] differences ignores relative strengths in various domains of cognitive functioning among minority groups. (p. 13)

Thus, they conclude,

The view that minorities have lower abilities is clearly wrong. IQ score differences between younger African-American and White children and Hispanic and White samples with only gross matches on SES are much less than a standard deviation, and the index scores are even smaller. What would the difference be if even more refined variables had been controlled for, such as household income, home environment (e.g., parental time spent with children), per-pupil school spending, medical and nutritional history, and exposure to toxins? (p. 15)

Information regarding profile differences on the WISC-III between groups also indicates that the percentages of Verbal Scale IQ versus Performance Scale IQ differences vary depending on racial/ethnic group. For the African American sample of the WISC-III, 57% of the children obtained a Verbal > Performance profile. Prifitera et al. (1998) noted that African American children had scores approximately 2 points higher on the Verbal Scale IQ than on the Performance Scale IQ. Discrepancies of more than 19 points were found for less than 10% of the African American sample. By comparison, most Hispanic children scored higher on Performance subtests than on Verbal subtests. On the WISC-III, 69% of the Hispanic sample had higher Performance Scale IQs. More specifically, a 20-point Performance > Verbal profile was noted for 14% of the Hispanic sample. Findings also indicated that for the Hispanic sample, the lower the level of parental education, the larger the Performance > Verbal IQ discrepancy. The reasons for this occurrence were not provided by the authors. One might hypothesize, however, that more educated parents might reinforce verbal abilities and provide more verbal stimulation, leading to a smaller discrepancy.

Stanford-Binet (Fourth Edition)

Mean scores on the fourth edition of the Stanford-Binet (SB-FE) by racial/ethnic group are reported in four age ranges: 2 to 6 years, 7 to 11 years, 12 to 17 years, and 18 to 23 years. Age groupings were selected given that individuals in these age ranges usually are administered the same subtests. Scores are delineated according to area scale means and standard deviations that are provided in the technical manual (Thorndike et al., 1986b). Examination of profile differences based on area scores obtained for Hispanics, Asians, and Blacks yield a distinct pattern of scores consistently found across all age groups (Thorndike et al., 1986b):

Hispanics had higher mean scores in Abstract/Visual Reasoning, Quantitative Reasoning, and Short-Term Memory than in Verbal Reasoning. Hispanic examinees, at all ages, had their highest mean score in Abstract/Visual Reasoning and Quantitative Reasoning. . . . In two of the three age groups, Asian examinees also had higher mean scores in Abstract/Visual Reasoning and Quantitative Reasoning than in Short-Term Memory. The black examinees had their highest scores in Short-Term Memory. (p. 37)

These findings appear consistent with research indicating ability profile differences by racial/ethnic group (see also Chapter 2 of the present book).

In examining the data between racial/ethnic groups, it also is clear that discrepancies between groups remain similar to the past literature in terms of the overall composite score. Whites (ages 12 to 23 years) score approximately 1 standard deviation higher than do Blacks in this same age range (note that the SB-FE mean = 100, standard deviation = 16). Native Americans, Hispanics, and Asians score somewhere between these two groups on the overall composite score at this age level. Whites scored higher than all groups in the Verbal Reasoning area. Relatively higher scores are noted for Whites and Asians on Abstract Visual Reasoning and Quantitative Reasoning in comparison to those scores for Blacks and Hispanics. Discrepancies like these continue to remain a concern to researchers and clinicians given the use of instruments such as the SB-FE in educational decision making.

Kaufman Assessment Battery for Children

Means and standard deviations by racial/ethnic group on the K-ABC Global Scales are noted in the test manual (Kaufman & Kaufman, 1983a). It appears that Black preschool children scored at approximately the standardization average on the Global Scales, with only small deviations from the mean. Black children's scores dropped at higher grade levels as they scored 3.8 to 8.9 points lower, with the largest difference being 0.5 standard deviation on the Nonverbal Scale. Overall, Blacks scored highest on the Sequential Processing Scale. The test authors commented that Black-White differences continue to be noted, especially on the Simultaneous Processing Scale. They pointed out, however, that the discrepancy still is less than that with other traditional IQ measures.

The Hispanic sample appears to do equally well on the Simultaneous Processing Scale and Sequential Processing Scale, although higher scores (5 to 7 points) are noted at the preschool level. This represents a similar pattern to that with the Black sample. At the school-age level, Mental Processing Scale and Nonverbal Scale scores appear to be close to the mean of 100. Achievement is 0.5 standard deviation below the mean; the test authors noted that this might be attributed to linguistic or cultural differences.[2]

The Sioux and Navajo samples obtained higher scores at the school-age level than at the preschool level, although the differences were somewhat negligible. The Navajo group scored 12 points higher on the Simultaneous Processing Scale than on the Sequential Processing Scale. The test authors, however, noted that this might be "illusory" (Kaufman & Kaufman, 1983b, p. 154). Both Navajo and Sioux children obtained higher Mental Processing Scale scores than Achievement Scale scores, "indicating the influence of cultural and linguistic factors in their K-ABC profile" (p. 154); this is similar to what is found in Hispanics. Again, score differences between means for these samples are less than what typically are found on other intelligence measures (i.e., 15-point discrepancy, 1 standard deviation).

Universal Nonverbal Intelligence Test

The UNIT (Bracken & McCallum, 1998) is one of the newest measures in the nonverbal reasoning area (for more information about the norming and standardization of the measure, refer to Tables 9.1 and 9.2). The test manual states, "The UNIT measures a broad range of complex memory and reasoning abilities, including those lending themselves to internal processes of verbal (symbolic) mediation as well as those that are less conducive to such mediation (nonsymbolic)" (p. 1). Subtests include Symbolic Memory, Cube Design, Spatial Memory, Analogic Reasoning, Object Memory, and Mazes. Overall group discrepancies and profile differences also are noted on nonverbal measures such as the UNIT. Mean discrepancies between the African American, Asian/Pacific Islander (A/PI), Native American, Hispanic, and bilingual/English as a second language groups with matched (i.e., sex, age, parental education) White comparison groups are provided in the test manual. A/PIs tended to score higher across areas of the Abbreviated Battery (two subtests), Standard Battery (four subtests), and Extended Battery (six subtests). African Americans scored relatively lower across all areas, although still within the average range (mean = 100, standard deviation = 15). In reference to their findings, Bracken and McCallum (1998) noted the following:

> According to the findings from the comparison studies, it is apparent that some group differences in intelligence do occur, at least as assessed by the UNIT; it is also apparent that differences do not occur in other groups. Moreover, when group differences on the UNIT do occur, those differences are smaller than the differences typically reported in the literature for many language-loaded tests. The UNIT tasks were designed to reduce the influence of culture, and that goal appears to have been met. Other noncognitive influences may also produce mean differences, and it is impossible to remove all of those. (pp. 195-196)

For example, Bracken and McCallum noted that groups were equated on the basis of SES, with parent education level the basis of SES. They cited literature by Valencia, Rankin, and Henderson (1986) suggesting that parent occupational status and schooling attainment may act differently as predictors of children's intelligence between White and minority groups. Thus, there is a degree of complexity in understanding the influence of particular variables based on how they are defined (e.g., SES). (See Chapter 3 of the present book for a more comprehensive description of the role of SES.)

Based on examination of the WISC-III, the SB-FE, the K-ABC, and the UNIT, it appears that the issues of racial/ethnic group discrepancies and profile differences continue to be evidenced despite increased sophistication in test development strategies. Smaller discrepancies are noted on nonverbal measures, and profile differences appear commensurate with past research. However, the meaning of such differences based on cultural, educational, and socioeconomic explanations remains unclear as to what specific variables are accounting for the differences.

Current Procedural Modifications of Traditional Cognitive Tests

Given the popularity of cognitive tests and the controversies surrounding their use with racial/ethnic minorities, researchers have attempted to provide procedures to adjust scores and modify administration of these measures for diverse populations. These include the early development of the System of Multicultural Pluralistic Assessment (SOMPA), the biocultural model of assessment, and the Gf-Gc Cross-Battery Assessment model. The SOMPA represents an example of modification of scores based on variables known to affect intelligence test performance, whereas the biocultural model illustrates ways of modifying test administration practices and collection of additional information on various aspects of intelligence. This practice often involves the incorporation of qualitative information (i.e., medical history, multiple intelligences). The Gf-Gc Cross-Battery Assessment model provides a means of expanding the breadth of ability areas assessed.

System of Multicultural Pluralistic Assessment

One of the first examples of score adjustment procedures was made with regard to the WISC-R. This was the development of the SOMPA (Mercer, 1979; for a description of the SOMPA, see also Chapter 8 of the present book). Questions have arisen, however, as to the contribution of the SOMPA procedures. For example, one evaluation of the SOMPA (Figueroa & Sassenrath, 1989) indicated that the estimated learning potential does not appear to predict or correlate with academic achievement as accurately as do other standardized IQ scores (e.g., Verbal Scale IQ).[3]

Biocultural Model of Assessment

Armour-Thomas and Gopaul-McNicol (1998) provided information regarding a four-tier assessment system incorporating preassessment activity areas that must be addressed prior to testing given the potential impact of these variables on intelligence test performance. These areas are health assessment, linguistic assessment, prior experiences (educational or psychosocial experiences [e.g., learning style]), and family issues (e.g., recent divorce might affect test performance).

The biocultural model consists of three dimensions: biological, cognitive processes, and culturally coded experience and cultural contexts. The four tiers of the biocultural model encompass these processes. The first tier is the psychometric assessment itself. This represents a score-oriented approach and constitutes the biological component of the model. The second tier is represented by psychometric potential assessment. This tier consists of the several components used to derive potential and estimated intellectual functioning in relation to the actual test administration. For example, incorporating "testing the limit" procedures by suspending time, contextualizing vocabulary, paper-and-pencil use, and test-teach-retest strategies are modifications that are recommended. The third tier involves ecological assessment. The components are family/community support assessment; observations to determine performance in the school, home, and community; stage of acculturation; and data from a teacher questionnaire. The fourth tier, other intelligences, consists of assessing musical, bodily kinesthetic, interpersonal, and intrapersonal areas. According to Armour-Thomas and Gopaul-McNicol (1998), each tier is equally weighted at 25%.

Armour-Thomas and Gopaul-McNicol (1998) provided extensive information and scales to facilitate a clinician's use of the biocultural model. The procedures are illustrated in the case of the WISC-III. To begin, the authors suggested that the examiner divide the test protocol into two sections: one for the "actual testing scores" and the other for the "potential testing scores" (p. 139). Actual testing scores are obtained based on the standardized administration of each subtest. The potential testing scores are derived based on the following modifications in test administration: (a) suspension of time constraints, (b) matching of items to culturally relevant questions and symbols (contextualize vocabulary), and (c) permitted teaching of item tasks (on the Arithmetic subtest, the student is allowed to use paper and pencil to solve the problems). Issues such as cultural/item equivalence and linguistic equivalence are raised throughout the discussion of the subtests. In addition, a test-teach-retest model is encouraged. The impact of speed is eradicated given that timing is not required for those who take the test.

The biocultural model represents an effort to integrate qualitative and quantitative data into the understanding of the cognitive abilities of students from diverse cultural backgrounds. In many ways, this blending of test data might represent the type of flexible and clinically oriented measurement procedure needed to identify the "true" abilities of students in various assessment contexts.

Gf-Gc Cross-Battery Assessment Model

The Gf-Gc Cross-Battery Assessment approach provides a structural schema to enable examiners to select the best battery of measures to tap into a wide range of ability areas. As described by McGrew and Flanagan (1998),

> The Gf-Gc Cross-Battery approach is a time-efficient method of intellectual assessment that allows practitioners to measure validly a wider range or a more in-depth but selective range of cognitive abilities than that represented by any one intelligence battery in a way consistent with contemporary psychometric theory and research on the structure of intelligence. (p. 357)

The Cross-Battery Assessment model is based on Horn (1985, cited in McGrew & Flanagan, 1998), and Cattell (1941, cited in McGrew & Flanagan, 1998), and Carroll's Gf-

Gc model of abilities (1997, cited in McGrew & Flanagan, 1998). McGrew and Flanagan (1998) noted that there are a number of broad classifications (Stratum II) that can be made regarding cognitive ability tests. In addition to the broad classifications, McGrew and Flanagan noted the importance of the narrow classifications (Stratum I) of cognitive ability tests. Based on these two classifications, the examiner can provide a more valid examination of the range of abilities being examined in a particular test or battery.

Based on these classifications, McGrew and Flanagan (1998) identified broad and narrow stratum abilities, which they related to particular measures of cognitive ability. Broad Stratum II abilities include Fluid Intelligence, Quantitative Knowledge, Crystallized Intelligence, Reading/Writing, Short-Term Memory, Visual Processing, Auditory Processing, Long-Term Storage and Retrieval, Processing Speed, and Decision/Reaction Time or Speed. Narrow Stratum I abilities address *particular* broad Stratum II abilities (e.g., Fluid Intelligence is made up of narrow Stratum I abilities including General Sequential Reasoning, Induction, Quantitative Reasoning, Piagetian Reasoning, and Speed of Reasoning). Based on a review of some of the most popular tests, the authors noted that most intellectual assessment focuses on examination of dichotomous abilities and on an incomplete examination of multiple intelligences. They noted that the Woodcock-Johnson Psychoeducational Battery–Revised represents the most advanced test in terms of progress in the applied measurement area given that it attends to the interaction of cognitive and noncognitive factors (i.e., coverage of all Gf-Gc factors).

In addition to examining the most popular cognitive measures in relation to the Gf-Gc ability model, McGrew and Flanagan (1998) provided information regarding cultural content and linguistic demands on a variety of test batteries. Measures identified by McGrew and Flanagan as having low cultural content and low linguistic demand include particular nonverbal subtests of the Leiter-R, the UNIT, the DAS, the Test of Nonverbal Intelligence (third edition), Raven's Progressive Matrices, the K-ABC, the Kaufman Adolescent and Adult Intelligence Test, the SB-FE, the Wechsler Test of Memory and Learning, the Wechsler Memory Scale–Revised, and so forth. Specific citations for many of these measures may be found in Table 9.1 (see also McGrew & Flanagan, 1998).

The Gf-Gc Cross-Battery Assessment model represents a structured framework to determine the abilities being assessed using a particular instrument. Based on the Gf-Gc model, one may select different measures to address the specific abilities of importance to the student being tested. In addition, McGrew and Flanagan (1998) rate the strengths and limitations of various instruments in relation to cultural content that might be helpful to examiners who are not familiar with particular batteries or subtests.

Based on the preceding discussion, it appears that our knowledge base of increasing the multicultural sensitivity toward cognitive and intellectual assessment is expanding. Such attention is occurring in a number of ways including modifications in scoring procedures, changes in administration practices, and additions of measures to tap into a wider range of ability areas based on a validated model. The effectiveness and efficiency of such procedures, however, remain to be seen in terms of how they translate into actual assessment practice.

Conclusions

This chapter has attempted to provide the reader with a multicultural overview of intelligence and cognitive testing through the examination of reviews of numerous tests. The focus was primarily on test development given the importance of planning for appropriate examination of racial/ethnic differences and other relevant variables. Tests are being developed to address a variety of areas related to cognition and intelligence. However, given current test development practices (i.e., establishment of concurrent and predictive validity), many tests appear to assess the same constructs with little variation. Contemporary norming procedures address critical issues related to race/ethnicity, gender, community size, geographic region, and so forth. However, the complexity of the issues that can potentially affect test performance remains problematic. As suggested by Helms (1997), there is a great need to be more specific regarding what variables we control, how we measure these variables, and what assumptions and implications are present in our test development practices. Too often, assumptions are made regarding "equivalence" without the depth of understanding to address cultural issues. For example, as noted in this chapter, the use of expert panels and Rasch models is not adequate to provide the necessary information needed to identify item bias. There is a lack of standardization in terms of how expert panels are selected and used. Procedures for evaluation often are not specified, and the process appears to rely on subjective judgments. Item difficulty levels examined using Rasch models also are limited given that an individual's performance on a particular test item may be based on a number of issues, not merely ability. Often in our reliance on these procedures, we do not look beyond the scores that we obtain on a particular measure.

Discrepancies in scores between racial/ethnic groups and profile differences need to be examined more closely to understand the meanings and interpretive value of these findings within a cultural context. In addition, all of the tests cited in this chapter report racial/ethnic group differences without regard to potential within-group variability. There appear to be important differences within racial/ethnic groups that research virtually ignored in the empirical literature.

As discussed in previous chapters, SES, home environment, and cultural bias are important to take into account for possibly explaining these persistent differences. All of these complex variables must be considered in our understanding of the results of our administered ability tests. Also, modifications in the interpretation of test scores, administration of particular measures, and frameworks for test selection must be studied empirically to determine their effectiveness in measuring the intellectual performance of racial/ethnic minority group members.

10

Future Directions and Best-Case Practices:
Toward Nondiscriminatory Assessment

The preceding chapters of this text have highlighted important foundations, performance factors, and assessment issues in relation to intelligence testing with minority populations. From its earliest historical roots, intelligence as a definitive construct, as well as its measurement, has been challenged regarding its applicability in a multicultural context.

Over the years, a number of changes have occurred in the measurement field due, in part, to these concerns. For example, some researchers have recommended movement away from the use of the term "intelligence" because of the negative connotations historically associated with this construct (e.g., Reschly, 1981). Other writers have begun referring to "cognitive ability tests" (e.g., Helms, 1992) as a conceptual umbrella inclusive of intelligence tests. Measures theoretically based on various dynamic processes of abilities (e.g., simultaneous and successive processing) have been developed (e.g., Cognitive Assessment System [Naglieri & Das, 1997]; Kaufman Assessment Battery for Children [Kaufman & Kaufman, 1983b]), as opposed to those that focus on overall IQ scores. It is important to note, however, that even these new measures maintain formats similar to those of more "traditional" tests given that the major purpose of using these tests within an educational setting is to identify students with learning problems. Other scholars have begun a movement toward more biologically based measures such as "average evoked potentials" (e.g., Eysenck, 1998).

Despite these changes, cognitive ability tests (including those purporting to measure intelligence) have grown in use to become gatekeepers and tracking devices in many educational settings. Although Vance and Awadh (1998) reported that "assessment is in a state of flux and that in many educational and clinical settings psychological assessment has fallen into disrepute during the past 20 years" (p. 6), actual use of various testing practices has increased. The focus of the present book has been on intelligence tests, but the reader should be aware that group-based achievement tests (norm-referenced and criterion-referenced tests) are widely used in the United States. As part of the movement to require basic levels of skill competency, a number of school systems currently are moving toward greater accountability for the achievement of students and are using group-administered tests as a major indicator of student, school, and district success. For example, many states have implemented standardized examinations as a requirement for grade promotion and high school graduation. In addition, various aptitude tests continue to be used as major criteria for college admission.

It is clear that the issues surrounding the use of intelligence tests are complex and controversial, with many minority communities expressing discontent due to the discrepancies in scores found repeatedly between particular racial/ethnic groups. Although test developers note that current measures withstand empirical scrutiny with regard to psychometric definitions of test bias, other writers continue to point to issues of cultural loading and test fairness issues (see Chapter 5 of the present book). Intelligence tests measure what an individual has learned within a particular cultural context (e.g., U.S. culture; Kaufman, 1990). Given the increasing racial/ethnic diversity of the U.S. population, it appears that this issue might continue to adversely affect the outcome of such intelligence measures with community members whose experiences are not reflective of the dominant culture. The focus of this final chapter is on this major concern. In addressing this, we offer our thoughts on how nondiscriminatory assessment can be advanced through preprofessional and in-service training.

Research has focused a great deal on racial/ethnic *group* differences in measured intelligence. Indeed, the present book has focused on this area, which is reflective of the literature. Individual differences within groups, however, often exceed between-group racial/ethnic differences on intelligence tests. Therefore, our current knowledge base is somewhat limited in terms of clinical applicability. We do not test groups, nor are we particularly interested in these group differences if they are not clinically meaningful to the diverse individuals we are designated to serve. How often have clinicians bemoaned research findings that do not directly apply to the cases they currently see in practice? This is not to say that knowledge in this area is unimportant. Understanding group differences can, when used appropriately, provide a more culturally sensitive context in which to interpret scores of minority group members. For example, a finding that an American Indian student's Performance IQ exceeds his or her Verbal IQ on the Wechsler Intelligence Scale for Children (third edition; WISC-III) might be reflective of profile differences found within this population and not necessarily indicative of a "problem." Hence, cultural explanations of this occurrence can be integrated into the interpretation of the test scores.

Clinicians also have raised concerns due to the lack of availability of measures that can be used with individuals who do not speak English as their first language. The use of translators and interpreters has raised serious questions regarding issues of equivalence, not to mention problems in using inappropriate norms. Inclusion of non-native speakers of English in the development of various measures (e.g., Universal Nonverbal Intelligence Test; Bracken & McCallum, 1998) and instruments such as the Bilingual Verbal Ability Test (Muñoz-Sandoval, Cummins, Alvarado, & Ruef, 1998) reflect movement toward addressing these concerns. Clearly, more is needed in this area given the increasing growth of language-minority students in the U.S. population. In addition, the complexity of assessing language proficiency is noted by many authors (e.g., Sandoval & Duran, 1998):

> Psychologists' sensitive interpretation of the meaning of communicative assessments of non-native speakers of English requires attention to the background characteristics of examinees, the nature of the criterion, the institutional and social contexts of the assessment, and the value of combining formal assessment strategies with informal but informative alternatives. (p. 208)

As such, there is an increasing need for greater attention to language issues as they pertain to the use of various intelligence instruments. Despite these concerns, American-based psychometric technology (i.e., tests) has been exported to other countries, and it is not unusual to find translated and renormed versions of our most popular instruments—with little variation—being used in other countries. For example, numerous versions of the Wechsler scales currently are available (e.g., the Japanese WISC–Revised; Kodama, Shinagawa, & Motegi, 1978).

As noted in the preceding chapters of this book, our research has not always kept pace with new developments in the testing arena. Each year, it is not uncommon for new tests to enter the market with little to support their use except reliance on correlations with other "traditional" measures (known as concurrent validity evidence). By contrast, often few validity studies are available to support the use of particular instruments with minority populations. Attempts to modify traditional scoring procedures for minority populations based on our knowledge of variables affecting the assessment of intelligence (e.g., System of Multicultural Pluralistic Assessment [SOMPA]) have not become part of mainstream clinical practice. In fact, the effectiveness and efficiency of alternative procedures, such as the SOMPA, have been heavily criticized (see Chapter 8 of the present book).

Despite increasing numbers of cognitive measures available, the Wechsler tests continue to reign as the most frequently used scales. Although updated versions of the test incorporate more sophisticated and seemingly culture-oriented test development strategies (e.g., racial/ethnic oversampling, expert review panels), results of the tests continue to yield findings similar to those in earlier versions (i.e., mean discrepancies between racial/ethnic groups remain at similar levels).

In the educational domain, different assessment practices of a more qualitative nature directly tied to instructional objectives (Chittooran & Miller, 1998) have been suggested. Examples include informal assessment practices (e.g., behavioral observations, criterion-referenced tests, curriculum-based assessment), authentic assessment (e.g., performance-based tasks and exhibitions, portfolios, formative evaluation), and computer-assisted assessment (Maddux & Johnson, 1998). Although these assessment practices provide additional information regarding a child's academic performance, they have done little to replace the use of standardized tests in the overall evaluation of intellectual functioning. The focus on intelligence tests might be due, in part, to the collegiate training that testing personnel receive in their cognitive assessment courses that focus primarily on instrumentation.

Writers in this area have indicated that the focus needs to be changed from the tests to the training of evaluators (Kaufman, 1990; Sandoval, 1998a, 1998b). What are the best practices to be used by consumers of intelligence tests? Any attempts to quantify and standardize procedures and to make testing into a psychometric technician role underestimate the importance of the clinical aspects of assessment. A good clinician obtains more than just an IQ score (or some other indicator of global cognitive functioning) in the process of administering an intelligence test. Qualitative observations of how individuals process information, deal with frustration, apply problem-solving strategies, and so forth can be identified during the assessment. In addition, procedures for allowing the testing of limits also can facilitate data gathering about an individual's abilities.

There is a need for assessment practices to be located within a multicultural context where theoretical knowledge, performance factors, and state-of-the-art psychometric testing practices are integrated and merged. This is a major premise of this book. All too often, clinicians and educators have operated out of sole reliance on tests without consideration of the numerous factors influencing the outcome of these measures. As noted previously, controversies have arisen when we focus only on test scores without consideration of the multiple variables that affect obtained results.

Assessment Competencies

Although this book has focused primarily on the importance of quality in the applied practice of intellectual assessment, it is important to preface this discussion of assessment competencies to include foundational concepts related to measurement theory and test development practices. It appears that as training programs have invested more in applied clinical assessment courses, this might have been done at the expense of important training in psychometrics. Aiken et al. (1990) conducted a survey of all Ph.D. programs in the United States and Canada investigating the current status of statistics, measurement, and methodology training. Findings indicated that training in measurement has declined substantially. One fourth of the departments reported that

> most or all of their students are competent in methods of reliability and validity assessment; over one third indicated that few or none of the students are. Even smaller percentages of their students are judged to be competent in classical test theory or item analysis. (p. 725)

Newer measurement techniques and test development strategies often are not covered in psychology programs. The information obtained in the survey by Aiken et al. leads to concerns in terms of applied clinical competencies. If recently trained testing personnel are unaware of issues related to basic test theory, reliability, and validity, then they might be unable to effectively evaluate the strengths and limitations of various measures given different testing populations. Clearly, the current state of training programs needs to address these notable shortcomings in measurement training.

Figure 10.1 presents our conceptual schema of the areas related to best-case practices in the intellectual assessment of minority students. Graduate training and continuing education courses should incorporate information in each of these areas to promote nondiscriminatory assessment procedures.

Understanding of Historical Issues Pertaining to Minority Students and Intelligence Testing and of Theoretical Foundations of Intelligence

As we have discussed, any current understanding of intelligence testing and minority students is incomplete without understanding historical developments and issues. Many contemporary concerns (e.g., test bias, the misuse of test results, racial/ethnic group underrepresentation in standardization samples) have historical roots and, thus, need to be considered in that context. To understand the complexity of intelligence as a construct, it is important for evaluators to integrate the theoretical foundations of intelligence. Most texts in the area of intelligence devote entire chapters to the multiple perspectives of this area (e.g., Flanagan, Genshaft, & Harrison, 1997). It also is imperative that evaluators understand the relation between these theories and intelligence tests. Many factor theories and process theories have had an impact on the development of particular testing practices. Other measures remain somewhat atheoretical, as noted earlier in the present book.

In addition, being able to translate the multiple facets of intelligence (i.e., ability areas) and how they relate to an individual's school achievement and overall functioning is invaluable. A good examiner is able to relate and make sense of test findings in relation to the individual examinee's life and educational experience. Test scores are merely one part of the complex puzzle of understanding an individual's intelligence. It should be noted that a major criticism of intelligence tests is their apparent lack of applicability to the actual treatment of learning problems.

Understanding of Empirical Literature

Examiners must be aware of and understand historical and contemporary information from the empirical literature. Without this knowledge, examiners are practicing unethically because they have failed to be aware of a vital knowledge base with respect to this area of expertise. This includes understanding the performance factors noted in Part II of this book (i.e., socioeconomic status, home environment, and test bias). In addition, testing personnel need to be aware and knowledgeable of the nature-nurture controversy given that minority populations have been central to this debate for decades. This

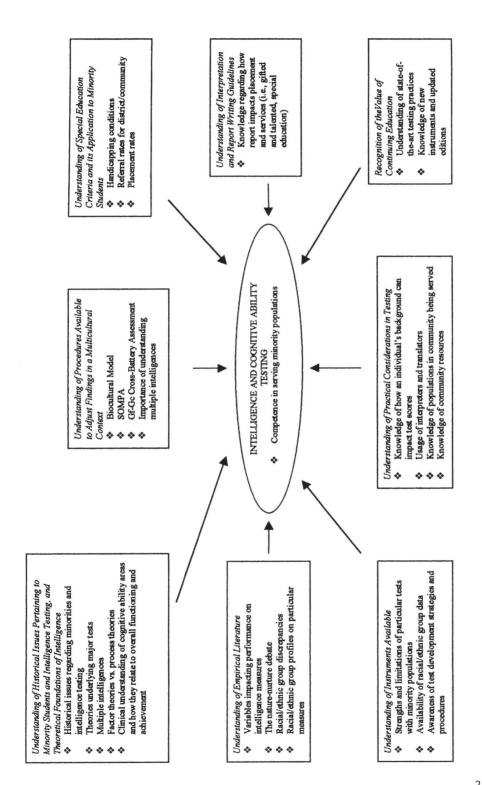

Figure 10.1. Training Model for Multicultural Competence in Intellectual Assessment

NOTE: SOMPA = System of Multicultural Pluralistic Assessment.

book has provided important updated overviews and the historical context of these areas.

Furthermore, it is imperative that examiners have knowledge of the racial/ethnic group differences literature to gain an understanding of what might be potentially "normative" within different populations (i.e., overall group discrepancies, ability profile differences). This is not to say that one should assume that all members of a particular group will reflect these same score differences. However, as noted repeatedly in this book, minority intellectual discrepancies and differences from the majority (i.e., dominant White group) in the past frequently were interpreted as "deficits."

Understanding of Instruments Available

Examiners should understand the strengths and limitations of the tools of our profession. This pertains especially to the availability of data regarding the use of a particular scale with racial/ethnic minority samples. Instruments should be used with caution in the absence of this information. In addition, gaining knowledge of the procedures involved in the test development process of the instruments used is imperative.

It also is important that examiners be aware of the most current testing practices. This includes the use of the most recent versions of particular scales. Over the years, we recall meeting clinicians who were continuing to use old versions of scales because they had not been trained to administer and interpret the newer editions or because they were more "comfortable" with the older versions. Thus, they were using outdated measures and comparing their clients to norming samples that were, in some cases, more than 20 years old.

Understanding of Procedures Available to Adjust Findings in a Multicultural Context

Given our understanding of the variables that can affect intelligence test performance, examiners should be aware of state-of-the-art assessment procedures to obtain the broadest range of intellectual indicators during a testing session. "For the effective testing of many groups in American society, test developers and test users must adapt their testing procedures to some extent" (Geisinger, 1998, p. 29). This includes, for example, the biocultural model of assessment and an understanding of multiple intelligences. In addition, texts such as the *Intelligence Test Desk Reference* (McGrew & Flanagan, 1998) can provide information to assist clinicians and examiners in thinking more broadly about the ability areas being assessed by particular instruments and the potential need to use supplementary measures. Modifications of testing procedures as recommended by procedures, such as the SOMPA, also require an understanding of how these scoring procedures affect the psychometric characteristics of a particular measure (e.g., reliability, validity; Geisinger, 1998).

Understanding of Practical Considerations in Testing

Of critical consideration is the integration of individual background information with testing outcomes. The performance factors discussed in Part II of this book highlight some of the areas of information that should be obtained prior to the testing. This information can influence the decision as to whether testing is needed as well as the appropriate selection of instruments.

Another consideration pertains to the use of translators and interpreters. Many evaluators have test items translated into the examinee's native language. It is important to consider how translation of items (either by individual evaluators or by test publishers) may affect the standardization of the testing procedures and whether the items still maintain psychological equivalence in the target language (for a discussion of a number of concerns regarding test translation, see Geisinger, 1994). Attention also needs to be given to the training that translators and interpreters receive so that the process and goals of the assessment are familiar to them. With proper training, they can provide invaluable information regarding observations of an individual's performance in his or her native language.

In a similar vein, given the numerous cultures potentially represented within any city or school district, it is imperative that evaluators become familiar with the cultures of the communities they are primarily serving. This might involve having cultural "informants" or "experts" in the community to provide consultation on cases. Knowledge of other resources found within a community also might prove beneficial in that evaluators can work collaboratively with other professionals to obtain knowledge regarding the best services for the individual being tested.

Understanding of Special Education Criteria and Their Application to Minority Students

Given the major role that tests play within our educational system, it is essential that evaluators have knowledge of special education criteria as well as referral and placement rates of minority students within their schools and districts. Disproportionate representation (i.e., over- or underrepresentation in the various special education categories) might be indicative of systemic problems, as noted in Chapter 7 of this book. Our school systems currently are mandated to keep records regarding racial/ethnic group representation in the various special education classifications. This information may provide valuable clues in assessing the quality, effectiveness, and appropriateness of both referrals from regular education and outcomes of special education programs.

Understanding of Interpretation and Report Writing Guidelines

The documents that represent the outcome of the testing procedures become a vital part of an individual's record and should be written with the utmost care and consideration. Evaluators need to be aware of the impact that their reports will have on the examinee's life (e.g., school placement, employment opportunities). As noted earlier, it is critical that background information, as well as information regarding other qualitative information that might affect the test scores (i.e., performance factors), be integrated throughout the report. Familiarity with alternative testing procedures (e.g., biocultural model, multiple intelligences) should be reflected in the reports. Finally, special attention should be paid by the evaluators regarding how their interpretations and report writing might influence whether or not a minority student is identified as a candidate for placement in a gifted and talented program. This is one particular situation in which multiple data sources should be considered (see Chapter 8 of this book).

Recognition of the Value of Continuing Education

As in all disciplines, the importance of remaining up-to-date with current assessment measures and strategies is imperative. Evaluators must maintain a commitment to their profession to continue to do justice to the important task for which they have been designated. Involvement in workshops and ongoing research projects, attendance at conferences, and so forth will go a long way toward ensuring state-of-the-art assessment practices.

Obtaining knowledge in each of the areas described will promote greater competence in the assessment of minority populations. Training programs in psychology and measurement should facilitate professional development in these areas as new practitioners set the standards for future generations.

In addition, Sandoval (1998a) noted the importance of examiners engaging in their own self-assessment procedures: "Clinicians' cognitive limitations and biases are certainly going to affect them in working with clients from diverse populations" (p. 45). Thus, Sandoval recommended that evaluators "identify preconceptions . . . , develop complex schemas or conceptions of client groups . . . , actively search for disconfirmatory evidence . . . , resist a rush to judgment . . . , seek supervision . . . , [and] distrust memory" (p. 45). To this, Frisby (1998) added the importance of members of our profession being "socially conscious advocates" for the communities we serve. Regarding the assessment of language-minority students, Damico and Hamayan (1991) asserted that evaluators need to become "advocacy oriented" and "change agents" so as to implement nondiscriminatory assessment in the "real world."

Geisinger and Carlson (1998) noted the responsibility of academic programs to ensure the quality of training that their students receive. Their recommendations include providing courses on individuality, field placements, consultation, and self-study with external program reviews to ensure quality of training.

A Final Note

Unlike the preceding chapters of this book, these final thoughts represent our thinking regarding the future of intelligence testing and the growing need to address concerns in the assessment of minority students. It is our hope that in the new millennium, we will be able to address the issue of nondiscriminatory assessment appropriately and move on to new challenges. Many have noted that the profession has repeatedly focused on the same issues regarding test bias and test fairness, even as newer measures are developed (Armour-Thomas, 1999; Vraniak, 1999). Perhaps as our testing strategies and procedures become more sophisticated, we will be better able to address the complexities of the assessment process with minority populations. As Vraniak (1999) reported, we need to move beyond issues of test bias and fairness and instead focus on issues of individual beneficence and greater effectiveness in addressing the learning difficulties of those individuals we are attempting to serve. For example, would minority communities be as concerned if our special education programs were effective in alleviating the learning difficulties of students and if movement within the system (e.g., mainstreaming, declassification) were more commonplace and less restrictive settings were used? Would we be as concerned about disproportionate rates of referrals and placements in special education categories if all children eventually were attaining high achievement in school? The answer is that we probably would not be focused on testing to the degree that we currently find our level of discourse.

The complexity of issues that face evaluators using current intelligence tests is overwhelming. Sandoval (1998b) noted that the situation will become increasingly more challenging in the future due to progress in a number of areas. For example, he cited factors such as cultural blending (i.e., biracial children, intermarriage), progress in medical technology, increased globalization, increased testing technology and use of computers, increased sophistication of psychometric theory, increased neuroscience technology, race norming, and regulation of testing practices as areas of major change.

At the 1999 convention of the National Association of School Psychologists, a panel of intelligence test authors discussed how their measures fit with the agenda for school psychology training and practice (Harrison, 1999). The presenters were in agreement that intelligence tests often have been used inappropriately in school settings. They also noted, however, that the tests themselves are not the primary issue. Their comments focused, instead, on the need for professionals to address issues related to special education requirements and school policies. Their remarks advocated for the continued use of intelligence tests as a means to obtain important information about students when the tests are used appropriately and in conjunction with other tools. Discussants cited the possibility that intelligence testing will move toward a more neuropsychological foundation in the future and that computer-assisted administration procedures would increasingly be implemented. In lieu of these changes, the need for well-trained and knowledgeable evaluators will continue in the new millennium.

Notes

Introduction

1. Our population of focus is kindergarten through Grade 12 (K-12) students. At times, however, the literature we examine has various age levels as its focal groups (e.g., infants, toddlers, preschoolers, adults). Regarding the term "racial/ethnic minority," as well as the referent "White," we use the descriptions provided in the Office for Civil Rights biennial surveys of elementary and secondary schools. These racial/ethnic descriptions, as listed in Chinn and Hughes (1987), are as follows:

> *American Indian.* American Indian or Alaskan Native—A person having origins in any of the original peoples of North America and who maintains cultural identification through tribal affiliation or community recognition.
>
> *Asian.* Asian or Pacific Islander—A person having origins in any of the original peoples of the Far East, South Asia, the Pacific Islands, or the Indian subcontinent. This area includes, for example, China, India, Japan, Korea, the Philippine Islands, and Samoa.
>
> *Hispanic.* A person of Mexican, Puerto Rican, Cuban, Central or South American, or other Hispanic culture or origin regardless of race.
>
> *Black.* Black not of Hispanic origin—A person having origins in any of the black racial groups of Africa.
>
> *White.* White not of Hispanic origin—A person having origins in any of the original peoples of Europe, North Africa, or the Middle East. (p. 42)

There is one other point about the concept of race that needs to be mentioned. In the study of racial (or ethnic) differences in intelligence, how one conceptualizes race is a thorny issue. Loehlin, Lindzey, and Spuhler (1975) commented, "In the biological sciences, races are customarily considered as subspecies, defined by gene frequencies" (p. 13). By contrast, in the social sciences (especially cultural anthropology), race typically is conceptualized as a social concept (i.e., a social construction) in which people assign meaning and value to "racial characteristics" (see, e.g., Omi & Winant, 1994). Still, there is another perspective. Population geneticists Cavalli-Sforza, Menozzi, and Piazza (1994) noted, "The classification into races has proved to be a futile exercise. . . . From a scientific point of view, the concept of race has failed to obtain any consensus" (p. 19). So, what does one make of all this? To us, the case that Cavalli-Sforza et al. forge about the meaninglessness of race as a biological concept makes much sense. Yet, given the race-conscious nature and racial stratification of U.S. society, the social concept of race also makes sense. Thus, in the present book, we prefer to view race as a social construction. In any event, the reader should be aware of the growing criticism of research on racial differences in intelligence in which the social concept of race is implied, yet inferences sometimes are made in a biological sense when conclusions are drawn about interpreting such group differences in intelligence (for more on this and related issues, see Helms, 1992, 1997).

2. We place quotation marks around the term "Negro" here because that was the nomenclature (along with "colored") typically used during this period. When reviewing the historical research on racial/ethnic differences in intelligence, we often report these racial/ethnic designations to allow the reader to get a better sense of race relations during those times.

3. Throughout this book, we frequently modify the term "intelligence" with the adjective "measured." We do so to underscore that obtained scores on intelligence tests are *samples of behavior.* That is, intelligence tests measure developed abilities, not innate ones.

4. There is strong evidence, however, that the IQ scores of American racial/ethnic groups are rising. Although the 15-point IQ gap between African American and White groups still remains, there are findings that the mean intelligence test performance of African Americans in 1990 was approximately equivalent to the mean IQ of Whites, as a group, in 1940 (Neisser, 1998a). On this, Neisser (1998a) noted that although the 1 standard deviation mean intellectual difference between African Americans and Whites continues, "it is now clear that a gap of this size can easily result from environmental influences, specifically, from the differences between the general American environments of 1940 and 1990" (p. 18). For further discussion on the "rising curve," see Neisser, 1998b).

5. This is not to say that these are the only factors involved that might help to account for the mean difference in measured intelligence (and, in some cases, achievement test scores) across racial/ethnic groups. A host of other variables have been discussed in the literature. These factors include, for example, intrauterine difficulties (as evidenced by premature birth [Storfer, 1990]), nutrition (Loehlin et al., 1975; Samuda, 1998; Sigman & Whaley, 1998), lead exposure (Brody, 1992), motivation (Cook & Ludwig, 1998; Fordham & Ogbu, 1986; Samuda, 1998), and test anxiety (Samuda, 1998; Steele & Aronson, 1998). Another factor that can be advanced to help explain the average difference in measured intelligence between minority group students (e.g., African Americans, Latinos, American Indians), as a whole, and their White peers is inequalities in schooling opportunities. Although we do not offer a separate chapter on schooling in Part II of this book, we do integrate the schooling inequalities factor in several chapters (e.g., Chapters 1 and 3). Scholars of minority education, such as for Mexican Americans (see, e.g., San Miguel & Valencia, 1998; Valencia, 1991), assert that Mexican American students have experienced, historically and contemporarily, a number of schooling conditions (e.g., segregation, tracking, school financing inequities) that have led to diminished achievement (e.g., low achievement test scores, high dropout rates) among these students. Given the moderate correlation (typically .50) between intelligence and scholastic achievement test scores (Jensen, 1980), it is not surprising that the mean intellectual performance of Mexican American students as a group is lower than the mean score of their White peer group. As such, there is strong evidence that inequalities in educational opportunities play some role in explaining the intellectual mean differences between Mexican American and White student groups. The same argument holds for African American-White and American Indian-White comparisons in mean intelligence test scores.

6. This paragraph is excerpted, with very slight modifications, from Suzuki and Valencia (1997, p. 1103).

7. To some extent, this assertion about stereotypical thinking regarding racial/ethnic group differences in intelligence is supported by empirical research. Results from a 1990 survey conducted by the National Opinion Research Center at the University of Chicago found that a majority of White respondents believed that Blacks and Hispanics were less intelligent than Whites (cited in Duke, 1991).

8. See Jensen (1980) for an empirical study clearly demonstrating that the sources of variance in intelligence largely reside within families.

Chapter 1

1. The launching of the psychological testing movement also is attributed, in part, to American psychologist James McKeen Cattell (Anastasi, 1988). Cattell studied under Wilhelm Wundt in Leipzig, Germany, during the late 1880s. Like Galton, Cattell believed that intellectual functions could be measured by sensorimotor techniques. "On his return to America, Cattell was active both in the establishment of laboratories for experimental psychology and in the spread of the testing movement" (p. 9). Cattell, in an 1890 article titled "Mental Tests and Measurements," used the term "mental tests" for the first time in the psychological literature (Cohen et al., 1992).

2. Although Terman often is given credit for developing the first widely used American version of the Binet-Simon scale, Goddard (1928) and Kuhlmann (1912) were the first to produce highly useful versions (Freeman, 1926). It is believed that Terman's success in having his Binet-Simon scale version more widely used and known was due to the assertiveness of the test's publisher (Houghton Mifflin) and Terman's reputation as a scholar and author of many books, articles, and tests (French & Hale, 1990).

3. William Stern was the developer of the construct of "mental quotient," that is, a child's mental age divided by his or her chronological age (Stern, 1914). For example, a 12-year-old child whose mental age was estimated to be 14 years would have a mental quotient of 1.17. Terman, in his development of the Stanford-Binet intelligence test, extended the construct of mental quotient by multiplying it by 100. In the preceding example, the IQ would be 117.

4. It should be noted, however, that Terman did test foreign-born children (and, we suspect, some minority children), but he did not incorporate their test results into the norming process. On this, Terman (1916) commented, "To avoid accidental selection, *all* the children within two months of a birthday were tested in whatever grade enrolled. Tests of foreign-born children, however, were eliminated in the treatment of results" (p. 53).

5. The most current version, the fourth edition of the Stanford-Binet (Thorndike, Hagen, & Sattler, 1986a, 1986b), contains a standardization sample in which race/ethnicity is one of the stratified variables (based on the 1980 U.S. census data).

6. To be sure, the development of the 1916 Stanford-Binet intelligence test led to a long domination in the field of individually administered intelligence tests. In 1949, however, David Wechsler published the Wechsler Intelligence Scale for Children (Wechsler, 1949). In this edition, the standardization sample also was exclusively White children. It was not until the 1974 revision (Wechsler, 1974) that minority children were represented in the norm group. Likewise, the most current edition, the Wechsler Intelligence Scale for Children: Third Edition (Wechsler, 1991), contains minority representation in the standardization sample.

7. Our assertion is not that *all* psychologists were ethnocentric when it came to test development and administration. Terman (1917), for example, appeared to be aware of the problems associated with social class, cultural, and language variables as possible confounds in intellectual assessment of children. Nevertheless, such awareness did not stop Terman from drawing racist conclusions about observed racial/ethnic differences in measured intelligence (see Terman [1916] quote in next section).

8. Terman (1916) defined borderline cases as "those which fall near the boundary between that grade of mental deficiency [70 to 80 IQ] which will be generally recognized as such and the

higher group usually classified as normal but dull. They are doubtful cases, the ones we are always trying (rarely with success) to restore to normality" (p. 87).

9. Cattell coined the term "individual differences" in a 1916 address commemorating the 25th anniversary of the American Psychological Association (Cattell, 1947, cited in Marks, 1981).

10. The prime mover of the development of the Army mental tests was Robert M. Yerkes, a Harvard University professor and president of the American Psychological Association. Yerkes's test development team consisted of the top hereditarians of the time including, for example, Goddard and Terman.

11. In addition to Terman and Yerkes, there were three other codevelopers of the NIT: M. E. Haggerty, E. L. Thorndike, and G. M. Whipple. It is interesting to note that the NIT team did not always see eye to eye on the test's development. For example, regarding the development of standard norms, Thorndike wanted "foreign" populations and rural districts represented in the standardization sample. Haggerty desired separate age norms for "colored" children. On the latter, "Yerkes replied, 'No, we don't want to use them. The political situation here will prevent us from going into colored schools' " (quoted in Chapman, 1988, p. 81). As Chapman (1988) noted, due to the pressure of time, the schools finally selected for the norming of the NIT were in the cities of Cincinnati, Pittsburgh, Kansas City, and New York. For a test that used "national" in its name, the final locations chosen for norming certainly were not representative of American culture. On this issue, Chapman's point is well taken: "The point to be drawn from this discussion is that the examination board, like Terman in his work on the Stanford-Binet, made rather broad assumptions about the nature of American culture and what constituted the 'normal' American experience. These standards were then used to judge the intelligence of many Americans whose cultural background was not adequately reflected in either the tests or the norms. These decisions would have important consequences, too, for in little more than a year, the NIT would be in wide use around the country" (p. 81).

12. The new corpus of group intelligence tests included the familiar "general" instrument (i.e., similar to the NIT) as well as nonverbal and performance tests. Some of these tests (with their respective publication dates) are the Terman Group Tests of Mental Ability (1920), Haggerty Delta I Test (1920), Detroit First Grade Intelligence Test (1921), Otis SA Test of Mental Ability (1922), Otis Classification Test (1923), Detroit Primary Intelligence Test (1924), Cole-Vincent Group Test (1924), and Goodenough Draw-a-Man Test (1926) (see Valencia, 1985a).

Hildreth (1933), in *A Bibliography of Mental Tests and Rating Scales* (cited in Chapman, 1988), reported that from 1920 to 1929, 70 "mental tests" were developed. From 1900 to 1932, 133 were available. Unfortunately, Hildreth used the term "mental test" in an omnibus fashion, including achievement and aptitude tests and tests measuring various types of ability. Thus, her analysis does not inform us of the number of *intelligence tests*. In any event, the mental tests she did identify were used, according to Hildreth, for large-scale classification of schoolchildren's abilities.

13. The second major area of race psychology interest stemmed from the massive database of the mental test scores of the Army Alpha and Beta testing of World War I recruits (Chapman, 1988; Gould, 1981; Valencia, 1997b). In his book, *A Study of American Intelligence*, Princeton University professor Carl C. Brigham provided a reanalysis of the Army mental test data (Brigham, 1923). Hard-hitting themes of his treatise were (a) the decline of American intelligence (imputed to racial admixture), (b) the racial (intellectual) inferiority of Negroes, (c) the racial superiority of the Nordic stock over men from Southern and Eastern Europe, and (d) the need to restrict immigration from Southern and Eastern Europe. Yerkes wrote a laudable foreword to Brigham's book, commenting that Brigham's treatment was fact-filled and not opinionated. Some scholars have noted, however, that Brigham's conclusions did not differ much from those of Madison Grant, the leading American theorist of racism of the period, who wrote *The Passing of the Great Race* (Grant, 1916).

14. Garth, as well as a number of other race psychologists of this period, used "race" in a biological sense. We know now that race is best viewed as a social concept. Also, Garth and others confused race with national origin and ethnicity.

15. The term "Mexican" typically was used during the first four decades of the 20th century to refer to Mexican-origin children living in the United States. A substantial proportion of these children were citizens of the United States.

16. Not all of these nearly 37,000 participants served in investigations of intellectual performance. A small percentage of studies involved psychological measurements, for example, of color preference, academic achievement, and personality self-ratings.

17. Garretson (1928) reported that the monthly mean incomes for the parents of the White and Mexican children were $250 and $110, respectively.

18. Garretson (1928) commented, "With few exceptions, Spanish is the language spoken in the homes and on the streets of the Mexican part of town. It is unusual for a Mexican child to be able to speak English when he enters the kindergarten or first grade" (pp. 32-33). Even after admitting this and presenting data to show that the IQ differences in Grade 1, favoring the White sample, on the verbal Pintner-Cunningham test and the NIT were substantial, and after commenting that the Mexican children's language difficulty operated to their disadvantage, Garretson still suggested a genetic explanation for group differences in intellectual performance.

19. Furthermore, on several other grade levels, the IQ differences, although favoring Whites, were negligible and, it appears, statistically nonsignificant. For example, on the Pantomime test in Grade 5, the mean IQs were 92.0 and 91.5 for the White and Mexican American groups, respectively.

20. In Goodenough's (1926b) study, there were 79 American Indian participants (Hoopla Valley). Their mean IQ was 85.6.

21. The other two management-of-instruction functions that Resnick (1979) discusses are (a) "monitoring" (i.e., when tests are administered *during* the course of the instructional process to provide feedback on students' strengths and weaknesses and to make curricular adjustments to improve teaching/learning) and (b) "grading" (i.e., when tests are given at the *end* of the instructional process to evaluate students' performance).

22. Today, curriculum differentiation is commonly referred to as "ability grouping" at the elementary school level and as "tracking" at the middle and high school levels (Valencia, 1997c). Another framework to describe curriculum differentiation was presented by Slavin (1997). "Tracking" refers to curricular assignments in high school (e.g., college preparatory, general). "Between-class ability grouping" is in reference to student groupings by ability in separate classes in junior high and middle schools. "Within-class ability grouping" refers to student groupings (e.g., mathematics) of similar levels in elementary school classes.

23. Chapman (1988) commented that Terman's use of "homogeneous class groups" was one of the first uses of the notion, which subsequently became a common term in education.

24. It is interesting to note that Terman's five percentage breakdowns coincide with how measured intelligence on later intelligence tests was distributed (by design). The breakdowns are strikingly similar to the areas (i.e., cases) under the normal curve. For example, about 2.5% of the area is contained in 2 standard deviations above the mean (IQ of 130 or Terman's "gifted" group), about 15.0% of the area is contained between the 1st and 2nd standard deviations above the mean (IQs between 115 and 130 or Terman's "bright" group), and so on. The point here is that Terman's a priori groupings were manufactured given his knowledge of the normal distribution. Such groupings are reflections of probability theory through a human-made construction, which subsequently became a social construction of reality.

25. It appears that Terman had considerable influence on some of his students' thinking about race and intelligence. For example, Kimball Young, a former student of Terman's, wrote the following in his autobiography: "I was pretty well fed up with Terman's doses of intelligence being inherited as a biological [racial] trait. But he made a tremendous impression on the educational world. So I played it cool and didn't say much about this in my dissertation, though I wanted to. I had to get that union card, as we all know" (quoted in Lindstrom, Hardert, & Johnson, 1995, p. 16). This admission by Young might explain why he made the racist conclusions in his dissertation (published as Young, 1922) about the intellectual performance of Mexican-origin children. Later

publications of Young demonstrated that environmental factors were more important than genetics in influencing intelligence (Lindstrom et al., 1995). In any event, Young went on to have an esteemed career in sociology. He taught at a number of universities (e.g., University of Oregon, Clark University, University of Wisconsin, Northwestern University). In 1945, Young was elected president of the American Sociological Association.

26. In the preface to the book, Terman praised the book's applied implications for actual classroom practice and school organization. He also lauded the author: "Dr. Dickson's experience in the educational use of test results has probably been more extensive than that of any other living educator" (Dickson, 1923, p. xiii). In *Mental Tests and the Classroom Teacher*, which was written in nontechnical prose and geared for teachers, principals, and administrators, Dickson's views on the nature of intelligence, individual differences, equality of opportunity, educability of children, curriculum differentiation, and social engineering are clone-like in Termanian thought (for a discussion, see Valencia, 1997b).

27. González (1974b) drew this median IQ data from McAnulty (1929), whose study in turn drew on a compilation of IQ test data surveys in Los Angeles from 1926, 1927, and 1928. The observed median IQ of 91.2 of Mexican American children in Los Angeles during the 1920s was about two thirds of 1 standard deviation below the typical White median (and mean) of 100. This difference of two thirds of 1 standard deviation frequently was found in other race psychology studies of the time. Furthermore, Valencia (1985a), in a comprehensive review of research on intelligence testing and Mexican American school-age children, estimated an aggregated mean IQ of 87.3. His analysis was based on 10,739 Mexican American children in 78 studies spanning six decades. Valencia noted, however, that this aggregated mean IQ of 87.3 is entirely misleading and greatly confounded by the failure of researchers to control critical variables.

28. For further discussion of González's research on the role of intelligence testing in Los Angeles during the 1920s, see Valencia (1997b).

29. The failure of race psychologists to assess (prior to intelligence testing) the language status of children and youths who were most likely to be limited English proficient was a frequent occurrence during the hereditarian period. For example, in Valencia's analysis of intelligence testing research of Mexican American children during the 1920s (Valencia, 1985a), he found that of the eight identified studies, none of the investigators attempted to assess the children's language for dominance and fluency. In addition, in the same review of the literature, Valencia found that in the 124 intelligence testing instances identified in 106 studies (spanning six decades), the language status of the Mexican American children *was not even reported* in 63 (50.8%) instances. Furthermore, in those studies that did report language status of the participants, about two thirds assessed language (predominantly through indirect means and not through the more precise direct procedures), but one third of the studies still *failed* to assess language. Thus, one can conclude that taking into account all pertinent studies of Mexican American intellectual performance from 1922 to 1984, the vast majority of investigators failed to assess language status.

30. Klineberg's finding that the various groups studied scored higher on performance tests than on linguistic tests certainly was the case. Support for this is seen in Valencia (1985a), where he employed the same procedure as Klineberg (1935b) except that Valencia used the more appropriate weighted means analysis. In 22 investigations (spanning decades) involving 2,380 Mexican American children, their mean IQs on verbal and performance tests were 85.6 and 94.7, respectively, a difference of 9.1 points (very close to Klineberg's observed difference of 10 points in a much smaller number of studies involving Mexican American children).

31. In his review of pertinent research by African American scholars who criticized early mental testing investigations, Valencia (1997b) identified a study by an African American woman, Martha MacLear (MacLear, 1922).

32. For other discussions of these African American scholars' contributions, see Guthrie (1976).

33. As Chapman (1988) noted, group achievement tests were about as widely used as their sister tests, group intelligence instruments. Achievement tests were used in an accountability role to

evaluate the performance of local schools as well as for homogeneous grouping. Achievement tests were produced in far greater numbers than were intelligence tests. Hildreth (1933) reported that between 1920 and 1929, 741 achievement tests were developed (cited in Chapman, 1988). From 1900 to 1932, 1,298 such tests were produced.

34. It should be noted that the administrative practice of ability grouping in public schools did not follow one uniform organizational plan. Chapman (1988), commenting on a 1929 survey by Otto (1931) of 395 superintendents in cities ranging in size from 2,500 to 25,000, wrote that "he [Otto] discovered 20 different plans to classify students within classrooms and 122 different promotion schemes" (p. 159).

35. Regarding intelligence testing research in which minority students served as participants, a similar trend noted by Haney (1981) has been reported. Valencia (1985a), in a literature review of Mexican Americans and intelligence testing, identified 18 journal articles published between 1920 and 1939. Only 1 such article was identified for the period from 1940 to 1949.

36. This paragraph is excerpted, with minor modifications, from Valencia (1999, pp. 125-126).

37. There are some critics, however, who argue that these cases (especially *Diana* and *Larry P.*) missed the point regarding the major reason for overrepresentation of African Americans and Mexican Americans in EMR classes. For example, MacMillan and Meyers (1980) contended that "the Court was led into an erroneous conclusion that IQ was the principal determiner of EMR status. . . . The single critical and unexceptional variable that leads to EMR placement is academic failure" (p. 146).

Chapter 2

1. For further discussion of these issues of reification of intelligence tests and hereditarianism, see Chapter 1 of the present book.

2. The 1921 symposium (Terman et al., 1921) was attended by 17 top psychologists who were invited to share their views on how they conceptualize intelligence. This symposium sparked controversy (e.g., over the constancy of IQ) and set the climate for further controversy during the 1920s (see, e.g., Chapman, 1988; Valencia, 1997b).

3. This is a major issue that we also discuss in Chapter 8 of this book.

Chapter 3

1. It must be kept in mind, however, that in terms of *absolute* numbers, there are more Whites (compared to people of color) who live in poverty. Based on 1990 census data, about 65% of all children who lived in poverty (defined as an urban family of four with an income of less than $13,359) were White (Macionis, 1994).

2. It should be noted that reporting SES data for only the Latino aggregate can be misleading because subgroups vary. For example, based on 1990 census data (cited in Chapa & Valencia, 1993), the following Latino child poverty rates were reported (in descending order): Puerto Rican (48.4%), Mexican origin (37.1%), other Latino (28.4%), Central and South American (26.1%), and Cuban (23.8%). The Latino aggregate was 36.2%.

3. Regarding the opposite end of the continuum, only 2% of African American families and 2% of Latino families had incomes of more than $75,000. By contrast, 11% of White families had incomes of more than $75,000. Interestingly, Asian American families were close behind at 10% (Chapa & Valencia, 1993).

4. The reality that minority groups (particularly African Americans, Mexican Americans, Puerto Ricans, and American Indians) face economic disadvantagement and are positively skewed on SES measures is long-standing. Numerous studies in the past have confirmed this (e.g., Garth, 1928; Grebler, Moore, & Guzman, 1970; Klineberg, 1935b; Pettigrew, 1964b; Reynolds, 1933).

5. In Chapter 4 of the present book, we argue that home atmosphere variables as reported in White (1982) are conceptually different from conventional SES variables and, therefore, are poor choices of SES indicators in research studies.

6. Unfortunately, Binet did not have an opportunity to pursue, with vigor, the question of potential class bias in the Binet-Simon scale. He died at an early age in 1911 (Jensen, 1980).

7. Furthermore, these 12 studies frequently were cited in the reviews of the time. For example, in Neff's (1938) literature review, he examined "a list of the more important studies" (p. 729). Of our own 12 studies selected for analysis, 9 are on Neff's list.

8. In some studies (e.g., Bridges & Coler, 1917; Chapman & Wiggins, 1925; Stroud, 1928), the authors mentioned that their participants were White; in other studies, they did not. In any event, we are confident that all participants were White. When minority (e.g., African American) children did serve in "race psychology" studies during this historical era, it was explicitly communicated in the titles of the journal articles or in the texts of the studies.

9. Actually, Goodenough (1928) administered the Kuhlman-Binet scale twice to the participants. The data that we present are from the second testing. As Goodenough commented, "The second examination is the more valid of the two measurements" (p. 288).

10. The favored school is described as being located in a very good residential area near the university, having a modern building, and having well-qualified teachers. By sharp contrast, the unfavored school is located near the railroad in a poor factory district of Columbus, is old and without up-to-date equipment, and has young teachers who have not taught for very long.

11. Regarding children who scored *above* the norm (third quartile), 44.2% of the students who attended the favored school did so; only 8.1% of the students from the unfavored school scored above the norm.

12. These eight studies, in alphabetical order, are Bridges and Coler (1917), Chapman and Wiggins (1925), Chauncey (1929), Dexter (1923), Goodenough (1928), Haggerty and Nash (1924), Pressey and Ralston (1919), and Stroud (1928). Of these eight studies, only two (Goodenough, 1928; Pressey & Ralston, 1919) interpreted their results as genetic as we have categorized; the other six investigations had environmental, equivocal, or "none" interpretations.

13. Although Pintner (1931) proffered a hereditarian perspective, he did discuss the subject of overlap: "We must remember, of course, that the occupational status of an individual is by no means a sure guide to his mentality. It is only in a general sense that occupational status correlates with mental ability. The distribution of children in all occupational groups runs from very low to very high intelligence. Since it is the duty of education to make the most of all the mental ability of all the pupils, educational classification according to social status is not justified. The brighter children in the lower social groups should be given just as much opportunity as they can profit by. The road through high school and college should be open to all who have the intelligence and interest to travel along it" (pp. 518-519). Furthermore, Pintner did appear to show some understanding of the problems that non-English-speaking persons have when being tested on English-language intelligence tests. Earlier in his career, Pintner (1919) commented, "The best known group tests at the present time depend largely, if not entirely, upon the knowledge and use of language. They presuppose the ability to read and write the English language. . . . It is very desirable to have tests which do not involve any language" (p. 199). To remedy this, he developed the Pintner Non-Language Group Test. Subsequently, research using his test demonstrated its efficacy. For example, Garth, Elson, and Morton (1936) administered the Otis Classification Test (a verbal intelligence test) and the Pintner Non-Language Group Test to 455 Mexican (i.e., Mexican American) students in Grades 4 to 8. Garth et al. concluded, "Age for age and grade for grade, the Mexican children are inferior to American whites in verbal test results. But in the non-language test results, the Mexicans are practically equal in performance to the American whites" (p. 58). Nonetheless, we find it curious that Pintner, on the one hand, would show sensitivity to the mental testing of non-English-speaking individuals, but on the other, would assert that "mental ability is inherited" (Pintner, 1931, p. 465) and observe, "The general picture [mental testing results of Mexicans and Mexican Americans] is one of low mentality. And this seems to be true for verbal, non-verbal, non-

language, and performance tests. Undoubtedly, the recent Mexican immigration into this country is bringing in individuals of poor mental caliber" (p. 456).

14. For reviews of within- and between-racial/ethnic group studies of the relation between SES and intellectual performance among African American students, see Dreger and Miller (1960, 1968), Klineberg (1935b), Osborne and McGurk (1982b), Pettigrew (1964b), and Shuey (1958, 1966). Shuey (1966), who reviewed 42 studies, is the most comprehensive.

15. It is not clear why Beckham (1935) did not use the more recent Taussig (1920) classification.

16. The respective mean IQs for the four groups, in descending order, were New York (104.7), Washington (97.3), Baltimore (95.7), and delinquents (86.0). The higher mean IQ of the New York group might have been due to "educational selection" (Beckham, 1935, p. 89). "In Washington and Baltimore, the schools for white and colored children are separate" (p. 71). This suggests that the New York participants attended schools that were racially/ethnically mixed, probably indicating better schooling.

17. The mean IQ differences that Beckham (1935) described in this quote are slightly in error given that he used the unweighted means (weighted means analysis is called for here). Nonetheless, the patterns still hold.

18. Shuey (1966) criticized Beckham's (1935) study, arguing that the Washington sample was not randomly selected. Shuey (1966) asserted that all of the Washington participants were brought to Howard University for testing by Howard students as part of a psychology course requirement (actually, the Baltimore participants also were brought to Howard for testing by psychology students or parents or were referred by teachers). Thus, Shuey apparently contended that the grand mean IQ (95.2) was likely spurious (inflated) because "the subjects were heavily biased in favor of siblings, neighbors, and friends of the *Howard* students" (p. 35). Shuey failed to mention, however, that Beckham had 100 delinquents (mean IQ = 86.0) in his total sample. Valencia (1997b) deleted the mean of the delinquent group, recalculated the grand mean for the three metropolitan groups, and observed a new grand mean of 97.2 (2.0 points higher than Beckham's grand mean of 95.2 for the four groups). Valencia contended that the grand mean of 95.2 was spurious, that is, pulled down by the delinquents' mean, a mean of an unrepresentative African American adolescent group. In short, Shuey's (1966) criticism and Valencia's (1997b) recalculation could suggest that there is a canceling out effect and that, therefore, Beckham's initial grand mean is valid.

19. For a more precise comparison, we calculated the mean IQ for the adolescent Negro sample in Kennedy et al. (1963) and compared it to Beckham's (1935) mean IQ (95.2). The mean Stanford-Binet IQ for the Kennedy et al. (1963) adolescent sample (ages 12 to 16 years [same age interval reported in Beckham, 1935], $n = 150$) was 72.6, a difference (22.6 points) even larger than the mean difference of 14.5 IQ points between Beckham's grand mean and the Kennedy et al. grand mean.

20. The discussion of Arlitt's (1921) study is excerpted, with minor modifications, from Valencia (1997b, pp. 84-85, 103).

21. Arlitt, a psychology professor at Bryn Mawr in Pennsylvania, collected the data with the assistance of four graduate students. Thus, it is very likely that the location of the study was in the Bryn Mawr area.

22. Arlitt (1921) did not mention which edition of the Taussig classification she used. We believe, however, that it was Taussig (1920).

23. It appears that Arlitt (1921) relabeled the occupational groupings because of the home environments of the children. "The home conditions followed closely the divisions by occupation" (p. 180).

24. According to Arlitt (1921), "All of the Italians spoke English without difficulty" (p. 180).

25. Degler (1991), in his own analysis of Arlitt's (1921) study, asserted that she captured a central issue in the field of race psychology, but the implications of controlling for SES in cross-racial/ethnic investigations were not recognized quickly by researchers: "An indication of how slowly social scientists recognized the implications of findings such as those published by Arlitt is provided by some research Arlitt [1921] herself reported the following year. In this study, Arlitt found that black children of ages five and six [years] scored above white children of the same age and

social class but that, as the students grew older, the whites surpassed the blacks. That shift Arlitt attributed "to a genuine race difference." On the basis of her article the previous year, one would have thought she would have asked how it came about that black children fell below whites as they grew older. Instead, she proceeded to explain why the superior scores of the black children at an earlier age may have been influenced by factors other than intelligence. In short, she fell back to the traditionalist position of assuming that race was controlling even when scores clearly called it into question" (p. 173).

26. Bird et al. (1952) did not mention the name of the intelligence test. We do suspect, however, that it was a verbally loaded group-administered test (see Shuey's [1966, pp. 182-183] listing; furthermore, the test data were provided to the researchers by school administrators, who typically administered verbal group IQ tests).

27. In their study, Bird et al. (1952) also compared the intellectual performance of White and Negro children (3rd- to 5th-graders) from another district, one of lower-SES background. Thus, between-SES group differences among Negro children are of interest here. As expected, the Negro children of higher SES background had a higher mean IQ (103.6) than did their lower-SES peers (92.3).

28. For example, in the Boston sample (high-SES group), the White and Negro mothers had completed an average of about 12.0 years of schooling. By contrast, in the Baltimore and Philadelphia sample (low-SES group), the average maternal schooling attainment was just 10.0 years for the Negro mothers and 9.5 years for the White mothers.

29. We also calculated the aggregated grand mean WISC IQ for Whites (mean IQ = 103.3) and for Negroes (mean IQ = 90.3). The mean IQ difference was 13 points, identical to what we found for the Stanford-Binet comparison.

30. The high-SES Negro group showed a slight departure from this pattern. The mean IQs were 100.7, 100.3, and 103.1 for the lowest, intermediate, and highest-SES intervals, respectively.

31. Although González (1932) offered no explanation, it appears that she selected Mexican American participants from a broad range of neighborhoods so as to avoid the range restriction problem that Ellis (1932) experienced in her similar study in San Antonio.

32. González (1932) suggested that the higher mean IQ on the Non-Language Mental Test might have been due to (a) practice effect (the Pantomime was administered first) and (b) inequivalent norms on the two tests.

33. González (1932) also reported a related analysis in which she compared the IQs of the highest-SES group (n = 33) and the lowest-SES group (n = 34). Although she did not report the r values for the groups, González found that the mean IQ for the highest-SES group was 98.6, more than 5 points higher than the mean IQ (93.2) for the lowest-SES group. The author also noted that there was considerable overlapping of IQs in the two groups.

34. González (1932) commented that when she visited the participants' homes to gather the SES data, the subject of lack of employment frequently was raised by the parents. "One of the most pitiful cases observed was that of an elderly man, who for seven years had been working as a tailor in one of the largest department stores in San Antonio and who without any warning was dismissed. They had been paying on a nice comfortable home, but after several months of non-payment, they were forced to give it up. [Incidentally], his daughter, who was included in this research, made an IQ of 120" (p. 30).

35. We assume that the term "Spanish American" refers to Mexican American. The term Spanish American is commonly used in New Mexico to refer to people of Mexican descent. Furthermore, the term "Anglo-American," which is used by Christiansen and Livermore (1970), is a common referent for Whites in New Mexico and other southwestern states.

36. Children of white-collar workers (e.g., professional, managerial) were considered middle class. Children of blue-collar workers (e.g., semiskilled, laborers) were deemed lower class.

37. As can be seen in Table 3.6 and in the preceding paragraph, SES was particularly powerful in differentiating participants on Verbal Scale IQ. Regarding the lower Verbal Scale IQ scores of the Mexican American participants, Christiansen and Livermore (1970) commented that the

depressed scores, "regardless of social class, were probably due to the bilingual nature of the typical Spanish American home which makes it more difficult for these children to acquire many of the verbal skills needed in an Anglo culture" (p. 12). It is not clear from this conjecture whether the authors were suggesting that bilingual homes thwart English-language development (a contentious point, by the way) or whether there simply were limited opportunities to learn English in these homes. Nonetheless, the authors should have assessed all the Mexican American participants' fluency, proficiency, and dominance in English before undertaking this investigation.

38. Oakland (1978) did not report the *n*'s for the six participant groups. However, based on some data about sample sizes he did report (p. 575, Table 1), we surmise that the *n*'s for the six groups range somewhere between the high 20s and high 50s.

39. The administration of the Slosson Intelligence Test was just one aspect of Oakland's (1978) study. His main purpose was to investigate the predictive validity of six tests of academic readiness across the three racial/ethnic groups.

40. The investigation by Valencia et al. (1995) did have SES as a control variable of concern. The major purpose, however, was to investigate content bias on the K-ABC (Kaufman & Kaufman, 1983a). This study is discussed in Chapter 5 of the present book.

41. Because the population of White 5th-grade students was smaller than that of the Mexican American population, it was necessary to select 24 White 6th-graders who met the SES criterion.

42. The Mexican American participants (50 boys, 50 girls) attended public schools in a highly segregated school district in central California. Of the 100 White children, 37 also attended the high-enrollment Mexican American district, and 63 attended schools in a contiguous, fairly segregated White district in the same community. For the total White group, there were 45 and 55 boys and girls, respectively. The mean ages of the two racial/ethnic groups were quite similar—11.3 and 11.1 years for the White and Mexican American groups, respectively.

43. The K-ABC also contains a totally separate Achievement Scale that focuses on children's knowledge of acquired facts and skills learned in the school and home environments. Given that our focus in this chapter is on intellectual performance and not academic achievement, the results of the participants' performance on the Achievement Scale in Valencia et al. (1995) are not of interest here. Such results are of importance, however, in our coverage of test bias in Chapter 5 of this book.

44. It needs to be noted that Jensen (1973a) was referring to intellectual differences *between* social classes but *within* racial groups. Later, however, Jensen suggested that such differences also are found *between* racial groups (see his Chapters 7 and 11).

45. Shuey (1966) also noted, "Twenty five of the 41 studies were located in the North, and in at least fourteen of the [research studies,] the colored and white children were not only attending the same schools but were living in the same district or neighborhood. The combined mean difference in IQ between the 2,760 colored subjects' tests in the North and the whites of comparable socioeconomic status or occupation was 7.6. Nearly all of these *S*s in the eighteen studies were of school age, [with] the whites and Negroes attending the same school and living in the same areas, many with large Negro populations (p. 519).

46. These investigations are Coleman et al. (1966), Scarr-Salapatek (1971), and Wilson (1967).

47. This discussion of Herrnstein and Murray (1994) is excerpted, with minor modifications, from Valencia and Solórzano (1997, pp. 174, 201).

48. Herrnstein and Murray (1994), based on the normal distribution of IQ scores, broke the continuum into five "cognitive classes." The respective classes (with names given, IQ intervals, and percentages of cases) are as follows: Class I (very bright, 125 and above, 5%), Class II (bright, 110 to 124, 20%), Class III (normal, 90 to 109, 50%), Class IV (dull, 75 to 89, 20%), and Class V (very dull, below 75, 5%). These classifications by IQ are very similar to the interval breaks that testing specialists have used since the time of Lewis Terman. Herrnstein and Murray noted, however, that they substituted more "neutral" terms (e.g., "very dull") for "less damning terms" (e.g., "retarded").

49. The issue of caste in U.S. society has a long scholarly history. Feagin (1984) pointed out that the "caste school" of race relations developed during the late 1940s under the leadership of

W. Lloyd Warner (Warner, 1941, 1949; Warner & Srole, 1945). The caste school of race relations focused on the post-Civil War era during which structured patterns of discrimination replaced the now defunct system of slavery in the South. The new social system, based on caste, led to severe economic and social inequality (Feagin, 1984). Another early analyst of the caste school of race relations was Oliver C. Cox (1948).

50. Ogbu's scholarship is not without its critics. See, for example, Foley (1991) and Trueba (1991).

51. Blau (1981) operationalized "middle-class exposure score" as "the proportion of close neighbors in white-collar occupations, of college-educated neighbors, and of neighbors with a child who has gone on to college . . . [as well as] the mean education of mother's three closest friends" (p. 18).

52. Valencia, Rankin, and Henderson (1986), in a path-analytic investigation of White ($n = 632$), African American ($n = 488$), and Mexican American ($n = 558$) schoolchildren (ages 5 to 12 years), found weaker status consistency (intercorrelations of parental occupational status and educational attainment) among the two minority groups, compared to that among the White group.

53. Parents' "demographic origins" was a composite score based on four measures (e.g., location of community where parent grew up, size of home community).

54. For discussions of how this is supported mathematically, see Knapp (1977) and Robinson (1950).

55. Although White (1982), in this particular analysis, collapsed studies across dependent measures (achievement and intelligence), the observed r of .35 can be fairly generalizable to studies that examined only the relation between SES and intelligence. That is, achievement and intelligence tests share much in common. This assertion is based on the observation that achievement and intelligence tests show a typical range of correlations of .60 to .70 for elementary school students and of .50 to .60 for high school students (Jensen, 1980).

56. See, for example, Boocock (1972), Charters (1963), and Coleman et al. (1966).

Chapter 4

1. For reviews of this literature and reports of empirical studies, see, for example, Gottfried (1984), Henderson (1981), Laosa and Sigel (1982), Marjoribanks (1979), and Walberg and Marjoribanks (1976).

2. The Whittier Scale evaluated five aspects of the home environment: (a) necessities, (b) neatness, (c) size, (d) parental conditions, and (e) parental supervision.

3. Correlations between the various predictor variables and children's IQ for the control group were of higher magnitudes. In descending order, the correlations (corrected for attenuation) were mother's mental age (.57), father's mental age (.55), father's vocabulary (.52), cultural index (.49), mother's vocabulary and Whittier Scale (both .48), and family income (.26).

4. As noted by Wachs and Gruen (1982), Van Alstyne also reported that the home environment variables were correlated with maternal IQ. Wachs and Gruen suggested that this relation might reflect the action of "Type 1 organism-environment covariance," which refers to "parents transmitting not only genes but also rearing environments to their offspring" (p. 43).

5. The construct of a press variable stems from Murray's (1938) theory of personality. He proposed that an environment can be classified by the ways in which an individual is harmed or benefited (Marjoribanks, 1979). According to Murray (1938), if the environment has an effect that is potentially harmful, then the individual will avoid the environment or defend himself or herself. On the other hand, if the environment has an effect that is potentially beneficial, then the individual typically will select to interact with it. This directional tendency of avoidance/approach is conceived as the "press of the environment." In his theoretical framework, Murray distinguished between the "alpha press" (i.e., the environment as seen by observers) and the "beta press" (i.e., the individual's perception of the environment). Whereas Murray focused on analyzing the beta

press of the family environment, Bloom (1964) and his doctoral students concentrated on the alpha press of family environments (Marjoribanks, 1979).

6. In his dissertation investigation, Davé's (1963) home environment measure consisted of six press variables and 21 process characteristics (for the listing, see Marjoribanks, 1979, pp. 33-34). The criterion variable was the Metropolitan Achievement Tests (MAT), a measure of academic achievement. When the home environment ratings were correlated with MAT scores, Davé (1963) reported multiple correlations ranging from .56 to .79 on the various MAT subtests. The overall index of the home environment rating scale correlated .80 with the total score on the MAT (Marjoribanks, 1979, Table 2.3).

7. Weiss (1969), in his dissertation, examined the relation between the home environment and measures of achievement motivation and self-esteem for 53 boys and girls (M age = 11 years) attending public schools in an Illinois community. Weiss formulated three press variables and 15 process characteristics to define the subenvironment for achievement motivation as well as three press variables and 16 process characteristics to define the subenvironment for self-esteem (for the listing, see Marjoribanks, 1979, pp. 46-47). Weiss found multiple correlations ranging from .27 to .79 when the home environment rating scales were correlated with the criterion measures (Marjoribanks, 1979, Table 2.4).

8. Marjoribanks (1979) also spoke of the "British" school of family environment research. He commented, "As the research of the 'Chicago' school was undertaken, another set of studies, which may be loosely grouped together as the 'British' school, was being generated. Although these latter studies did not conspicuously adopt Murray's [1938] concept of the alpha press [or] generally use Bloom's [1964] model of sub-environments of press variables, they may be interpreted in relation to these constructs" (p. 31). The interested reader might want to inspect Marjoribanks (1979, Figure 2.2), who provided an informative flowchart of the origins and development of the Chicago and British schools of family environment research.

9. In their review article, Bradley and Caldwell (1978) referred to the impact of Barker and Wright (1954); Escalona and Corman (1971); Honzik (1967); Schaefer, Bell, and Bayley (1959); White and Carew (1973); Whiteman, Brown, and Deutsch (1967); and Yarrow, Rubenstein, and Pederson (1975).

10. For the complete set of 12 propositions, see Henderson (1981).

11. For a list of the 45 items, see Bradley and Caldwell (1984).

12. For a list of the eight subscales, see Gottfried and Gottfried (1984).

13. See, for example, Barnard, Bee, and Hammond (1984); Bradley and Caldwell (1976a, 1976b, 1977, 1980, 1981, 1984); Bradley, Caldwell, and Elardo (1977, 1979); Bradley et al. (1989); Elardo, Bradley, and Caldwell (1975, 1977); Gottfried and Gottfried (1984); Hawk et al. (1986); Johnson, Breckenridge, and McGowan (1984); Johnson et al. (1993); Ramey, MacPhee, and Yeates (1982); Ramey, Mills, Campbell, and O'Brien (1975); Siegel (1984); Wachs, Uzgiris, and Hunt (1971); and Wulbert, Inglis, Kriegsman, and Mills (1975).

14. This set of studies consists of Bradley and Caldwell (1976a, 1976b, 1977) and Elardo et al. (1975, 1977).

15. For a comprehensive and very readable discussion of nonshared and shared environments, see Plomin et al. (1997, Chapter 14).

16. Plomin et al. (1997) asserted that phenotypic variance for cognition is "caused" by environmental variance as well as genetic variance (about 50%). The environmental variance is overwhelmingly of the nonshared variety. According to Plomin et al., environmental factors (both nonshared and shared) are not restricted to the home. Peer groups and educational experiences, for example, also constitute one's environment.

17. Stoolmiller (1999) noted that until more definitive data are made available, his 50% estimate should be viewed as an upper limit.

18. Two notions central to this finding are "genotype-environment correlation" and "genotype-environment interaction." For explication, see Plomin et al. (1997, Chapter 14).

19. For a discussion of "passive genotype-environment correlation," see Plomin et al. (1997).

20. These six studies are Bracken, Howell, and Crain (1993); Bradley and Caldwell (1976a, 1976b, 1977, 1980); and Elardo et al. (1975).

21. In two studies (Bradley & Caldwell, 1981; Trotman, 1977), an academic achievement test was used as a criterion measure. Although our focus is on the relation between home environment and measured intelligence, we listed those studies that used academic achievement as the criterion variable given the small number of total investigations that used African American participants.

22. The study by Trotman (listed in Table 4.1) also was published in 1977. That study's African American and White participants were in the 9th grade.

23. The ISC places families in social classes according to a total of weighted ratings based on occupation, source of income, house type, and dwelling area.

24. Trotman's (1977) investigation has been severely criticized on substantive and methodological lines. Longstreth (1978) called into question the reliability and validity of the Wolf (1964) home environment measure, which Trotman used. In our view, it appears that Longstreth did not think highly of maternal interview studies in general. Longstreth (1978) also criticized Trotman for not considering the role of "child-to-parent effects" on shaping the home environment. This criticism has merit, and we briefly discuss this issue later in this chapter. Wolff (1978) commented that Trotman's (1977) study is fatally flawed. First, Wolff (1978) noted that Trotman might be guilty of researcher bias (i.e., her "strong environmentalist leanings" [p. 473]). Second, Wolff commented that Trotman's observed correlations were anomalous (i.e., not showing greater attenuation, compared to existing research). For responses to her critics, see Trotman (1978).

25. These six sites are Seattle, Washington; Hamilton, Ontario; Chapel Hill, North Carolina; Houston, Texas; Fullerton, California; and Little Rock, Arkansas.

26. These investigations are Barnard et al. (1984); Bradley and Caldwell (1984); Caldwell, Elardo, and Elardo (1972); Gottfried and Gottfried (1984); Johnson et al. (1984); Ramey et al. (1982); and Siegel (1984).

27. Because of a confound between race/ethnicity and SES, White, African American, and Mexican American matched samples were constituted. Triads were matched on their 12-month HOME total scores. The result, in effect, was to eliminate a substantial proportion of the middle-class White participants. A total of 112 exact matches were made (±1 point) from the original total sample of more than 900 children.

28. For further discussion of this point, see Chapter 3 of the present book

29. The sample of 65 White children included 6 children "with Spanish surnames" (whom we assume to be Mexican American) and 1 Asian American child. Given the very small percentage (11%) of minority students in the White sample, it does not appear that there is confoundment of data.

30. We also located an unpublished investigation (doctoral dissertation) in which Mexican Americans served as participants (Wickwire, 1971).

31. These seven sociological factors contain 32 characteristics. For a listing, see Henderson and Merritt (1968, Table 1).

32. Henderson and Merritt (1968) is based on Henderson's (1966) doctoral dissertation. For further discussion of the Spanish vocabulary test, see Henderson (1966).

33. For a listing of the items used in the development and validation of the HELPS, see Henderson et al. (1972, pp. 191-192, Table 1).

34. The SEAT is an academic achievement test, and the BTBC is a school readiness test. Although neither instrument is considered a measure of intelligence (the criterion measure of focus in this chapter), we decided to include the Henderson et al. (1972) study for our review given the small number of investigations on the relation between home environment and Mexican American children's measured intelligence.

35. In Valencia et al. (1985), the authors made some changes in the HELPS on the basis of pilot test data suggesting that "(a) some of the items were culturally sensitive; (b) some items were

redundant with information from the family background information questionnaire; and (c) a few items were judged to be age inappropriate for this particular sample" (p. 326).

36. The criterion score of the MSCA used in Valencia et al. (1985) was the GCI, which is a global index of cognitive functioning.

37. The Spanish version of the MSCA was developed by Valencia and Cruz (1981). In Valencia et al. (1985), it was used for research purposes, and it is not recommended for use in the psychoeducational assessment of Mexican American children.

38. For a study that sought to investigate child (→) parent effects, see Bradley et al. (1979). For general readings on child effects on adults, see, for example, Bell (1974) and Bell and Harper (1977).

39. Ceci (1990) raised a very important point about the relation between schooling attainment and measured intelligence vis-à-vis African Americans, Latinos, and American Indians. Implicit in his conjecture is that limited access to schooling and subsequent diminishment in schooling attainment for minorities are associated with the persistence of lower IQs, compared to their White peers. Drawing from Sarason and Doris's (1979) review of literature on early immigrants' IQs, Ceci (1990) commented, "What Sarason and Doris [1979] show so nicely in their review is that as the levels of school completion increased among Italian-American children during the first five decades of this century, so did their IQs. I mention this review because nearly every social commentator and IQ researcher during the early part of this century was convinced that the low IQs of Italian-American children were due to nonenvironmental causes. As a matter of fact, no one (except Sarason and Doris) has bothered to ask what happened to the second- and third-generation offspring of these Italian families and what this development implies about the persistence of low IQs among blacks, Hispanics, and American Indians. Today, Italian-American students' IQs are not considered remarkably low (they are, in fact, slightly above average)." (p. 80)

40. For a comprehensive discussion of the construct of deficit thinking, see Valencia (1997a).

Chapter 5

1. For more discussion of these cases, see Chapter 1 of the present book.

2. Test validity investigations involving a single group sometimes are referred to as single population validity studies. For example, Valencia (1984), in a concurrent validity study, administered the Kaufman Assessment Battery for Children (K-ABC; Kaufman & Kaufman, 1983a) and the Wechsler Preschool and Primary Scale of Intelligence (WPPSI; Wechsler, 1967) to 42 English-speaking Mexican American preschool children. Statistically significant correlations between K-ABC Global Scale Standard scores and WPPSI IQ Scale scores were observed. In addition to validity research, single population studies can be undertaken on intelligence test reliability (Valencia & Rankin, 1983) as well as on stability (Valencia, 1985b). Although single population studies are useful in understanding a test's psychometric integrity with minority groups, such investigations are limited for assessing cultural bias given the absence of the White comparison group.

3. As we shall see later, in the case of content validity (i.e., item bias) studies, bias also can be detected against the major group.

4. After we had sent our book manuscript to Sage Publications for the production phase, we identified two test bias investigations pertaining to the WISC-III (Weiss & Prifitera, 1995; Weiss, Prifitera, & Roid, 1993). Due to this timing, we were not able to incorporate these studies into our comprehensive analysis of test bias findings provided in this chapter. These two studies (predictive validity bias) found that the WISC-III predicted achievement scores about equally well for African American, Hispanic, and White students. Weiss et al. (1993) did note, however, that using the common regression line *overpredicted* minority achievement. Thus, if one uses the common regression line (not the minority regression line), then minority students are likely to be *overidentified* as learning disabled. This contention also supports the findings of Olivárez, Palmer, and Guillemard

(1992, No. 34) and Palmer, Olivárez, Wilson, and Fordyce (1989, No. 35), investigations that we discuss later in this chapter.

5. This N of nearly 46,000 participants (as well as the n's for total major and minor group participants) is inflated. In some studies (e.g., Valencia & Rankin, 1986, Nos. 57 and 58), the researchers examined more than one psychometric property of one instrument. Therefore, we counted this as two separate studies, and the sample sizes were counted twice.

6. Some investigations used standard SES measures (e.g., Valencia, 1984, No. 55, used the Warner Revised Occupational Rating Scale [Warner, Meeker, & Eells, 1960]). Other investigations used indirect SES measures. For example, Valencia, Rankin, and Livingston (1995, No. 60) used participants' status in the school district's free or reduced-price lunch program (enrolled/not enrolled).

7. Carlson and Jensen (1981, No. 3) also computed reliability estimates using Hoyt's formula (analogous to the Kuder-Richardson Formula 20) and split-half correlations corrected for attenuation.

8. Valencia (1984, No. 55) and Valencia and Rankin (1986-2, No. 58) used the procedure developed by Feldt (1969), which results in a test statistic distributed as F with $N_1 - 1df$ (v_1) and $N_2 - 1df$ (v_2).

9. Investigators frequently use the coefficient of congruence (Harman, 1976; Levine, 1977) to determine the degree of similarity of the various factors across the major and minor groups. The coefficient of congruence is an index of factorial similarity that has interpretive aspects similar to the Pearson r.

10. Bias in content validity also can be investigated at the subtest level (see, e.g., Study Nos. 56 and 60). Our focus here, however, is on bias at the item level.

11. In two such studies we reviewed, the empirically defined and testable definition of content bias suggested by Reynolds (1982b) served as a guide. See Valencia and Rankin (1985, No. 56) and Valencia et al. (1995, No. 60).

12. The *PASE* (1980) case, which resulted in a landmark legal decision involving the issue of cultural bias in intelligence tests, was initiated by African American students in Chicago. Plaintiffs argued that they were misdiagnosed as mentally retarded based on intelligence tests that were racially biased. Although Grady concluded that the WISC contained some biased items, he eventually upheld that the WISC (as well as the WISC-R and Stanford-Binet) were valid. His method of bias detection and his decision have been severely criticized by some members of the measurement community. For example, Bersoff (1984) commented, "[Grady's analysis] can best be described as naive. At worst, it was unintelligent and completely devoid of empirical content... . At bottom, what it represented was a single person's subjective and personal judgment cloaked in the apparent authority of judicial robes. If submitted as a study to one of psychology's more respected refereed journals, rather than masquerading as a legal opinion, it would have been summarily rejected as an experiment whose sample size and lack of objectivity stamped it as unworthy of publication" (p. 100).

13. The seven items include three from the Information subtest and four from the Comprehension subtest.

14. The 15 biased items were distributed as follows: Vocabulary subtest ($n = 6$), Information subtest ($n = 5$), and Similarities subtest ($n = 4$).

15. This statistical method calls for the correlation between the item and group membership with the contribution of the general ability true score partialled out. The particular item under examination is not included in the true score. The test for bias was the significance of the item-partial correlation, with subgroup membership (English- and Spanish-speaking) controlling for the total score (MSCA General Cognitive Ability, an IQ analog), age, and sex.

16. The subtests are Word Knowledge II, Verbal Memory I, Draw-a-Child, Draw-a-Design, and Opposite Analogies.

17. There is a considerable amount of empirical research on the word length effect. As the length of the stimuli words to be recalled increases (as measured by syllable count), short-term

memory becomes taxed (see, e.g., Baddeley, Thomson, & Buchanan, 1975; Eriksen, Pollack, & Montague, 1970). Regarding the acoustic similarity effect, research has found that systematic confusion occurs when words are very similar in phonemic structure. Such confusion arises in short-term memory for item and order information; thus, limited recall occurs (see, e.g., Conrad, 1964; Baddeley, 1976).

18. The subtests in which biased items were detected are Hand Movements, Gestalt Closure, Number Recall, Triangles, Spatial Memory, and Photo Series.

19. The numbers and percentages of biased items in the respective Achievement Scale subtests are as follows: Faces and Places (15 of 26, 57.7%), Arithmetic (10 of 16, 62.5%), Riddles (16 of 18, 88.9%), Reading/Decoding (9 of 16, 56.3%), and Reading/Understanding (8 of 16, 50%).

20. Valencia et al. (1995, No. 60) noted, "An interesting question we asked about possible ethnic differences in achievement was, 'How did the 37 White children fare who attended Mexican American segregated schools?' To address this question, we ran a one-way ANOVA. The groups examined were Mexican Americans in segregated schools (Group 1, $n = 100$), Whites in Mexican American segregated schools (Group 2, $n = 37$), and Whites in White segregated schools (Group 3, $n = 63$). The criterion variable was the K-ABC Achievement Scale standard score. As expected, the observed F was significant, $F(2, 197) = 21.51$, $p = .00001$, $MS_e = 107.00$. Scheffé tests revealed that the mean K-ABC Achievement score for Group 1 ($M = 91.33$) was significantly lower than the mean K-ABC Achievement scores for Group 2 ($M = 98.41$) and Group 3 ($M = 101.92$). There was no significant difference between Groups 2 and 3. These findings suggest that the White students and the Mexican American students attending the Mexican American segregated schools were different populations. It is unknown, however, whether learning opportunities differed for the two ethnic groups in the Mexican American segregated schools" (pp. 166-167).

21. White (1982), in a review of literature that examined the relation between SES and academic achievement (and measured intelligence), reported a mean correlation of .33 between SES and IQ.

22. Halpern (1997) also noted that males, compared to females, are overrepresented at the low-ability end of distributions in mental retardation, dyslexia, stuttering, attention disorders, and delayed speech.

Chapter 6

1. It is beyond the scope of this chapter to present comprehensive discussions of the history, assumptions, models, and findings germane to the field of BGIA. We refer the reader to the following examples of literature in this area: Chipeur, Rovine, and Plomin (1990); Loehlin (1989); McArdle and Goldsmith (1990); McArdle and Prescott (1997); Plomin, DeFries, McClearn, and Rutter (1997); Plomin and Neiderhiser (1991); Plomin and Rende (1990); Scarr (1992, 1997); and Scarr and Carter-Saltzman (1982). Of these publications, McArdle and Prescott (1997) offer one of the best discussions (e.g., comprehensiveness, readability, balance) of the BGIA field.

2. Some examples of publications that voice diverse views pertaining to this debate include Baumrind (1993); Block and Dworkin (1976); Bouchard (1997); Brody (1992); Brody and Brody (1976); Ceci (1990); Ehrman, Omenn, and Caspari (1972); Fancher (1985); Jackson (1993); Kelves (1985); Lewontin, Rose, and Kamin (1984); Neisser (1998); Pearson (1991); Scarr (1997); Steen (1996); Storfer (1990); and Taylor (1980). One book that we particularly recommend is *Intelligence, Heredity, and Environment* (Sternberg & Grigorenko, 1997). This book offers a comprehensive, fairly balanced, and updated survey on the nature-nurture debate in relation to measured intelligence. We do, however, have some concerns about balance at the book's end (see Note 21 of the present chapter).

3. An example of advances in gene location can be seen in "gene mapping." McArdle and Prescott (1997) noted, "Rather than inferring genetic influence by studying resemblance among relatives, these methods attempt to locate specific locations in the genome that code for specific

functions. Some researchers have begun to apply this approach to intelligence by initiating a molecular genetic search for a small number of single genes of relatively large effect using a methodology termed 'quantitative trait loci' (QTL)" (pp. 426-427).

4. Genotype refers to the "sum total of the genetic material in a particular individual, i.e., the total genetic constitution, either at one specific locus or more generally" (Kaplan, 1976, p. 434). Phenotype refers to "the appearance of an individual that results from genotype and environment" (Plomin et al., 1997, p. 315).

5. A related, but distinct, term is "range of reaction" (or "reaction range"). Miller (1997) referred to reaction range "as the distribution of observed phenotypic reactions associated with a given genotype" (p. 293). For a discussion of the conceptual differences between norm of reaction and range of reaction, see Wahlsten and Gottlieb (1997).

6. Hirsch (1976) commented that the norm of reaction is of fundamental significance to understand "because it saves us from being taken in by glib and misleading textbook clichés such as 'heredity sets the limits but environment determines the extent of development within those limits' " (p. 163).

7. More recently, Wahlsten and Gottlieb (1997) made a similar assertion: "The reaction norm is something to be discovered through experimentation, and it allows us no prediction of what will happen when an individual is placed beyond a rearing environment that has already been observed in scientific studies" (p. 172).

8. There remains, however, differences in opinion about the empirical role of the norm of reaction in human development (Baumrind, 1993; Ceci, Rosenblum, de Bruyn, & Lee, 1997; Miller, 1997; Scarr, 1993). Miller (1997), arguing from a cultural psychology perspective, noted that the behavioral genetic perspective (Scarr, 1993) appears to downplay the notion of reaction norm and, by contrast, emphasizes genetically determined reaction range. Miller (1997) commented, "From this perspective [behavioral genetics], claims that development should be understood as an open process [e.g., cultural psychology] are rejected as purely hypothetical in nature since there appears to be no empirical evidence of environmental variation in nonimpaired intellectual outcomes. The indeterminate view of development implied by the concept of reaction norm is also rejected as incompatible with the predictive goals of science" (p. 294).

Critics of the behavioral genetics perspective that the reaction range is genetically determined assert that this view does not account for heredity by environment interaction. For example, Baumrind (1993) contended, "The norm of reaction for a particular individual or group can be ascertained only by attempting to expand it, and prevention or intervention efforts are limited by extant scientific knowledge. Change an individual's experience and, within indeterminate limits, the individual can be changed" (p. 1302).

9. For a discussion of the distinctions between broad- and narrow-sense heritability, see Plomin et al. (1997). It is beyond the scope of this chapter to describe the mathematical models and procedures for estimating the heritability of intelligence. The interested reader might want to refer to the following examples of sources for such explications: Bouchard (1997); Cancro (1971); Falconer (1990); Jensen (1969); Loehlin, Lindzey, and Spuhler (1975); McArdle and Prescott (1997); Plomin et al. (1997); Plomin and Loehlin (1989); Scarr and Carter-Saltzman (1982); and Schönemann (1994). For critiques of how the heritability of intelligence has been estimated and conclusions have been drawn from such work, see Farber (1981), Kamin (1974), Lewontin et al. (1984), and Taylor (1980). Furthermore, there are discussions that assert that calculations of heritability estimates of intelligence frequently violate critical assumptions of the statistical models used and use statistical techniques that are insensitive to identifying heredity-environment interaction (see Wahlsten, 1990, and those who concur with his position: Bullock, 1990; Chiszar & Gollin, 1990; and Schönemann, 1994). Finally, Lewontin (1975) commented that to have a proper study of heritability, there are six requirements that must be met (e.g., sample sizes need to be large [at least several hundred families], the various relationship pairs must be representative of the

population to which the heritability estimate is applied such as representativeness in schooling attainment or SES). Lewontin concluded, "In fact, no study of the genetics of IQ has even come close to fulfilling the six requirements, and most studies fail all tests in a significant way" (p. 394).

10. A similar point was made by Kamin (1974): "They [MZ twins] presumably have identical genes. They are necessarily of the same sex, and their physical similarities are typically very striking" (p. 33). Kamin continued, "Evidence [from studies of separated MZ twins] seems especially powerful because it is based on fewer—and simpler—arbitrary assumptions that must be made to interpret other forms of data. . . . MZ twin pairs seem uniquely equipped to serve as subjects in experiments concerned with heredity" (p. 33).

11. Critics of the research on MZ twins separated and reared apart are not without their countercritics. For critical examinations of Farber (1981), see Bouchard (1982a), Brody (1992), Locurto (1991), and Loehlin (1981), all cited in Bouchard (1997). For criticisms of Kamin (1974), see Bouchard (1982b), Fulker (1975), and D. N. Jackson (1975), all cited in Bouchard (1997).

12. Taylor (1980) pointed out that there actually is a fourth condition: Prenatal (intrauterine) environmental effects on each twin pair are equal and negligible in magnitude.

13. In addition to his reanalysis of the three investigations, Taylor's (1980) conclusion was informed by additional analyses throughout his book.

14. A polygenetic theory of inheritance of intelligence is "the higher the proportion of genes two family members have in common, the higher the average correlation between their IQ's" (Bouchard & McGue, 1981, p. 1055).

15. McArdle and Prescott (1997) commented that the earlier studies commonly sampled individuals from a restricted range of SES backgrounds. Thus, this sampling procedure likely resulted in environmental variance that was limited and probably resulted in a greater percentage of the observed variance being ascribed to genetic sources.

16. For these possible explanations of why newer research, in comparison to older research, has resulted in conclusions of lower estimates of genetic influence on measured intelligence, McArdle and Prescott (1997) referred to Chipeur et al. (1990); Loehlin (1989); Loehlin, Horn, and Willerman (1989); and Plomin and DeFries (1980).

17. It needs to be pointed out that there are many scholars (including a number of leading geneticists) who assert that the heritability construct itself is a misguided notion (for a discussion, see Tucker, 1994, pp. 224-225). For example, Otto Kempthorne, an expert in biostatistics, contended that "heritability does not even exist in the human context" (Kempthorne, 1978, p. 19).

18. The authors of *Not in Our Genes* (Lewontin et al., 1984) make an interesting team regarding their academic fields. Lewontin is an evolutionary geneticist, Rose is a neurobiologist, and Kamin is a psychologist.

19. For further discussion and critiques of biological determinism, see *The Dialectical Biologist* (Levins & Lewontin, 1985).

20. Suffice it to say, Scarr's (1992) theorizing did not escape challenges. Baumrind (1993), for example, argued that considerable evidence has accrued from child development and socialization literature to support the position that "what normal parents do or fail to do crucially affects their children's development" (p. 1299). Jackson (1993), who also disagreed with Scarr's (1992) position, focused on rival hypotheses that could explain findings from empirical research on human behavioral genetics, thus leading to environment (→) phenotype interpretations. For her reply to Baumrind (1993) and Jackson (1993), see Scarr (1993).

21. The subject of balance is, in our opinion, a bit uneven at the book's end. We found it curious that Sternberg and Grigorenko (1997) did not participate in the "Integration and Conclusions" section. Rather, they allowed two authors who did not contribute to the core chapters to provide closing chapters. The chapter by Hunt (1997), who appeared to place himself as the referee in the nature-nurture "Super Bowl," called it a "lopsided victory. . . . Nature wins 48-6" (p. 531). If this metaphor is not graphic enough, Hunt also likened the nature-nurture debate to "a stomping match between Godzilla and Bambi. For those who did not see the movie, Godzilla wins. Yet we all

love Bambi. Why do people keep looking for cultural explanations of intelligence?" (p. 546). It is this latter point about "cultural explanations" that signals Hunt's theoretical bias. To Hunt, the various "culturalist" (his term) arguments in the book by Sternberg and Grigorenko (1997) are mere "worldviews" of "humanistic arguments," whereas the "nativist side" advances "scientific arguments." To us, this manufactured distinction between the genetic and cultural/environmental perspectives is harsh and inaccurate because it sets up the latter as a straw person. That is, by relegating the cultural/environmental perspective to a location outside the realm of scientific inquiry, Hunt (1997) created a weak opposition whose arguments are easily refuted. Although Hunt did call for scholars to "go beyond statistics to investigate the mechanisms of genetic action, including the interaction between genetic predisposition and the environment" (p. 549), most of what he presented in his chapter is biased in favor of the behavioral genetics field. At least, Sternberg and Grigorenko (1997) should have allowed rejoinders from the so-called "culturalists" (e.g., Ceci et al., 1997; Gardner et al., 1997; Miller, 1997). This would have provided the reader with a more stimulating and balanced conclusion.

22. The symbol systems approach described by Gardner et al. (1997), which we introduced in Chapter 2 of this book, "holds that the human experience—the thought and behavior of our species—is most felicitously analyzed in terms of the symbol and systems of symbols [e.g., language, mathematics] in which human beings traffic from early in life" (p. 246). The authors asserted that their approach can be seen as a "third perspective" that integrates the nature-nurture debate.

23. The notion of "proximal processes," which Ceci et al. (1997) defined as "reciprocal interactions between the developing child and other persons, objects, and symbols in [his or her] immediate environment" (p. 311), is somewhat similar to what we referred to as "proximal variable" in Chapter 4 of the present book. For a brief introduction to the bioecological model of intellectual development, see Chapter 2 of the present book.

24. Miller (1997) noted that cultural psychology is an emerging tradition that can be characterized as having a variety of shared assumptions and goals. Its various perspectives include, for example, psychological anthropology, cultural-social psychology, and sociolinguistics. Theory and research from cultural psychology, Miller asserted, challenges a number of assumptions underlying research in behavioral genetics.

25. For a similar point of view on the invalidity of the separation of nature and nurture effects, see Wahlsten and Gottlieb (1997).

26. For examples of other scholars whose writings are part of this growing alternative framework, see Bronfenbrenner and Ceci (1993), Gottlieb (1970, 1992), Lerner (1980, 1991), and Oyama (1985), all cited in Bidell and Fischer (1997).

27. The following sections on Shuey, Garrett, Jensen, Dunn, and the team of Herrnstein and Murray are excerpted, with some revisions (deletions and expansions) from Valencia and Solórzano (1997, pp. 160-183).

28. These criticisms are in reference to Shuey's (1958) first edition of *The Testing of Negro Intelligence*. Her genetic hypothesis of racial differences in intelligence was presented in both the 1958 and 1966 editions.

29. Estimated ω^2 is a measure of the association between the independent and dependent variables, calculated after a statistical test of significance is run. Estimated ω^2 has the element of "practical significance." That is, it estimates the proportion of variance in the dependent variable that can be accounted for by specifying the independent variable. Estimated ω^2 is more meaningful than statistical significance because it can be used to inform decisions about practical matters.

30. For a discussion on how hereditarianism was debunked during the 1920s, see Valencia (1997b) and Chapter 1 of the present book. For coverage of what led to the continuing demise of hereditarianism during the 1930s and 1940s, see Foley (1997). For a slightly different interpretation of the vicissitudes of hereditarian thought during these decades, see Provine (1986).

31. After Shuey's death, and in her memory, Osborne and McGurk (1982b) published an edited book, *The Testing of Negro Intelligence: Volume 2*. The book covered new material on Black-White studies of intellectual performance, spanning 1966 to 1979. Osborne and McGurk (1982a),

echoing (word by word) Shuey's earlier conclusion, commented that Black-White differences in intelligence "inevitably point to the presence of native differences between Negroes and Whites as determined by intelligence tests" (p. 297).

32. For interesting discussions of the events that led to the publication of Jensen's (1969) monograph and accounts of what transpired shortly after publication, see Fancher (1985), Jensen (1972), and Pearson (1991). These accounts attest to the highly politicized climate surrounding the publication of Jensen's (1969) treatise.

33. It appears that Jensen's current understanding of the magnitude of heritability of intelligence has changed from years past. "The broad heritability of IQ is about .40 to .50 when measured in children, [is] about .60 to .70 in adolescents and young adults, and approaches .80 in later maturity" (Jensen, 1998, p. 169).

34. Jensen (1998) noted that, based on citation data from the Institute for Scientific Information, his 1969 *HER* article soon became a "citation classic"—a publication that has an unusually high frequency of being cited in the scientific literature. Based on a list of the 100 most frequently cited articles in the social science literature (1969-1978) compiled by Tucker (1994), Jensen's (1969) article ranked number 6. Tucker (1994) did comment, however, that the other articles tended to be "seminal works" in their fields, whereas Jensen's (1969) article was frequently cited because it was a subject of controversy.

35. Whether Jensen wished to embrace this accolade is, perhaps, another question.

36. For examples of publications in response to Jensen (1969) and writings about the debate that ensued, see Block and Dworkin (1976), Eysenck (1971), Fancher (1985), Flynn (1980), Herrnstein (1973), Jensen (1972), Kamin (1974), Modgil and Modgil (1987), Pearson (1991), Snyderman and Rothman (1988), and Tucker (1994). See also "Bibliography of Articles About 'How Much Can We Boost IQ and Scholastic Achievement? by Arthur R. Jensen" (in Jensen, 1972).

37. Although Jensen (1969) did not cite the Westinghouse-Ohio study, it appears that he was referring to the preliminary report by Cicirelli et al. (1969).

38. During recent years, however, the case of fraud against Burt appears to have weakened. Fletcher (1991) and Joynson (1990) presented unabashed defenses of Burt's work and reputation. Part of these authors' cases are criticisms of the biography by Hearnshaw (1979), who wrote, in part, a comprehensive exposé of Burt's alleged fraudulent publications on MZA twins. Fletcher (1991) and Joynson (1990) claimed that Hearnshaw (1979) made numerous errors, questionable assumptions, and false conclusions. For book reviews (pro, con, and neutral) of Fletcher's (1991) and Joynson's (1990) works, see Apple (1992a, 1992b), Beloff (1990), Blinkton (1989), Eysenck (1989), Fancher (1991), Fletcher (1990), Skrabanek (1989), and Zenderland (1990). What the "Burt affair" truly means in the final analysis is not clear. Drawing from Mackintosh's (1995) book, *Cyril Burt: Fraud or Framed?*, Plomin et al. (1997) observed, "Although the jury is still out on some of the charges . . ., it appears that some of Burt's data are dubious" (p. 137).

39. There is one study that presents strong evidence questioning the assumed invariance of Jensen's (1969) Level I-Level II theory. See the Taylor and Skanes (1976) investigation with Canadian Inuit Indian and Canadian White children.

40. The subtitle of the monograph is *Emphasizing Studies Involving the English- and Spanish-Language Versions of the Peabody Picture Vocabulary Test–Revised.*

41. Dunn (1988) contended that American Guidance Service (AGS) "did not *publish* my monograph and did not *sell* it. They only reproduced and mailed it, upon request, as a free service" (p. 302). It appears that Dunn attempted to clarify matters in that some critics of Dunn's piece advocated the blacklisting of AGS. Although it might be true that Dunn had such an agreement with AGS, the title page of the monograph does read "published by AGS."

42. Notwithstanding the consensus that the PPVT series does not represent general measures of intelligence, the PPVT-R still is listed under "assessment of intelligence" in texts on testing (Anastasi, 1988; Cohen et al., 1992; Salvia & Ysseldyke, 1988). Valencia and Solórzano (1997) opined that for authors to do such listings not only is misleading but also helps to perpetuate the

misconception that the PPVT-R is an appropriate measure of intelligence. We believe that the test best fits under the "assessment of language" category.

43. Dunn (1988), in his rejoinder to Berliner (1988), asserted, "I did not say that the PPVT/TVIP [Spanish version of the Peabody test] should be used as a measure of verbal intelligence/scholastic aptitude for immigrant Hispanic youth. . . . For such children, the [tests] are only achievement tests of vocabulary for single words spoken in isolation" (p. 315). Yet, in his monograph, Dunn (1987) relied on data from two sources to draw his conclusions of heritability of intelligence. The first source included studies with "instruments other than the Peabody tests" (e.g., WISC-R), and the second cluster included research involving the original PPVT (which used an IQ index) and the PPVT-R. Thus, Dunn's (1988) response to Berliner (1988) made little sense, based on the overall approach he took in reviewing studies germane to his partial genetic interpretation.

44. These reform ideas are not really new. For the most part, these ideas stem from discourse beginning during the early 1980s and are viewed, by critics, as agenda items of the current conservative school reform movement.

45. Herrnstein and Murray (1994) argued that the funding priority for the gifted and the economically disadvantaged "turned 180 degrees" with the passage of the Elementary and Secondary Education Act of 1965, a federal program designed to serve the educational needs of students in low-income areas.

46. For an earlier discussion of the argument for the custodial state, see Murray (1988).

47. Of these five books, two (Fraser, 1995; Kinchloe et al., 1996) consist of collections of short reviews, all of which are unfavorable to *The Bell Curve*. The book by Jacoby and Glauberman (1995) also contains brief reviews and commentaries, but of diverse views. The books by Dickens et al. (1996) and Fischer et al. (1996), by contrast, are based in large part on reanalysis of the National Longitudinal Survey of Youth study data used by Herrnstein and Murray (1994) in *The Bell Curve*. Both books' conclusions about class structure formation are very different from those presented by Herrnstein and Murray.

48. Our search for book reviews of *The Bell Curve*, which was conducted in August 1999, was confined to the *Periodical Abstracts* database. As such, there probably are more book reviews and commentaries than the nearly 70 we identified.

49. This range of responses to *The Bell Curve* also was observed by Murphey (1995), who did a survey of approximately 125 book reviews, articles, columns, and television commentaries that appeared from late 1994 to early 1995.

50. For Murray's response to the critics of *The Bell Curve*, see Murray (1995).

51. As Gould (1995) noted, a correlation of .40 yields an R^2 of only .16. R^2, a statistic known as the coefficient of determination, is the square of the observed correlation coefficient and is useful in explaining how much of the variance in the dependent variable (e.g., dropping out of high school) can be accounted for by the independent variable (e.g., IQ). Gould commented that the values of R^2 presented in Appendix 4 of *The Bell Curve* are very low.

52. One example of this type of study is by Mercer and Brown (1973).

53. For more than a decade, J. Phillipe Rushton, a Canadian developmental psychologist at the University of Western Ontario, has been advancing a theory that the three "races" he has studied have evolved to different levels of intellectual development. According to Rushton, "Mongoloids" are most advanced, "Caucasoids" are intermediate, and "Negroids" are least advanced. In that such an ordering of the races stems from an evolutionary basis, the engine must be genetics, he asserted (see, e.g., Rushton, 1995). One of the many variables in his analysis is brain size. Rushton claims that, given that the "Negroid race" has the smallest mean brain size and the lowest mean IQ, such findings are evidence of a genetic hypothesis about racial differences in intelligence. Suffice it to say, Rushton's overall theory (and his work on brain size) has been criticized frequently (Cain & Vanderwolf, 1990; Horowitz, 1995; Lynn, 1989; Zuckerman & Brody, 1988). For discussions of other controversial scholars (and their connections with the Pioneer Fund, a right-

wing organization) on whom Herrnstein and Murray (1994) relied, in part, for source material, see Lane (1995), Linklater (1995), Miller (1995), Sedgwick (1995), and Tucker (1994).

54. For a discussion of some of the earlier transracial adoption studies, see Loehlin et al. (1975). Loehlin et al. pointed out that these studies are not that informative because the focus was mainly on the adopting parents.

55. Scarr and Weinberg (1976) also found that Black adopted children who had two natural Black parents, compared to Black adopted children who had one White natural parent and one Black natural parent, tended to have lower IQs. The authors pointed out, however, that this finding should not be interpreted as genetic given that the differences in IQs were associated with maternal schooling attainment and preplacement history. Once these variables were controlled, the mean IQ difference between the two groups was substantially less.

56. Brody (1992) commented, however, that the data from Scarr and Weinberg (1976) should not be construed as definitive evidence in favor of an environmental interpretation because the children were quite young at the time of testing. Research has shown that heritability of intelligence increases from childhood to adulthood (Plomin et al., 1997). See also McCartney, Bernieri, and Harris (1990).

57. For a discussion of research problems associated with racial admixture studies, see Loehlin et al. (1975), Mackenzie (1984), and Scarr et al. (1977).

58. For several other studies on the degree of White ancestry in the African American population, see the references provided in Scarr et al. (1977, p. 70).

59. It must be pointed out, however, that in *Race Differences in Intelligence* (Loehlin et al., 1975), then considered the definitive work on the topic, the authors concluded that the studies they reviewed "provided no unequivocal answer to the question of whether the differences in ability-test performance among U.S. racial-ethnic populations do or do not have a substantial component reflecting genetic differences among the subpopulations" (p. 133). Provine (1986) observed that, based on the studies reviewed by Loehlin et al. (1975), such a conclusion is considered an objective assessment of the available studies. Provine (1986) did comment, however, that "the social and public-policy implications of the *possible* genetic differences between racial-ethnic groups in the U.S. were, however, far less objective" (p. 881). Note the following statement by Loehlin et al. (1975): "We consider it quite likely that *some* genes affecting *some* aspects of intellectual performance differ appreciably in frequency between U.S. racial-ethnic groups—leaving open the issue of which groups, which aspects, and which direction of difference. Thus, we consider it most unwise to base public policy on the assumption that no such genetic differences exist" (p. 240). Provine (1986) asserted that, based on the evidence reviewed by Loehlin et al. (1975), the first sentence in the preceding quote cannot be supported. Provine (1986) also found that the public policy conclusion presented by the authors was disturbing. He opined that objective scientific ignorance of genetic differences between races cannot be used to guide social policy. In the final analysis, the ambiguous timbre of the conclusions drawn by Loehlin et al. (1975) left a specter of openness to the question of genetic bases in mean intelligence differences between races. Scarr and Carter-Saltzman (1982) commented, "Their [Loehlin et al.'s (1975)] equivocal conclusion led many social scientists to take the position that the possibility of genetic differences between the races was an open issue" (p. 863).

Chapter 7

1. Based on OCR survey data presented by Chinn and Hughes (1987) and from the U.S. Department of Education, OCR (1991), we calculated overrepresentation rates of Black students in EMR classes from 1978 to 1988 and juxtaposed these data with the most recent figure (for 1994) provided in this chapter. These disparities are as follows: 1978, +141.8%; 1980, +125.7%;

1982, +109.3%; 1984, +97.0%; 1988, +94.4%; 1994, +86.7%. Although the overrepresentation rate for Blacks in the EMR category (national level) has steadily declined over the data points, it still remains high.

2. Using the same sources mentioned in Note 1 (Chinn & Hughes, 1987; U.S. Department of Education, OCR, 1991), we calculated underrepresentation rates of Hispanic students in EMR classes from 1978 to 1988 and compared these data to the most recent figure (for 1994) provided in this chapter. Our calculations are as follows: 1978, –31.7%; 1980, –46.4%; 1982, –56.8%; 1984, –23.5%; 1988, –23.6%; 1994, –35.1%. Notwithstanding fluctuations, the underrepresentation of Hispanics in EMR classes (national level) has been holding steady.

Chapter 8

1. Searches were done electronically. ERIC served as the database for *CIJE*, and PsycINFO served as the database for *PA*. The time frame for the searches was from 1967 to 1999. We also conducted searches using "gifted" and "talented" as well as "children," "youth," and "students." There were 1,836 and 2,358 citations listed in the *CIJE* and *PA* databases, respectively. Of these citations, only a handful pertained to gifted minority children and youths.

2. Ford and Harris (1990) did not describe how they undertook their bibliographic search.

3. Good (1960) noted, "The girls were given domestic education by the mother in the home" (p. 25).

4. See, for example, Cleveland (1920), Gosling (1919), Race (1918), and Specht (1919), all cited in Newland (1976). According to Margolin (1994), the earliest use of the term "gifted child" that he could find was in Van Sickle (1910).

5. As a supplementary test, the Goddard form board also was administered as a basis for entrance.

6. It appears that Hollingsworth (1929) suggested a causal link between the racial origins of successful American adults and distinction in life. Citing Woods (1914), she commented that Anglo-Saxons are 3 to 10 times more likely than other races to achieve positions of distinction nationally. Hollingsworth failed to consider that it might not be racial origin that is the driving force of distinction but rather that socially transmitted privilege, status, and power are keys to mobility and distinction among Anglo-Saxon people.

7. For our discussion here, we refer to Terman's second edition (Terman, 1926).

8. Group II participants were selected by volunteer assistants who administered the Stanford-Binet to students not covered in the main study. Group II cities included, for example, Santa Barbara, Fresno, San Jose, Pasadena, and Burbank.

9. For exact nomination instructions given to teachers, see Terman (1926, pp. 21-22).

10. See Terman (1926, p. 55, Table 8). As we noted in Chapter 1 of the present book (Note 14), many scholars of this era confused race with national origin and ethnicity.

11. The *n* values provided are our own calculations given that Terman provided only percentages. Our estimates for the various *n* values are very accurate. We calculated *n* values for each of the 37 racial stocks. Our total *N* was 636 cases, very close to Terman's *N* of 643 cases. The difference between our *N* and Terman's *N* is due to rounding error.

12. For the total Jewish group, Terman (1926) also reported nine subgroupings (e.g., Russian Jewish, German Jewish, Polish Jewish).

13. Although four Japanese children surfaced in Terman's main study, they and Chinese students were not canvassed. These children were required, by state law, to attend segregated schools. Terman (1926) acknowledged that he intentionally did not canvass these schools. He did not, however, provide a reason. We suspect that he did not because of the atypical nature of these segregated schools. This rationale, if true, makes little sense given that African American and Mexican American children also attended, for the most part, segregated schools in California (see, e.g., González, 1990).

14. Whether or not the teachers exhibited social class bias in their nominations is unknown. It is clear, however, that the gifted children identified in Terman's (1926) study came from backgrounds of privilege. Based on occupational data from 560 fathers in the main study, Terman reported that 31.4% of children's fathers were classified as professional in occupational status, 50.0% as semiprofessional and business, 11.8% as skilled labor, and 6.8% as semiskilled or slightly skilled labor. This skewed distribution in favor of high-SES fathers was quite different from the distribution in the general population in the target cities. For example, Terman noted that, based on census data, the percentage of professional fathers in the population was 2.5%. Yet, in his study, 31.4% of the fathers were considered part of the professional group (an overrepresentation of 1,156%). "Common laborers" constituted 15.0% of the population, yet fathers who were common laborers in Terman's study constituted 0.1% of the total group (an underrepresentation of 99%). We calculated these disparities based on the data described in Terman (1926, pp. 63-64).

15. For the larger quote from which this quote is taken, see Terman (1916, pp. 91-92).

16. Terman and his associates also followed up on his original sample in one of the most ambitious longitudinal studies of the gifted ever undertaken (see, e.g., Cox, 1926; Terman, Burks, & Jensen, 1930; Terman & Oden, 1947, 1959). One of Terman's major findings was that many of the children identified as gifted in his initial study went on to lives of distinction (e.g., a disproportionate representation in *Who's Who in America*). However, the argument that "IQ = intelligence = real-world success" (Ceci, 1990, p. 58) has been sharply challenged. For example, drawing from the analysis by Sorokin (1956), Ceci (1990) raised an interesting methodological issue that questions Terman's assertion about the relation between IQ and life success. Ceci commented that if the gifted children from the original study "are compared to children from their *same* childhood social class instead of to economically unselected children, as Terman had originally done, they do not appear to be nearly as exceptional in their later professional or personal lives. I became aware of this when I came across a relatively obscure paper by the sociologist P. Sorokin (1956), writing in *Fads and Foibles in Modern Sociology*. Sorokin conducted an informal comparison of Terman's "geniuses" with children from their same social group on everything from high school and college grades to marital dissolution rates and professional accomplishments. He concluded that the instances of superiority among the "geniuses" in Terman's study are within the expectation of superiority *irrespective of IQ*, in view of their family social background. Notwithstanding the effusive praise that psychologists have heaped on this study as evidence of the predictive power of IQ tests, the IQ scores of these "geniuses" appear to have added little to the prediction of their life outcomes over that gained simply from a consideration of their parental income, education, and occupational status" (p. 59).

17. It appears that little attention was given to the gifted even into the 1950s. The survey by Heck (1953, cited in Tannenbaum, 1983) revealed that of 3,203 cities with populations of 2,500 people or more, only 15 cities (0.5%) had special classes or schools for children with superior measured intelligence.

18. At first observation, this trend of increased interest in gifted children might seem at odds with the earlier quote by Tannenbaum (1983, p. 15), who commented that interest likely hit a low during and after World War II. It appears, however, that Tannenbaum was speaking about the actual practice of "gifted education," whereas Albert (1969) was focusing on the general research/scholarly publication base.

19. The survey was funded under a grant from the Office of Gifted and Talented in the U.S. Office of Education.

20. See *California Administrative Code* (1971, cited in Zettel, 1979).

21. It must be pointed out, however, that the use of intelligence tests in more recent times has not been the exclusive means to identify gifted students. In a national survey of elementary and secondary school programs for gifted students in 1,108 school districts, Cox, Daniel, and Boston (1985) found a variety of procedures that were used to identify gifted students. In descending order, the percentages of districts reporting these procedures were teacher nomination (91%), achievement tests (90%), IQ tests (82%), grades (50%), peer nomination (25%), "other" (22%), par-

ent nomination (21%), self-nomination (6%), creativity measure (6%), and "none" (3%). Although not discussed by Cox et al., we assume that the categories were not mutually exclusive.

22. Renzulli (1986) drew a distinction between "motivation" and "task commitment" (a focused form of motivation): "Whereas motivation is usually defined in terms of a general energizing process that triggers responses in organisms, task commitment represents energy brought to bear on a particular problem (task) or specific performance area" (p. 69).

23. See also Renzulli (1978).

24. For a listing of research studies providing evidence of the effectiveness of the Schoolwide Enrichment Model, see Renzulli and Reis (1997). For a discussion of this research, see Renzulli and Reis (1994).

25. For a full discussion of his triarchic theory of intelligence, see Sternberg (1988).

26. As a case in point regarding deficit thinking views held toward minority children, Margolin (1994) referred to a list of "common descriptors affected by cultural diversity, socioeconomic deprivation, and geographical location" described in *Educational Psychology of the Gifted* by Khatena (1982, pp. 247-249), who in turn drew from Baldwin (1977). Khatena (1982) listed a number of "external and internal deficits" of children that have "possible environmental causality." For example, one deficit is "inability to attend to task without supervision." The likely environmental cause is that "tradition dictates strict adherence to directions" and that "discipline does not encourage inner locus of control" (p. 247). Another deficit is "inability to trust or consider 'beauty' in life." A possible cause: "Environment dictates need to survive. Anger and frustration increase animalistic desire to survive" (pp. 247-248). Similar deficit views toward minority children are seen in Clark's (1997) *Growing Up Gifted.* Although she provided some positive descriptions of minority students, the negative characteristics that she listed can be thought of as being barriers in identifying gifted children. Low-SES African American students, for example, are said to have "unsupportive home environments," "low expectations from family," "manipulative behavior," and "limited experience with varied or extended language patterns" (p. 508). Clark noted that such problems "are typical of the poverty culture" (p. 508; for a critique of the "culture of poverty" model, a deficit thinking perspective, see Foley, 1997). Clark also described other minority group students in negative ways. For example, American Indians have a "thinking style" that is "nonanalytic" (p. 509). Asian Americans have "little experience with independent thinking" (p. 510). Finally, Hispanics hold a "belief in school as primary educator; value of parent as educator is not recognized" (p. 511). It is noteworthy that Clark's negative descriptions of minority students are unchanged from an edition of *Growing Up Gifted* published 14 years earlier (Clark, 1983).

27. Margolin's (1994) reference to a "pedagogy of the oppressed" is in regard to Freire's (1970) *Pedagogy of the Oppressed.*

28. The first paragraph of this section is excerpted, with minor modifications, from Reyes and Valencia (1993, pp. 258-259).

29. Goals 2000: The Educate America Act was signed into law on March 31, 1994. At that time, two additional goals had been added: parental involvement and professional development (Gallagher, 1993).

30. The excellence versus equity debate also is active abroad (e.g., in Russia, Taiwan, South Africa, Israel, France, and Sweden; Gallagher, 1984, 1986).

31. For the position that the elimination of grouping practices would be harmful to U.S. education, see Kulik and Kulik (1997).

32. By underrepresentation, we mean that a minority student group (e.g., African Americans) is disproportionately represented (percentage-wise) in a gifted program relative to the group's percentage in the school population.

33. See, for example, Chinn and Hughes (1987); Baldwin (1977, 1987); Fatemi (1991); Hartz, Adkins, and Sherwood (1984); High and Udall (1983); McKenzie (1986); Passow (1989); Richert (1985a, 1987); Riley (1993); U.S. Department of Education (1991b); Van Tassel-Baska (1991); and Van Tassel-Baska, Patton, and Prillaman (1989).

34. See, for example, Clark (1997); Colangelo and Davis (1997); Davis and Rimm (1998); Khatena (1982); Maker and Schiever (1989a); and Pendarvis, Howley, and Howley (1990).

35. Based on the most recent OCR survey (U.S. Department of Education, OCR, 1997), other categories included, for example, "corporal punishment," "suspension," and "AP [advanced placement] mathematics."

36. The "loss" of 502 school districts (5,173 minus 4,671) is a result of school districts being removed from the sample for a number of reasons (e.g, the school district no longer exists, the school district was consolidated with another district or districts, the school district was found to be an administrative school district rather than a district with students). The 502 figure also represents nonresponding school districts (personal communication, Peter McCabe, July 26, 1999).

37. Disparities were calculated by comparing each racial/ethnic group's percentage in total school enrollment to each group's percentage in the gifted and talented category. The OCR data allow users to access projected or reported data. Projected data, which we selected to use for our disparity analysis, represent the total values that would have resulted if every district in the universe had been sampled.

38. The top 10 states in combined minority K-12 enrollments (American Indian, A/PI, Hispanic, and Black) are as follows:

| | Minorities | | Whites | |
State	Number of Students	Percentage Total Enrollment	Number of Students	Percentage Total Enrollment
1. California	2,924,564	58.9	2,044,070	41.1
2. Texas	1,785,398	52.5	1,614,118	47.5
3. New York	1,160,160	42.1	1,594,022	57.9
4. Florida	847,219	41.3	1,205,685	58.7
5. Illinois	691,816	36.9	1,184,043	63.1
6. Georgia	533,874	42.7	717,555	57.3
7. New Jersey	423,493	35.9	754,797	64.1
8. North Carolina	415,626	35.6	752,929	64.4
9. Virginia	371,486	34.2	715,971	65.8
10. Louisiana	368,527	46.9	417,934	53.1

SOURCE: U.S. Department of Education, Office for Civil Rights (1997).

39. Here, a predominantly White school is defined as having an enrollment of 70% or greater White students. A predominantly minority school is defined as having an enrollment of 70% or greater combined minority enrollment.

40. The year-to-year fluctuations probably are due to methodological factors (e.g., different size samples, nonrandom samples, the OCR not checking on the accuracy of the school or district reports; Finn, 1982).

41. See, for example, National/State Leadership Training Institute, Office of Ventura County Superintendent of Schools (1981); Baldwin, Gear, and Lucito (1978); Callahan and McIntire (1994); Kitano and Chinn (1986); and Maker and Schiever (1989a). See also sections/chapters on gifted minorities in major texts on the gifted and their education (citations listed in Note 34 of the present chapter).

42. Bond (1927) reported that his 30 participants included children from "professional . . . . , middle-class . . . . , and laboring homes" (p. 259). He did not, however, report how the children were selected, nor did he provide their ages.

43. Bond's (1927) participants were indeed exceptional. Of the 30 children, nearly half ($n = 14$) scored IQs higher than 125.

44. See Jenkins (1936, p. 178, Table 1). In most of the 15 studies, the focus was not on gifted African American children but rather on racial comparisons in intelligence (e.g., Goodenough, 1926b; Hewitt, 1930). All of the investigations, however, included some African American children with IQs higher than 120.

45. Examples of other publications on gifted African American children and youths during this early period are Bond (1959), Long (1935), and Valien and Horton (1954).

46. Notwithstanding this praise of Jenkins's (1935) dissertation, Wilkerson (1936) also offered some concerns (e.g., representativeness of the seven elementary schools, possible experimenter bias).

47. This extremely bright child mentioned in Guthrie (1976) was discussed in "The Case of 'B': A Gifted Negro Girl" (Witty & Jenkins, 1935). This child (9 years, 4 months of age), first identified in Jenkins (1935), had a Stanford-Binet IQ of 200 and was described by Witty and Jenkins (1935) as follows: "Undoubtedly . . ., [B] is one of the most precocious and promising children in America" (p. 118). B, a student in the Chicago public schools, was described as being quite precocious in language development during her toddler years. As a student, "B's preferred leisure-time activity is reading; she spends about 28 hours weekly in this activity" (p. 123). B's parents were college educated; her mother was a schoolteacher, and her father was an electrical engineer. Regarding B's racial ancestry, "The mother reports B to be of pure Negro stock. There is no record of any white ancestors on either the maternal or paternal side" (p. 123).

48. We urge the reader to examine the article by Ford and Webb (1994) for details germane to these reasons that help to explain the underrepresentation of African American students. For example, regarding the "misunderstanding of and inattention to cultural differences in learning" (p. 363) of African American students, Ford and Webb drew from several works (i.e., Dunn & Griggs, 1990; Hale-Benson, 1986; Hilliard, 1992). Based on these studies, Ford and Webb (1994) asserted, "Several researchers have noted differences in learning styles among African American and White students. . . . Specifically, African Americans tend to be relational, visual, mobile/kinesthetic, concrete/global tactile learners, while school success in the United States is heavily dependent upon abstract, auditory, less mobile, tactile, and kinesthetic learning" (p. 363).

49. Our assessment of the literature on gifted African American students, as well as on other gifted minorities, is that the predominant proportion of these writings focuses on (a) factors that hinder the identification of gifted minority students and (b) principles and strategies that can lead to increased identification of such students (Baldwin, 1987; Ford, 1995; Ford & Harris, 1990; Ford & Webb, 1994; Frasier, 1987; Gay, 1978; Hamilton, 1993; Harris & Ford, 1991; Patton, 1992; Rhodes, 1992). By sharp contrast, there is a small proportion of empirically based investigations that actually are designed to demonstrate the effectiveness of implementing innovative procedures that lead to increased minority representation in gifted programs. We are not suggesting, however, that gifted African American students and other gifted minority students cannot be identified using traditional standardized intelligence tests. For example, regarding Mexican American students, as a case in point, Ortiz and Volloff (1987) reported that the Wechsler Intelligence Scale for Children–Revised (WISC-R) was quite effective in identifying migrant Mexican American students (Grades 3 to 6) in California who were referred by their teachers for possible placement in a gifted program. Compared to a number of group-based intelligence, achievement, and self-esteem measures, the WISC-R Performance Scale IQ and Full Scale IQ were the most effective in the identification of high-IQ students. If an IQ of 130 or higher was the standard used for identification of gifted students, then 32% and 23% of the students would have been identified based on their Performance Scale IQ and Full Scale IQ, respectively. By stark contrast, 0% of the students would have been identified based on the Otis-Lennon School Ability Test (OLSAT; Otis & Lennon, 1982), a group test. Ortiz and Volloff (1987) contended that because the children in their sample had reading skills only somewhat higher than the average range, these students' "performance on school ability tests like the OLSAT may be more of a function of reading skills . . . than of school ability" (p. 51). The authors suggested that group-based screening tests emphasizing reading skills might be inappropriate for identifying migrant Mexican American children and that indi-

vidually administered intelligence tests (e.g., the WISC-R) might be more appropriate instruments. On a final note, given the time required to administer the WISC-R, Ortiz and González (1989) undertook a study in which they provided validity data for a short form of the WISC-R. The abbreviated version was found to be effective as a screening tool in a sample of migrant Mexican American students referred for possible gifted program placement. The authors did comment, however, that the WISC-R short form should not be the only criterion used for the selection of these students.

50. The group tests were the California Achievement Test and the Test of Cognitive Skills. "Students who scored ≥ 85th percentile in a designated number of areas were eligible to participate in the target group" (Woods & Achey, 1990, p. 22).

51. The group meetings with parents were comprehensive in information dissemination (e.g., how children became eligible, evaluation procedures used, invitation to visit the schools).

52. The group-administered achievement test was the Comprehensive Test of Basic Skills. The group-administered intelligence test was the OLSAT. Regarding the individually administered intelligence tests, the WISC-R and the Kaufman Assessment Battery for Children (K-ABC, both the Mental Processing Composite and Achievement Scales) were used.

53. The number of White students increased 12% (n = 653 to n = 732).

54. The second most frequently occurring combination of tests was the WISC-R and the K-ABC Achievement scale.

55. For the non-SOMPA group, the mean IQs of children who were given the WISC-R and Stanford-Binet were 131.6 and 133.4, respectively. For the SOMPA group, the initial mean WISC-R IQ was 119.0. After SOMPA sociocultural adjustment procedures were done, the adjusted mean was 135.7.

56. The Ross Test of Higher Cognitive Processes (no author, date, or publisher provided by Matthew et al., 1992) was normed on gifted students (Grades 4 to 6) in Washington State and is designed to measure a student's ability to understand verbal abstractions (e.g., deductive reasoning, analogies).

57. For criticisms of the SOMPA, see Brown (1979a, 1979b), Clarizio (1979a, 1979b), Goodman (1979a, 1979b), and Oakland (1979a, 1979b). For a response to these criticisms, see Mercer (1979).

58. The SOMPA also has been shown to be quite effective in identifying gifted Mexican American children. See, for example, Mercer (1977).

59. Two explanations can be advanced to explain the historical inattention given to gifted Latino students. First, in sharp contrast to early African American scholars who were interested in mental testing (small in number but very active in research and publications), there was only one Latino scholar, George Sánchez (Valencia, 1997b; see also Chapter 1 of the present book). Second, scholarly work on Latino intellectual performance largely focused on the issue of White-Latino differences in measured intelligence (see also Chapter 1).

60. Examples of other earlier studies in which some Mexican American students attained exceptionally high IQs are, chronologically, Sheldon (1924), Terman (1925), Goodenough (1926b), Randalls (1929), Keston and Jiménez (1954), and Ratliff (1960).

61. By contrast, White students made up 71.0% of the total K-12 population but constituted 87.5% of the gifted population, an overrepresentation of 23.2%.

62. Examples of such publications are Bernal (1979), DeLeon (1983), Gratz and Pulley (1984), Melesky (1985), Perrine (1989), Riojas Clark (1981), and Zappia (1989).

63. For other empirical investigations in which the SRBCSS has been used in the identification of gifted Mexican American students, see Argulewicz, Elliott, and Hall (1982); Argulewicz and Kush (1984); Elliott and Argulewicz (1983); and High and Udall (1983).

64. These studies referred to by High and Udall (1983) are Jensen and Rosenfeld (1974), Leacock (1969), and Yee (1968).

65. For selection to the gifted program, the district used a nonweighted matrix consisting of several indicators—grades, parents' and teachers' nominations, standardized achievement tests, and group- and individually administered intelligence tests.

66. Examples of such publications are Chan and Kitano (1986), Chen (1989), Chen and Goon (1976), Gallagher (1989), Hasegawa (1989), Kaplan (1989), Kitano (1986, 1989), Larson (1989), Plucker (1996), Tanaka (1989), Woliver and Woliver (1991), Wong and Wong (1989), and Woo (1989).

67. Plucker (1996) referred to "Asian Americans." We believe that his assertions also can generalize to Pacific Islanders.

68. Much of the literature cited in Note 66 discusses this diversity among and within A/PI people.

69. For literature that has attempted to debunk the model minority myth, see Chen (1989), Gallagher (1989), Hartman and Askounis (1989), Hu (1989), Kim and Hurh (1983), Lesser (1985-1986), Ong (1984), and Plucker (1996).

70. For data to support this statement, see Chapa and Valencia (1993, p. 177, Table 10). See also Hu (1989).

71. This is not to suggest, however, that low-SES A/PI students cannot be identified as gifted. See, for example, Chen and Goon's (1976) study of gifted low-SES Asian American children in Manhattan's Lower East Side (greater Chinatown area) in New York City.

72. One must be very cautious in concluding that parents of students who are underrepresented among the gifted (e.g., Latinos) do not value education. The perception that Latinos do not value the importance of education is false and can be rebuked by the available evidence (see, e.g., Valencia & Solórzano, 1997).

73. Examples of other investigations from the 1920s in which some American Indian children and youths performed at relatively high levels on standardized intelligence tests are Fitzgerald and Ludeman (1926); Garth, Schuelke, and Abell (1927); Garth, Serafini, and Dutton (1925); and Hunter and Sommermier (1922).

74. Examples of publications on gifted American Indian students are Bradley (1989), Brittan and Tonemah (1985), Brooks (1989), Callahan and McIntire (1994), Christensen (1991), Daniels (1988), Garrison (1989), George (1989), Hartley (1991), Kirschenbaum (1988, 1989), Maker and Schiever (1989b), Montgomery (1989), Pfeiffer (1989), Robbins (1991), Romero (1994), Scruggs and Cohn (1983), Sisk (1989), Throssel (1989), and Tonemah (1987).

75. A common (and false) perception is that many American Indian children and youths attend separate reservation schools. The vast majority (85%) of American Indian students are enrolled in regular public schools. Another 12% of the students are enrolled in schools funded by the Bureau of Indian Affairs (i.e., bureau-operated, grant, or contract), and 3% attend private or mission schools (Callahan & McIntire, 1994).

76. See Ross (1989), who asserts this "right brain" position. For a refutation of the "right-brained Indian" thesis, see Chrisjohn and Peters (1989).

77. There are three distinct language families among the New Mexico Pueblos: Keresan, Zunian, and Tanoan. Romero's (1994) study focused on the seven Keresan-speaking communities of the Pueblos (e.g., Acoma, Laguna, Zia, Cochiti).

78. It appears that the study by Scott et al. (1992) was conducted in Miami, Florida. As such, it is very likely that the vast majority of the Hispanic students were Cuban American.

Chapter 9

1. "Racial/ethnic bias" is what we referred to as "cultural bias" in Chapter 5.

2. Independent investigations by Valencia and associates have provided some confirmations that Hispanic (i.e., Mexican American) children tend to score at normative means on the K-ABC Mental Processing Scale and Nonverbal Scale and below the normative mean on the K-ABC Achievement scale (see, e.g., Valencia, Rankin, & Livingston, 1995).

3. For other criticisms of the SOMPA, see references cited in Note 57 of Chapter 8 of the present book.

References

Aaron, P. G. (1997). The impending demise of the discrepancy formula. *Review of Educational Research, 67*, 461-502.

Adler, M. (1967). Reported incidence of giftedness among ethnic groups. *Exceptional Children, 34*, 101-105.

Aiken, L. S., West, S. G., Sechrest, L., Reno, R. R., Roediger, H. C., III, Scarr, S., Kazdin, A. E., & Sherman, S. J. (1990). Graduate training in statistics, methodology, and measurement in psychology: A survey of Ph.D. programs in North America. *American Psychologist, 45*, 721-734.

Albert, R. S. (1969). Genius: Present-day status of the concept and its implications for the study of creativity and giftedness. *American Psychologist, 24*, 743-753.

Alley, G., & Foster, C. (1978). Nondiscriminatory testing of minority and exceptional children. *Focus on Exceptional Children, 9*, 1-14.

American Association on Mental Retardation. (1999). *The definition of mental retardation* [on-line]. Available: http://www.aamr.org/policies/faqmentalretardation.html

American Educational Research Association, American Psychological Association, & National Council on Measurement in Education. (1985). *Standards for educational and psychological testing.* Washington, DC: Author.

Anastasi, A. (1958a). *Differential psychology.* New York: Macmillan.

Anastasi, A. (1958b). Heredity, environment, and the question "how." *Psychological Review, 65*, 197-208.

Anastasi, A. (1988). *Psychological testing* (6th ed.). New York: Macmillan.

Anastasi, A., & D'Angelo, R. V. (1952). A comparison of Negro and White preschool children in language development and Goodenough Draw-a-Man IQ. *Journal of Genetic Psychology, 81*, 147-165.

Angoff, W. H. (1982). Use of difficulty and discrimination indices for detecting item bias. In R. A. Berk (Ed.), *Handbook of methods for detecting test bias* (pp. 96-116). Baltimore, MD: Johns Hopkins University Press.

Angoff, W. H. (1988). The nature-nurture debate, aptitudes, and group differences. *American Psychologist, 43,* 713-720.

Apple, M. W. (1992a). [Review of the book *The Burt Affair*]. *Isis, 83,* 699-700.

Apple, M. W. (1992b). [Review of the book *Science, Ideology, and the Media: The Cyril Burt Scandal*]. *Isis, 83,* 699-700.

Argulewicz, E. N., Elliott, S. N., & Hall, R. (1982). Comparison of behavioral ratings of Anglo-American and Mexican-American gifted children. *Psychology in the Schools, 19,* 469-472.

Argulewicz, E. N., & Kush, J. C. (1984). Concurrent validity of the SRBCSS Creativity Scale for Anglo-American and Mexican-American gifted students. *Educational and Psychological Research, 4,* 81-89.

Arlitt, A. H. (1921). On the need for caution in establishing race norms. *Journal of Applied Psychology, 5,* 179-183.

Armour-Thomas, E. (1992). Intellectual assessment of children from culturally diverse backgrounds. *School Psychology Review, 21,* 552-565.

Armour-Thomas, E. (1999, August). [Discussant.] In L. Suzuki (Chair), *Intelligence 2000: A multicultural mosaic* (symposium conducted at the meeting of the American Psychological Association, Boston).

Armour-Thomas, E., & Gopaul-McNicol, S. (1998). *Assessing intelligence: Applying a bicultural model.* Thousand Oaks, CA: Sage.

Artiles, A. J., & Trent, S. C. (1994). Overrepresentation of minority students in special education: A continuing debate. *Journal of Special Education, 27,* 410-437.

Ayres, L. P. (1911). The Binet-Simon measuring scale for intelligence: Some criticisms and suggestions. *Psychological Clinic, 5,* 187-196.

Azuma, H., & Kashiwagi, K. (1987). Descriptors for an intelligent person: A Japanese study. *Japanese Psychological Research, 29,* 17-26.

Baca, L., & Chinn, P. C. (1982). Coming to grips with cultural diversity. *Exceptional Education Quarterly, 2,* 33-45.

Bacharach, S. B. (1991). *Education reform: Making sense of it all.* Boston: Allyn & Bacon.

Baddeley, A. D. (1976). *The psychology of memory.* New York: Basic Books.

Baddeley, A. D., Thomson, N., & Buchanan, M. (1975). Word length and the structure of short-term memory. *Journal of Verbal Learning and Verbal Behavior, 14,* 575-589.

Bagley, W. C. (1922). Educational determinism: Or democracy and the I.Q. *School and Society, 15,* 373-384.

Baldwin, A. Y. (1977). Tests do underpredict: A case study. *Phi Delta Kappan, 59,* 620-621.

Baldwin, A. Y. (1987). I'm Black but look at me, I am also gifted. *Gifted Child Quarterly, 31,* 180-185.

Baldwin, A. Y., Gear, G. H., & Lucito, L. J. (Eds.). (1978). *Educational planning for the gifted: Overcoming cultural, geographic, and socioeconomic barriers.* Reston, VA: Council for Exceptional Children.

Barkan, J. H., & Bernal, E. M., Jr. (1991). Gifted education for bilingual and limited English proficient students. *Gifted Child Quarterly, 35,* 144-147.

Barker, R., & Wright, A. (1954). *Midwest and its children.* Evanston, IL: Row, Peterson.

Barnard, K., Bee, H., & Hammond, M. (1984). Home environment and cognitive development in a healthy, low-risk sample: The Seattle Study. In A. W. Gottfried (Ed.), *The home environment and early cognitive development: Longitudinal research* (pp. 117-150). Orlando, FL: Academic Press.

Barona, A. (1989). Differential effects of WISC-R factors on special education eligibility for three ethnic groups. *Journal of Psychoeducational Assessment, 7,* 31-38.

Baumrind, D. (1991). To nurture nature. *Behavioral and Brain Sciences, 14,* 386-387.

Baumrind, D. (1993). The average expectable environment is not good enough: A response to Scarr. *Child Development, 64,* 1299-1317.

Bayley, N. (1969). *Bayley Scales of Infant Development: Birth to two years.* New York: Psychological Corporation.

Beckham, A. S. (1935). A study of the intelligence of colored adolescents of different social-economic status in typical metropolitan areas. *Journal of Social Psychology, 4,* 70-90.

Beckwith, L., & Cohen, S. E. (1984). Home environment and cognitive competence in 46 preterm children during the first 5 years. In A. W. Gottfried (Ed.), *The home environment and early cognitive development: Longitudinal research* (pp. 235-271). Orlando, FL: Academic Press.

Bell, R. Q. (1974). Contributions of human infants to caregiving and social interaction. In M. Lewis & L. Rosenblum (Eds.), *The effect of the infant on its caregiver* (pp. 1-20). New York: John Wiley.

Bell, R. Q., & Harper, L. V. (1977). *Child effects on adults.* Hillsdale, NJ: Lawrence Erlbaum.

Belmont, J. M. (1994). Mental retardation, organic. In R. J. Sternberg (Ed.), *The encyclopedia of human intelligence* (pp. 717-724). New York: Macmillan.

Beloff, H. (1990). [Review of the book *The Burt Affair*]. *British Journal of Psychology, 81,* 395-397.

Bergeman, C. S., & Plomin, R. (1988). Parental mediation of the genetic relationship between home environment and infant mental development. *British Journal of Developmental Psychology, 6,* 11-19.

Berk, R. A. (Ed.). (1982). *Handbook of methods for detecting test bias.* Baltimore, MD: Johns Hopkins University Press.

Berliner, D. C. (1988). Meta-comments: A discussion of critiques of L. M. Dunn's monograph, *Bilingual Hispanic Children on the U.S. Mainland. Hispanic Journal of Behavioral Sciences, 10,* 273-299.

Bernal, E. M., Jr. (1974). Gifted Mexican-American children: An ethno-scientific perspective. *California Journal of Educational Research, 25,* 261-273.

Bernal, E. M., Jr. (1978). The identification of gifted Chicano children. In A. Y. Baldwin, G. H. Gear, & L. J. Lucito (Eds.), *Educational planning for the gifted: Overcoming cultural, geographic, and socioeconomic barriers* (pp. 14-17). Reston, VA: Council for Exceptional Children.

Bernal, E. M., Jr. (1979). The education of the culturally different and gifted. In H. Passow (Ed.), *The gifted and talented: Their education and development—The seventy-fifth yearbook of the National Society for the Study of Education, Part I* (pp. 395-400). Chicago: University of Chicago Press.

Bernal, E. M., Jr., & Reyna, J. (1975). Analysis and identification of giftedness in Mexican American children: A pilot study. In B. O. Boston (Ed.), *A resource manual of information on educating the gifted and talented* (pp. 53-60). Reston, VA: Council for Exceptional Children.

Berry, J. W. (1976). Radical cultural relativism and the concept of intelligence. In J. W. Berry & P. R. Dasen (Eds.), *Culture and cognition* (pp. 225-240). London: Methuen.

Berry, J. W., & Irvine, S. H. (1986). Bricolage: Savages do it daily. In R. J. Sternberg & R. K. Wagner (Eds.), *Practical intelligence: Nature and origins of competence in the everyday world* (pp. 271-306). New York: Cambridge University Press.

Bersoff, D. N. (1982). The legal regulation of psychology. In C. R. Reynolds & T. B. Gutkin (Eds.), *The handbook of school psychology* (pp. 1043-1074). New York: John Wiley.

Bersoff, D. N. (1984). Social and legal influences on test development and usage. In B. S. Plake (Ed.), *Social and technical issues in testing: Implications for test construction and usage* (pp. 87-109). Hillsdale, NJ: Lawrence Erlbaum.

Bidell, T. R., & Fischer, K. W. (1997). Between nature and nurture: The role of human agency in the epigenesis of intelligence. In R. J. Sternberg & E. Grigorenko (Eds.), *Intelligence, heredity, and environment* (pp. 193-242). New York: Cambridge University Press.

Binet, A., & Henri, V. (1895). La psychologie individuelle [Psychology of the individual]. *Anneé Psychologique, 2,* 411-463.

Bird, C., Monachesi, E. O., & Burdick H. (1952). Infiltration and the attitudes of White and Negro parents and children. *Journal of Abnormal and Social Psychology, 47,* 688-699.

Blau, Z. S. (1981). *Black children/White children: Competence, socialization, and social structure.* New York: Free Press.

Blinkton, S. (1989). [Review of the book *The Burt Affair*]. *Nature, 340,* 439-440.

Block, N. (1995). How heritability misleads about race. *Cognition, 56,* 99-128.

Block, N. J., & Dworkin, G. (Eds.). (1976). *The IQ controversy: Critical readings.* New York: Pantheon Books.

Bloom, B. S. (1964). *Stability and change in human characteristics.* New York: John Wiley.

Blum, J. M. (1978). *Pseudoscience and mental ability: The origins and fallacies of the IQ controversy.* New York: Monthly Review Press.

Bodner, F. W., & Cavalli-Sforza, L. L. (1970). Intelligence and race. *Scientific American, 223,* 19-29.

Boehm, A. E. (1971). *Boehm Test of Basic Concepts.* New York: Psychological Corporation.

Bond, H. M. (1924). What the army "intelligence" tests measured. *Opportunity, 2,* 197-202.

Bond, H. M. (1927). Some exceptional Negro children. *Crisis, 34,* 257-259, 278.

Bond, H. M. (1958). Cat on a hot tin roof. *Journal of Negro Education, 27,* 519-525.

Bond, H. M. (1959). Talent and toilets. *Journal of Negro Education, 28,* 8-14.

Boocock, S. S. (1972). *An introduction to the sociology of learning.* Boston: Houghton Mifflin.

Boring, E. G. (1923, June). Intelligence as the tests test it. *New Republic,* pp. 35-37.

Bouchard, T. J., Jr. (1982a). [Review of the book *Identical Twins Reared Apart: A Reanalysis*]. *Contemporary Psychology, 27,* 190-191.

Bouchard, T. J., Jr. (1982b). [Review of the book *The Intelligence Controversy*]. *American Journal of Psychology, 95,* 346-349.

Bouchard, T. J., Jr. (1983). Do environmental similarities explain the similarity in intelligence of identical twins reared apart? *Intelligence, 7,* 175-184.

Bouchard, T. J., Jr. (1997). IQ similarity in twins reared apart: Findings and responses to critics. In R. J. Sternberg & E. Grigorenko (Eds.), *Intelligence, heredity, and environment* (pp. 126-160). New York: Cambridge University Press.

Bouchard, T. J., Jr., & McGue, M. (1981). Familial studies of intelligence: A review. *Science, 212,* 1055-1059.

Bourdieu, P., & Passeron, J.-C. (1977). *Reproduction in education, society, and culture.* London: Sage.

Bowles, S., & Gintis, H. (1976). *Schooling in capitalist America: Educational reform and the contradictions of economic life.* New York: Basic Books.

Bracken, B. A., Howell, K. K., & Crain, R. M. (1993). Prediction of Caucasian and African-American preschool children's fluid and crystallized intelligence: Contributions of maternal characteristics and home environment. *Journal of Clinical Child Psychology, 22,* 455-464.

Bracken, B. A., & McCallum, R. S. (1998). *Universal Nonverbal Intelligence Test.* Itasca, IL: Riverside.

Bracken, B. A., & Walker, K. C. (1997). The utility of intelligence tests for preschool children. In D. P. Flanagan, J. P. Genshaft, & P. L. Harrison (Eds.), *Contemporary intellectual assessment: Theories, tests, and issues* (pp. 484-502). New York: Guilford.

Bradley, C. (1989). Give me the bow, I've got the arrow. In C. J. Maker & S. W. Schiever (Eds.), *Critical issues in gifted education: Defensible programs for cultural and ethnic minorities* (pp. 133-137). Austin, TX: PRO-ED.

Bradley, R. H. (1994). [Review of the book *The Limits of Family Influence: Genes, Experience, and Behavior*]. *Journal of Marriage and the Family, 56,* 779-780.

Bradley, R. H., & Caldwell, B. M. (1976a). Early home environment and changes in mental test performance in children from 6 to 36 months. *Developmental Psychology, 12,* 93-97.

Bradley, R. H., & Caldwell, B. M. (1976b). The relation of infants' home environments to mental test performance at fifty-four months: A follow-up study. *Child Development, 47,* 1172-1174.

Bradley, R. H., & Caldwell, B. M. (1977). Home Observation for Measurement of the Environment: A validation study of screening efficiency. *American Journal of Mental Deficiency, 81,* 417-420.

Bradley, R. H., & Caldwell, B. M. (1978). Screening the environment. *American Journal of Orthopsychiatry, 48,* 114-130.

Bradley, R. H., & Caldwell, B. M. (1980). The relation of home environment to cognitive competence and IQ among males and females. *Child Development, 51,* 1140-1148.

Bradley, R. H., & Caldwell, B. M. (1981). The HOME inventory: A validation of the preschool scale for Black children. *Child Development, 53,* 708-710.

Bradley, R. H., & Caldwell, B. M. (1982). The consistency of the home environment and its relation to child development. *International Journal of Behavioral Development, 5,* 445-465.

Bradley, R. H., & Caldwell, B. M. (1984). 177 children: A study of the relationship between home environment and cognitive development during the first five years. In A. W. Gottfried (Ed.), *Home environment and early cognitive development: Longitudinal research* (pp. 5-56). Orlando, FL: Academic Press.

Bradley, R. H., & Caldwell, B. M. (1991). Like images refracted: A view from the interactionist perspective. *Behavioral and Brain Sciences, 14,* 389-390.

Bradley, R. H., Caldwell, B. M., & Elardo, R. (1977). Home environment, social status, and mental test performance. *Journal of Educational Psychology, 69,* 697-701.

Bradley, R. H., Caldwell, B. M., & Elardo, R. (1979). Home environment and cognitive development in the first two years: A cross-lagged panel analysis. *Developmental Psychology, 15,* 246-250.

Bradley, R. H., Caldwell, B. M., Rock, S., Barnard, K., Gray, C., Hammond, M., Mitchell, S., Siegel, L., Ramey, C., Gottfried, A. W., & Johnson, D. L. (1989). Home environment and cognitive development in the first 3 years of life: A collaborative study involving six sites and three ethnic groups in North America. *Developmental Psychology, 28,* 217-235.

Brady, P. M., Manni, J. L., & Winikur, D. W. (1983). Implications of ethnic disproportion in programs for the educable mentally retarded. *Journal of Special Education, 17,* 295-302.

Bridges, J. W., & Coler, L. E. (1917). The relation of intelligence to social class. *Psychological Review, 24,* 1-31.

Brigham, C. C. (1923). *A study of American intelligence.* Princeton, NJ: Princeton University Press.

Brittan, M., & Tonemah, S. (1985). *American Indian Gifted and Talented Assessment Model (AIGTAM).* Norman, OK: American Indian Research and Development.

Brody, E. B., & Brody, N. (1976). *Intelligence: Nature, determinants, and consequences.* New York: Academic Press.

Brody, N. (1992). *Intelligence* (2nd ed.). San Diego, CA: Academic Press.

Bronfenbrenner, U., & Ceci, S. J. (1993). Heredity, environment, and the question "how": A first approximation. In R. Plomin & G. E. McClearn (Eds.), *Nature, nurture, and psychology* (pp. 313-324). Washington, DC: American Psychological Association.

Brooks, D. L. (1989). Administrative implications for administering programs for gifted and talented American Indian students. In C. J. Maker & S. W. Schiever (Eds.), *Critical issues in gifted education: Defensible programs for cultural and ethnic minorities* (pp. 138-141). Austin, TX: PRO-ED.

Brooks-Gunn, J., Klebanov, P. K., & Duncan, G. J. (1996). Ethnic differences in children's intelligence test scores: Role of economic deprivation, home environment, and maternal characteristics. *Child Development, 67,* 396-408.

Brosman, F. L. (1983). Overrepresentation of low-socioeconomic minority students in special education programs in California. *Learning Disability Quarterly, 6,* 517-525.

Brown, F. (1944). An experimental and critical study of the intelligence of Negro and White kindergarten children. *Journal of Genetic Psychology, 65,* 161-175.

Brown, F. G. (1979a). The algebra works—But what does it mean? *School Psychology Digest, 8,* 213-218.

Brown, F. G. (1979b). The SOMPA: A system of measuring potential abilities. *School Psychology Digest, 8,* 37-46.

Brown v. Board of Education of Topeka, 347 U.S. 483 at 494 (1954).

Bullock, D. (1990). Methodological heterogeneity and the anachronistic status of ANOVA in psychology. *Behavioral and Brain Sciences, 13,* 122-123.

Bunch, G. (1994). Mainstreaming. In R. J. Sternberg (Ed.), *The encyclopedia of human intelligence* (pp. 683-688). New York: Macmillan.

Burks, B. S. (1928). The relative influence of nature and nurture upon mental development: A comparative study of foster parent-foster child resemblance and true parent-true child resemblance. *Yearbook of the National Society for the Study of Education, 32,* 219-316.

Burrill, L. E. (1982). Comparative studies of item bias methods. In R. A. Berk (Ed.), *Handbook of methods for detecting test bias* (pp. 161-179). Baltimore, MD: Johns Hopkins University Press.

Burt, C. (1966). The genetic determination of differences in intelligence. *British Journal of Psychology, 57,* 137-153.

Burt, C. (1969). Intelligence and heredity: Some common misconceptions. *Irish Journal of Education, 3,* 75-94.

Burt, C. (1972). Inheritance of general intelligence. *American Psychologist, 27,* 175-190.

Butcher, J. N. (1982). Cross-cultural research methods in clinical psychology. In P. C. Kendall & J. N. Butcher (Eds.), *Handbook of research methods in clinical psychology* (pp. 273-308). New York: John Wiley.

Byrns, R., & Henmon, V. A. C. (1936). Parental occupation and mental ability. *Journal of Educational Psychology, 27,* 284-291.

Cain, D. P., & Vanderwolf, C. H. (1990). A critique of Rushton on race, brain size, and intelligence. *Personality and Individual Differences, 11,* 777-784.

Caldwell, B. M., & Bradley, R. (1984). *Home Observation for the Measurement of the Environment.* Little Rock, AR: Authors.

Caldwell, B. M., Elardo, P., & Elardo, R. (1972, Spring). *The observational and intervention study.* Paper presented at the Southeastern Conference on Human Development, Williamsburg, VA.

Caldwell, B. M., Heider, J., & Kaplan, B. (1966). *The inventory of home stimulation.* Paper presented at the meeting of the American Psychological Association, Washington, DC.

California Administrative Code, Chapter 6, Article 14, Section 6421 (1971).

Callahan, C. M. (1989-present). Review of the Gifted Evaluation Scale. In *Buro's mental measurements yearbook* (WinSPIRS database).

Callahan, C. M., & McIntire, J. A. (1994). *Identifying outstanding talent in American Indian and Alaska Native students.* Washington, DC: U.S. Department of Education, Office of Educational Research and Improvement.

Callahan, R. E. (1962). *Education and the cult of efficiency.* Chicago: University of Chicago Press.

Camilli, G., & Shepard, L. A. (1987). The inadequacy of ANOVA for detecting test bias. *Journal of Educational Statistics, 12,* 87-99.

Cancro, R. (Ed.). (1971). *Intelligence: Genetic and environmental influences.* New York: Grune & Stratton.

Carroll, J. B. (1993). *Human cognitive abilities: A survey of factor-analytic studies.* New York: Cambridge University Press.

Carroll, J. B. (1997). The three-stratum theory of cognitive abilities. In D. P. Flanagan, J. L. Genshaft, & P. L. Harrison (Eds.), *Contemporary intellectual assessment: Theories, tests, and issues* (pp. 122-130). New York: Guilford.

Carspecken, P. F. (1996). The set-up: Crocodile tears for the poor. In J. L. Kinchloe, S. R. Steinberg, & A. D. Gresson (Eds.), *Measured lies: The bell curve examined* (pp. 109-125). New York: St. Martin's.

Cary, R. (1996). IQ as commodity: The "new" economics of intelligence. In J. L. Kinchloe, S. R. Steinberg, & A. D. Gresson (Eds.), *Measured lies: The bell curve examined* (pp. 137-160). New York: St. Martin's.

Cattell, J. McK. (1947). *James McKeen Cattell: Man of science* (Vol. 2). Lancaster, PA: Science Press.

Cattell, R. B. (1941). Some theoretical issues in adult intelligence testing. *Psychological Bulletin, 38,* 592.

Cattell, R. B. (1963). Theory of fluid and crystallized intelligence. A critical experiment. *Journal of Educational Psychology, 54,* 1-22.

Cavalli-Sforza, L. L., Menozzi, P., & Piazza, A. (1994). *The history and geography of human genes.* Princeton, NJ: Princeton University Press.

Ceci, S. J. (1990). *On intelligence . . . more or less: A bio-ecological treatise on intellectual development.* Englewood Cliffs, NJ: Prentice Hall.

Ceci, S. J. (1991). How much does schooling influence general intelligence and its cognitive components? A reassessment of the evidence. *Developmental Psychology, 27,* 703-722.

Ceci, S. J. (1996). *On intelligence . . . more or less: A bioecological treatise on intellectual development* (Expanded ed.). Cambridge, MA: Harvard University Press.

Ceci, S. J., Rosenblum, T., de Bruyn, E., & Lee, D. Y. (1997). A bio-ecological model of intellectual development: Moving beyond h^2. In R. J. Sternberg & E. Grigorenko (Eds.), *Intelligence, heredity, and environment* (pp. 303-322). New York: Cambridge University Press.

Chan, K. S., & Kitano, M. K. (1986). Demographic characteristics of exceptional Asian students. In M. K. Kitano & P. C. Chinn (Eds.), *Exceptional Asian children and youth* (pp. 1-11). Reston, VA: Council for Exceptional Children.

Chan, S. (1986). Parents of exceptional Asian children. In M. K. Kitano & P. C. Chinn (Eds.), *Exceptional Asian children and youth* (pp. 36-53). Reston, VA: Council for Exceptional Children.

Chapa, J., & Valencia, R. R. (1993). Latino population growth, demographic characteristics, and educational stagnation: An examination of recent trends. *Hispanic Journal of Behavioral Sciences, 15*, 165-187.

Chapin, F. S. (1933). *The measurement of social status*. Minneapolis: University of Minnesota Press.

Chapman, J., & Wiggins, D. M. (1925). Relation of family size to intelligence of offspring and socio-economic status of family. *Pedagogical Seminary, 32*, 414-421.

Chapman, P. D. (1988). *Schools as sorters: Lewis M. Terman, applied psychology, and the intelligence testing movement, 1890-1930*. New York: New York University Press.

Charters, W. W., Jr. (1963). *The social background of teaching*. In N. L. Gage (Ed.), *Handbook of research on teaching* (pp. 715-813). Chicago: Rand McNally.

Chauncey, M. R. (1929). The relation of the home factor to achievement and intelligence test scores. *Journal of Educational Research, 20*, 88-90.

Chen, J. (1989). Identification of gifted Asian-American students. In C. J. Maker & S. W. Schiever (Eds.), *Critical issues in gifted education: Defensible programs for cultural and ethnic minorities* (pp. 154-162). Austin, TX: PRO-ED.

Chen, J., & Goon, S. W. (1976). Recognition of the gifted from among disadvantaged Asian children. *Gifted Child Quarterly, 20*, 157-164.

Chen, M. J., & Chen, H. C. (1988). Concepts of intelligence: A comparison of Chinese undergraduates from Chinese and English schools in Hong Kong. *International Journal of Psychology, 23*, 471-487.

Chinn, P. C., & Hughes, S. (1987). Representation of minority students in special education classes. *Remedial and Special Education, 8*, 41-46.

Chipeur, H. M., Rovine, M., & Plomin, R. (1990). LISREL modeling: Genetic and environmental influences on IQ revisited. *Intelligence, 14*, 11-29.

Chiszar, D. A., & Gollin, E. S. (1990). Additivity, interaction, and developmental good sense. *Behavioral and Brain Sciences, 13*, 124-125.

Chittooran, M. M., & Miller, T. L. (1998). Informal assessment. In H. Booney Vance (Ed.), *Psychological assessment of children: Best practices for school and clinical settings* (2nd ed., pp. 13-59). New York: John Wiley.

Chorover, S. L. (1979). *From genius to genocide: The meaning of human nature and the power of behavior control*. Cambridge, MA: MIT Press.

Chrisjohn, R. D., & Peters, M. (1989, August). The right-brained Indian: Fact or fiction? *Journal of American Indian Education, 29*, 77-82 [Special issue].

Christensen, R. A. (1991). A personal perspective on tribal-Alaskan Native gifted and talented education. *Journal of American Indian Education, 31*, 10-14.

Christiansen, T., & Livermore, G. (1970). A comparison of Anglo-American and Spanish-American children on the WISC. *Journal of Social Psychology, 81*, 9-14.

Cicirelli, V., Evans, J. W., & Schiller, J. S. (1969). *The impact of Head Start: An evaluation of the effects of Head Start on children's cognitive and affective development*. Report of a study undertaken by Westinghouse Learning Corporation and Ohio University (Contract B89-4536, Office of Economic Opportunity).

Clarizio, H. F. (1979a). Commentary on Mercer's rejoinder to Clarizio. *School Psychology Digest, 8*, 207-209.

Clarizio, H. F. (1979b). In defense of the IQ test. *School Psychology Digest, 8*, 79-88.

Clark, B. (1983). *Growing up gifted*. Englewood Cliffs, NJ: Prentice Hall.

Clark, B. (1997). *Growing up gifted* (5th ed.). Upper Saddle River, NJ: Merrill/Prentice Hall.

Cleveland, E. (1920). Detroit's experiment with gifted children. *School and Society, 12,* 179-183.

Cohen, R. J., Swerdlik, M. E., & Smith, D. K. (1992). *Psychological testing and assessment: An introduction to tests and measurement* (2nd ed.). Mountain View, CA: Mayfield.

Colangelo, N., & Davis, G. A. (1997). Introduction and overview. In N. Colangelo & G. A. Davis (Eds.), *Handbook of gifted education* (2nd ed., pp. 3-9). Boston: Allyn & Bacon.

Cole, N. S., & Moss, P. A. (1989). Bias in test use. In R. L. Linn (Ed.), *Educational measurement* (3rd ed., pp. 201-219). New York: Macmillan.

Coleman, J. S., Campbell, E. Q., Hobson, C. J., McPartland, J., Mood, A. M., Weinfeld, F. D., & York, R. L. (1966). *Equality of educational opportunity.* Washington, DC: Government Printing Office.

Coles, R. (1997). *The moral intelligence of children.* New York: Random House.

Conrad, R. (1964). Acoustic confusion in immediate memory. *British Journal of Psychology, 55,* 75-84.

Cook, P. J., & Ludwig, J. (1998). The burden of "acting White": Do Black adolescents disparage academic achievement? In C. Jencks & M. Phillips (Eds.), *The Black-White test score gap* (pp. 375-400). Washington, DC: Brookings Institution.

Cooke, G. J., & Baldwin, A. Y. (1979). Unique needs of a special population. In H. Passow (Ed.), *The gifted and talented: Their education and development—The seventy-fifth yearbook of the National Society for the Study of Education, Part I* (pp. 388-394). Chicago: University of Chicago Press.

Coon, H., Fulker, D. W., DeFries, J. C., & Plomin, T. (1990). Home environment and cognitive ability of 7-year-old children in the Colorado Adoption Project: Genetic and environmental etiologies. *Developmental Psychology, 26,* 459-468.

Covarrubias v. San Diego Unified School District, Civil Action No. 70-30d (S.D. Cal.) (1971).

Cox, C. M. (1926). *Genetic studies of genius: Vol. 2. The early mental traits of three hundred geniuses.* Stanford, CA: Stanford University Press.

Cox, J. M., Daniel, N., & Boston, B. O. (1985). *Educating able learners: Programs and promising practices.* Austin: University of Texas Press.

Cox, O. C. (1948). *Caste, class, and race.* Garden City, NY: Doubleday.

Cravens, H. (1978). *The triumph of evolution: American scientists and the heredity-environment controversy, 1900-1941.* Philadelphia: University of Pennsylvania Press.

Cronbach, L. J. (1980). Validity on parole: How can we go straight? In W. B. Schrader (Ed.), *New directions for testing and measurement, No. 5, Measuring achievement: Progress over a decade* (pp. 99-108). San Francisco: Jossey-Bass.

Cronin, J., Daniels, N., Hurley, A., Kroch, A., & Webber, R. (1975). Race, class, and intelligence: A critical look at the IQ controversy. *International Journal of Mental Health, 3,* 46-132.

Cross, W. E., Jr. (1996). *The Bell Curve* and transracial adoption studies. In J. L. Kinchloe, S. R. Steinberg, & A. D. Gresson (Eds.), *Measured lies: The bell curve examined* (pp. 331-342). New York: St. Martin's.

Crow, J. F. (1969). Genetic theories and influences: Comments on the value of diversity. *Harvard Educational Review, 39,* 338-347.

Csikszentmihalyi, M., & Robinson, R. E. (1986). Culture, time, and the development of talent. In R. J. Sternberg & J. E. Davidson (Eds.), *Conceptions of giftedness* (pp. 264-284). New York: Cambridge University Press.

Cummins, J. (1988). "Teachers are not miracle workers": Lloyd Dunn's call for Hispanic activism. *Hispanic Journal of Behavioral Sciences, 10,* 263-272.

Damico, J. S., & Hamayan, E. V. (1991). Implementing assessment in the real world. In E. V. Hamayan & J. S. Damico (Eds.), *Limiting bias in the assessment of bilingual students* (pp. 304-316). Austin, TX: PRO-ED.

Daniels, R. R. (1988). American Indians: Gifted, talented, creative, or forgotten. *Roeper Review, 10,* 241-244.

Das, J. P. (1973). Structure of cognitive abilities: Evidence for simultaneous and successive processing. *Journal of Educational Psychology, 65,* 103-108.

Das, J. P., Kirby, J. R., & Jarman, R. F. (1979). *Simultaneous and successive cognitive processes.* New York: Academic Press.

Das, J. P., Naglieri, J. A., & Kirby, J. R. (1994). *Assessment of cognitive processes: The PASS theory of intelligence.* Boston: Allyn & Bacon.

Dasen, P. R. (1984). The cross-cultural study of intelligence and the Baoule. *International Journal of Psychology, 19,* 407-434.

Davé, R. H. (1963). *The identification and measurement of environmental press variables that are related to educational achievement.* Unpublished doctoral dissertation, University of Chicago.

Davenport, E. L. (1931). *A comparative study of Mexican and non-Mexican siblings.* Unpublished master's thesis, The University of Texas at Austin.

Davis, G. A., & Rimm, S. B. (1998). *Education of the gifted and talented* (4th ed.). Boston: Allyn & Bacon.

Decroly, O., & Degand, J. (1910). La mesure de l'intelligence chez des enfantes normeaux d'après les tests de MM Binet et Simon [Home intelligence measures of children according to the Binet-Simon tests]. *Archives de Psychologie, 9,* 81-108.

Degler, C. N. (1991). *In search of human nature: The decline and revival of Darwinism in American social thought.* New York: Oxford University Press.

DeLeon, J. (1983). Cognitive style difference and the underrepresentation of Mexican Americans in programs for the gifted. *Journal for the Education of the Gifted, 6,* 167-177.

DeLeon, P. H., & Vandenbos, G. H. (1985). Public policy and advocacy on behalf of the gifted and talented. In F. D. Horowitz & M. O. O'Brien (Eds.), *The gifted and talented: Developmental perspectives* (pp. 409-435). Washington, DC: American Psychological Association.

Deutsch, M., & Brown, B. (1964). Social influences in Negro-White intelligence differences. *Journal of Social Issues, 20,* 24-35.

Dexter, E. S. (1923). The relation between occupation of parent and intelligence of children. *School and Society, 17,* 612-614.

Diana v. State Board of Education, Civil Action No. C-70-37 (N.D. Cal.) (1970).

Dickens, W. T., Kane, T. J., & Schultze, C. L. (1996). *Does the bell curve ring true?* Washington, DC: Brookings Institution.

Dickson, V. E. (1923). *Mental tests and the classroom teacher.* Yonkers-on-the-Hudson, NY: World Book.

Dollard, J. (1949). *Caste and class in a southern town* (2nd ed.). New York: Harper.

Donato, R. (1997). *The other struggle for equal schools: Mexican Americans in the civil rights era.* Albany: State University of New York Press.

Donato, R., Menchaca, M., & Valencia, R. R. (1991). Segregation, desegregation, and integration of Chicano students: Problems and prospects. In R. R. Valencia (Ed.), *Chicano school failure and success: Research and policy agendas for the 1990s* (pp. 27-63, Stanford Series on Education and Public Policy). London: Falmer.

Dorfman, D. D. (1995). Soft science with a neoconservative agenda [Review of the book *The Bell Curve: Intelligence and Class Structure in American Life*]. *Contemporary Psychology, 40,* 418-421.

Dreger, R. M., & Miller, K. S. (1960). Comparative psychological studies of Negroes and Whites in the United States. *Psychological Bulletin, 57,* 361-402.

Dreger, R. M., & Miller, K. S. (1968). Comparative psychological studies of Negroes and Whites in the United States: 1959-1965 [Monograph supplement]. *Psychological Bulletin, 70,* 1-58.

Du Bois, W. E. B. (1903). The talented tenth. In *The Negro problem: A series of articles by representative American Negroes of today* (pp. 33-75). New York: James Pott.

Duke, L. (1991, January 9). Racial stereotypes found to persist among Whites. *Austin American-Statesman,* pp. A1, A6.

Dunn, I. M. (1968). Special education for the mildly retarded: Is much of it justifiable? *Exceptional Children, 23,* 5-21.

Dunn, L. M. (1959). *Peabody Picture Vocabulary Test.* Minneapolis, MN: American Guidance Service.

Dunn, L. M. (1987). *Bilingual Hispanic children on the U.S. mainland: A review of research on their cognitive, linguistic, and scholastic development.* Circle Pines, MN: American Guidance Service.

Dunn, L. M. (1988). Has Dunn's monograph been shot down in flames? Author reactions to the preceding critiques of it. *Hispanic Journal of Behavioral Sciences, 10,* 301-323.

Dunn, L(loyd) M., & Dunn, L(eota) M. (1981). *Peabody Picture Vocabulary Test–Revised.* Minneapolis, MN: American Guidance Service.

Dunn, R., & Griggs, S. A. (1990). Research on the learning style characteristics of selected racial and ethnic groups. *Reading, Writing, and Learning Disabilities, 6,* 261-280.

Durost, W. N., Bixler, H. H., Wrightstone, J. W., Prescott, G. A., & Balow, I. H. (1975). *Metropolitan Achievement Tests.* New York: Psychological Corporation.

Duster, T. (1995). [Review of the book *The Bell Curve: Intelligence and Class Structure in American Life*]. *Contemporary Sociology: A Journal of Reviews, 24,* 158-161.

Dyer, P. B. A. (1967). *Home environment and achievement in Trinidad.* Unpublished doctoral dissertation, University of Alberta.

Early, G. (1995). The bell curve: As the meaning of the academic will [Book review]. *Black Scholar, 25,* 32-38.

Education of handicapped children: Regulations implementing Education for All Handicapped Children Act of 1975. (1977, August). *Federal Register,* pp. 42474-42518.

Educational Testing Service. (1994). *GRE: 1994-1995 guide to the use of the Graduate Record Examinations program.* Princeton, NJ: Author.

Eells, K. W., Davis, A., Havighurst, R. J., Herrick, V. E., & Tyler, R. W. (1962). *Intelligence and cultural differences.* Chicago: University of Chicago Press.

Ehrman, L., Omenn, G. S., & Caspari, E. (Eds.). (1972). *Genetics, environment, and behavior.* New York: Academic Press.

Elardo, R., Bradley, R., & Caldwell, B. M. (1975). The relation of infants' home environments to mental test performance from six months to thirty-six months: A longitudinal study. *Child Development, 46,* 71-76.

Elardo, R., Bradley, R., & Caldwell, B. M. (1977). A longitudinal study of the relation of infants' home environments to language development at age three. *Child Development, 48,* 595-603.

Elliott, C. D. (1990). *Differential Ability Scales.* San Antonio, TX: Psychological Corporation.

Elliott, S. N., & Argulewicz, E. N. (1983). Use of a behavior rating scale to aid in the identification of developmentally and culturally different gifted children. *Journal of Psychoeducational Assessment, 1,* 179-186.

Elliott, S. N., Argulewicz, E. N., & Turco, T. L. (1986). Predictive validity of the Scales for Rating the Behavioral Characteristics of Superior Students for gifted children from three sociocultural groups. *Journal of Experimental Education, 55,* 27-32.

Ellis, C. E. (1932). *The relation between socio-economic status to the intelligence and school success of Mexican children.* Unpublished master's thesis, The University of Texas at Austin.

Eriksen, C., Pollack, M., & Montague, W. (1970). Implicit speech: Mechanism in perceptual coding? *Journal of Experimental Psychology, 84,* 502-507.

Erlenmeyer-Kimling, L., & Jarvik, L. F. (1963). Genetics and intelligence: A review. *Science, 142,* 1477-1479.

Escalona, S., & Corman, H. (1971). The impact of mother's presence upon behavior: The first year. *Human Development, 14,* 2-15.

Evans de Bernard, A. (1985). Why José can't get in the gifted class: The bilingual child and standardized tests. *Roeper Review, 8,* 80-82.

Evans, B. J., & Whitfield, J. R. (1988). *Black males in the United States: Abstracts of the psychological and behavioral literature, 1967-1987.* (Bibliographies in Psychology, No. 1). Washington, DC: American Psychological Association.

Eysenck, H. J. (1939). Review of primary mental abilities. *British Journal of Educational Psychology, 9,* 270-275.

Eysenck, H. J. (1971). *The IQ argument: Race, intelligence, and education.* New York: Library Press.

Eysenck, H. J. (1982). Introduction. In H. J. Eysenck (Ed.), *A model for intelligence* (pp. 1-6). New York: Springer-Verlag.

Eysenck, H. J. (1989). Sensitive intelligence issues [Review of the book *The Burt Affair*]. *The Spectator, 263*, 26-27.

Eysenck, H. J. (1998). *A new look at intelligence.* New Brunswick, NJ: Transaction Books.

Falconer, D. S. (1990). *Introduction to quantitative genetics* (3rd ed.). Edinburgh, Scotland: Oliver & Boyd.

Fancher, R. C. (1985). *The intelligence men: Makers of the IQ controversy.* New York: Norton.

Fancher, R. C. (1991). The Burt case: Another foray [Review of the book *Science, Ideology, and the Media: The Cyril Burt Scandal*]. *Science, 253*, 1565-1566.

Farber, S. L. (1981). *Identical twins reared apart: A reanalysis.* New York: Basic Books.

Fass, P. S. (1980). The IQ: A cultural and historical framework. *American Journal of Education, 88*, 431-458.

Fatemi, E. (1991, May 27). Reports show minorities overrepresented in special education classes. *Education USA*, pp. 273, 276.

Feagin, J. R. (1984). *Racial and ethnic relations* (2nd ed.). Englewood Cliffs, NJ: Prentice Hall.

Feagin, J. R., & Booher Feagin, C. (1999). *Racial and ethnic relations* (6th ed.). Upper Saddle River, NJ: Merrill/Prentice Hall.

Feldt, L. S. (1969). A test of the hypothesis that Cronbach's alpha or Kuder-Richardson coefficient twenty is the same for two tests. *Psychometrika, 34*, 363-373.

Fernández, R. R. (Ed.). (1988a). Achievement testing: Science vs. ideology [Special issue]. *Hispanic Journal of Behavioral Sciences, 10*(3).

Fernández, R. R. (1988b). Introduction. *Hispanic Journal of Behavioral Sciences, 10*, 179-198.

Figueroa, R. A., & Sassenrath, J. M. (1989). A longitudinal study of the predictive validity of the System of Multicultural Pluralistic Assessment (SOMPA). *Psychology in the Schools, 26*, 5-19.

Fine, B. (1949, September 18). More and more, the IQ idea is questioned. *New York Times Magazine*, pp. 7, 72-74.

Finn, C. E., Jr. (1995). For whom it tolls [Review of the book *The Bell Curve: Intelligence and Class Structure in American Life*]. *Commentary, 99*, 76-80.

Finn, J. (1982). Patterns in special education placement as revealed by OCR surveys. In H. Keller, W. Holtzman, & S. Messick (Eds.), *Placement of children in special education: A strategy for equity* (pp. 322-381). Washington, DC: National Academy Press.

Fischbein, S. (1980). IQ and social class. *Intelligence, 4*, 51-63.

Fischer, C. S., Hout, M., Jankowski, M. S., Lucas, S. R., Swidler, A., & Voss, K. (1996). *Inequality by design: Cracking the bell curve myth.* Princeton, NJ: Princeton University Press.

Fitzgerald, J. A., & Ludeman, W. W. (1926). The intelligence of Indian children. *Journal of Comparative Psychology, 6*, 319-328.

Flanagan, D. P., Genshaft, J. L., & Harrison, P. L. (Eds.). (1997). *Contemporary intellectual assessment: Theories, tests, and issues.* New York: Guilford.

Fletcher, R. (1990). [Review of the book *The Burt Affair*]. *Society, 27*, 85-87.

Fletcher, R. (1991). *Science, ideology, and the media: The Cyril Burt scandal.* New Brunswick, NJ: Transaction Books.

Flynn, J. R. (1980). *Race, IQ, and Jensen.* London: Routledge & Kegan Paul.

Flynn, J. R. (1999). Searching for justice: The discovery of IQ gains over time. *American Psychologist, 54*, 5-20.

Foley, D. E. (1991). Anthropological explanations of minority school failure. *Anthropology and Education Quarterly, 4*, 60-86.

Foley, D. E. (1997). Deficit thinking models based on culture: The anthropological model. In R. R. Valencia (Ed.), *The evolution of deficit thinking: Educational thought and practice* (pp. 113-131, Stanford Series on Education and Public Policy). London: Falmer.

Ford, D. Y. (1995). Desegregating gifted education: A need unmet. *Journal of Negro Education, 64*, 53-62.

Ford, D. Y., & Harris, J. J., III. (1990). On discovering the hidden treasure of gifted and talented Black children. *Roeper Review, 13,* 27-33.

Ford, D. Y., & Webb, K. S. (1994). Desegregation of gifted educational programs: The impact of *Brown* on underachieving children of color. *Journal of Negro Education, 63,* 358-375.

Fordham, S., & Ogbu, J. U. (1986). Black students' school success: Coping with the burden of "acting White." *Urban Review, 58,* 54-84.

Frankenberger, W., & Fronzaglio, K. (1991). A review of states' criteria and procedures for identifying children with learning disabilities. *Journal of Learning Disabilities, 24,* 495-506.

Frary, R. B. (1989-present). Review of the Batería Woodcock-Muñoz–Revisada. In *Buro's mental measurements yearbook* (WinSPIRS database).

Fraser, S. (1995). *The bell curve wars: Race, intelligence, and the future of America.* New York: Basic Books.

Frasier, M. M. (1987). The identification of gifted Black students: Developing new perspectives. *Journal for the Education of the Gifted, 10,* 155-180.

Frasier, M. M. (1989). Identification of gifted Black students: Developing new perspectives. In C. J. Maker & S. W. Schiever (Eds.), *Critical issues in gifted education: Defensible programs for cultural and ethnic minorities* (pp. 213-225). Austin, TX: PRO-ED.

Freeman, F. N. (1926). *Mental tests: Their history, principles, and applications.* Boston: Houghton Mifflin.

Freire, P. (1970). *Pedagogy of the oppressed.* New York: Seabury Press.

French, J. L., & Hale, R. L. (1990). A history of the development of psychological and educational testing. In C. R. Reynolds & R. W. Kamphaus (Eds.), *Handbook of psychological and educational assessment of children: Intelligence and achievement* (pp. 3-28). New York: Guilford.

Frisby, C. L. (1998). Culture and cultural differences. In J. Sandoval, C. L. Frisby, K. F. Geisinger, J. D. Scheuneman, & J. R. Grenier (Eds.), *Test interpretation and diversity: Achieving equity in assessment* (pp. 51-73). Washington, DC: American Psychological Association.

Fulker, D. W. (1975). [Review of the book *The Science and Politics of IQ*]. *American Journal of Psychology, 88,* 505-537.

Gage, N. L. (1972a). IQ heritability, race differences, and educational research. *Phi Delta Kappan, 53,* 308-312.

Gage, N. L. (1972b). Replies to Shockley, Page, and Kaplan: The causes of race differences in IQ. *Phi Delta Kappan, 53,* 422-427.

Gallagher, J. J. (1984). Excellence and equity: A worldwide conflict. *Gifted International, 2,* 1-11.

Gallagher, J. J. (1986). Equity vs. excellence: An educational drama. *Roeper Review, 8,* 233-235.

Gallagher, J. J. (1991). Educational reform, values, and gifted students. *Gifted Child Quarterly, 35,* 12-19.

Gallagher, J. J. (1993). Current and historical thinking on education for gifted and talented students. In P. O. Ross (Ed.), *National excellence: A case for developing America's talent* (pp. 83-108). Washington, DC: U.S. Department of Education, Office of Educational Research and Improvement.

Gallagher, J. J., & Weiss, P. (1979). *The education of gifted and talented children and youth.* Washington, DC: Council for Basic Education.

Gallagher, R. M. (1989). Are we meeting the needs of gifted Asian-Americans? In C. J. Maker & S. W. Schiever (Eds.), *Critical issues in gifted education: Defensible programs for cultural and ethnic minorities* (pp. 169-173). Austin, TX: PRO-ED.

Galton, F. (1865). Heredity, talent and character. *Macmillian's Magazine, 12,* 157-166, 318-327.

Galton, F. (1870). *Hereditary genius: An inquiry into its laws and consequences.* New York: Appleton.

Galton, F. (1883). *Inquiries into human faculty and its development.* London: Macmillan.

Garber, L. O. (1961). *The yearbook of school law 1961.* Danville, IL: Interstate Printers and Publishers.

Gardner, H. (1983). *Frames of mind: The theory of multiple intelligences.* New York: Basic Books.

Gardner, H. (1987). The theory of multiple intelligences. *Annals of Dyslexia, 37,* 19-35.

Gardner, H. (1995). Cracking open the IQ box. In S. Fraser (Ed.), *The bell curve wars: Race, intelligence, and the future of America* (pp. 23-35). New York: Basic Books.

Gardner, H. (1999, February). Who owns intelligence? *Atlantic Monthly*, pp. 67-76.

Gardner, H., Hatch, T., & Torff, B. (1997). A third perspective: The symbol systems approach. In R. J. Sternberg & E. Grigorenko (Eds.), *Intelligence, heredity, and environment* (pp. 243-268). New York: Cambridge University Press.

Gardner, J. J. (1961). *Excellence: Can we be equal and excellent too?* New York: Harper & Row.

Garretson, O. K. (1928). Study of the causes of retardation among Mexican children. *Journal of Educational Psychology, 19*, 31-40.

Garrett, H. E. (n.d.). *Breeding down* [Pamphlet]. Richmond, VA: Patrick Henry Press.

Garrett, H. E. (1962, May). Rejoinder by Garrett. *SPSSI Newsletter*, pp. 1-2. (Society for the Psychological Study of Social Issues)

Garrett, H. E. (1966). Foreword to the second edition. In A. M. Shuey, *The testing of Negro intelligence* (2nd ed., pp. vii-viii). New York: Social Science Press.

Garrett, H. E. (1973). *IQ and racial differences* [Pamphlet]. Cape Canaveral, FL: Howard Allen.

Garrison, L. (1989). Programming for the gifted American Indian student. In C. J. Maker & S. W. Schiever (Eds.), *Critical issues in gifted education: Defensible programs for cultural and ethnic minorities* (pp. 116-127). Austin, TX: PRO-ED.

Garth, T. R. (1923). A comparison of the intelligence of Mexican and full blood Indian children. *Psychological Review, 30*, 388-401.

Garth, T. R. (1925). A review of race psychology. *Psychological Bulletin, 22*, 343-364.

Garth, T. R. (1928). The intelligence of Mexican school children. *School and Society, 27*, 791-794.

Garth, T. R. (1930). A review of race psychology. *Psychological Bulletin, 27*, 329-356.

Garth, T. R., Elson, T. H., & Morton, M. M. (1936). The administration of non-language intelligence tests to Mexicans. *Journal of Abnormal and Social Psychology, 31*, 53-58.

Garth, T. R., Schuelke, N., & Abell, W. (1927). The intelligence of mixed blood Indians. *Journal of Applied Psychology, 11*, 268-275.

Garth, T. R., Serafini, T. J., & Dutton, D. (1925). The intelligence of full blood Indians. *Journal of Applied Psychology, 9*, 382-389.

Gay, J. E. (1978). A proposed plan for identifying Black gifted children. *Gifted Child Quarterly, 22*, 353-360.

Gear, G. H. (1976). Accuracy of teacher judgment in identifying intellectually gifted children: A review of the literature. *Gifted Child Quarterly, 20*, 478-490.

Geisinger, K. F. (1994). Cross-cultural normative assessment: Translation and adaptation issues influencing the normative interpretation of assessment instruments. *Psychological Assessment, 6*, 304-312.

Geisinger, K. F. (1998). Psychometric issues in test interpretation. In J. Sandoval, C. L. Frisby, K. F. Geisinger, J. D. Scheuneman, & J. R. Grenier (Eds.), *Test interpretation and diversity: Achieving equity in assessment* (pp. 17-30). Washington, DC: American Psychological Association.

Geisinger, K. F., & Carlson, J. F. (1998). Training psychologists to assess members of a diverse society. In J. Sandoval, C. L. Frisby, K. F. Geisinger, J. D. Scheuneman, & J. R. Grenier (Eds.), *Test interpretation and diversity: Achieving equity in assessment* (pp. 375-386). Washington, DC: American Psychological Association.

George, K. R. (1989). Imagining and defining giftedness. In C. J. Maker & S. W. Schiever (Eds.), *Critical issues in gifted education: Defensible programs for cultural and ethnic minorities* (pp. 107-112). Austin, TX: PRO-ED.

Gersten, R., & Woodward, J. (1994). The language-minority student and special education: Issues, trends, and paradoxes. *Exceptional Children, 60*, 310-322.

Gilbert, H. B. (1966). On the IQ ban. *Teachers College Record, 67*, 282-285.

Gill, R. E., & Keats, D. M. (1980). Elements of intellectual competence: Judgments of Australian and Malay university students. *Journal of Cross-Cultural Psychology, 11*, 233-243.

Ginsburg, H. (1986). The myth of the deprived child: New thoughts on poor children. In U. Neisser (Ed.), *The school achievement of minority children: New perspectives* (pp. 169-189). Hillsdale, NJ: Lawrence Erlbaum.

Giroux, H. A., & Searls, S. (1996). The bell curve debate and the crisis of public intellectuals. In J. L. Kinchloe, S. R. Steinberg, & A. D. Gresson (Eds.), *Measured lies: The bell curve examined* (pp. 71-90). New York: St. Martin's.

Glennon, T. (1995). Race, education, and the construction of a disabled class. *Wisconsin Law Review, 1237.* (Published in *Westlaw,* 1997, pp. 1-80)

Goddard, H. H. (1928). *School training for gifted children.* Yonkers-on-the-Hudson, NY: World Book.

Golden, M., & Bridger, W. (1969). A refutation of Jensen's position on intelligence, race, social class, and heredity. *Mental Hygiene, 53,* 648-653.

Goldenberg, C. (1996). The education of language minority students: Where are we? And where do we need to go? *Elementary School Journal, 96,* 353-361.

Goleman, D. (1995). *Emotional intelligence.* New York: Bantam Books.

González, A. M. (1932). *A study of the intelligence of Mexican children in relation to their socio-economic status.* Unpublished master's thesis, The University of Texas at Austin.

González, G. G. (1974a). Racism, education, and the Mexican community in Los Angeles, 1920-1930. *Societas, 4,* 287-301.

González, G. G. (1974b). *The system of public education and its function within the Chicano communities, 1910-1930.* Unpublished doctoral dissertation, University of California, Los Angeles.

González, G. G. (1990). *Chicano education in the era of segregation.* Philadelphia: Balch Institute Press.

Good, H. G. (1960). *A history of Western education.* New York: Macmillan.

Goodenough, F. L. (1926a). *Measurement of intelligence by drawings.* Yonkers-on-the-Hudson, NY: World Book.

Goodenough, F. L. (1926b). Racial differences in the intelligence of school children. *Journal of Experimental Psychology, 9,* 388-397.

Goodenough, F. L. (1928). The relation of the intelligence of pre-school children to the occupation of their families. *American Journal of Psychology, 40,* 284-294.

Goodenough, F. L., & Harris, D. B. (1950). Studies in the psychology of children's drawings: II. 1928-1949. *Psychological Bulletin, 49,* 369-433.

Goodlad, J. J., & Oakes, J. (1988). We must offer equal access to knowledge. *Educational Leadership, 45,* 16-22.

Goodman, J. F. (1979a). Is tissue the issue? A critique of SOMPA's models and tests. *School Psychology Digest, 8,* 47-62.

Goodman, J. F. (1979b). "Ignorance" versus "stupidity": The basic disagreement. *School Psychology Digest, 8,* 218-223.

Gosling, T. W. (1919). A special academic class in the junior high school. *School Review, 27,* 241-255.

Gottfried, A. W. (Ed.). (1984). *Home environment and early cognitive development: Longitudinal research.* Orlando, FL: Academic Press.

Gottfried, A. W., & Gottfried, A. E. (1984). Home environment and cognitive development in young children of middle-socioeconomic-status families. In A. W. Gottfried (Ed.), *Home environment and early cognitive development: Longitudinal research* (pp. 57-115). Orlando, FL: Academic Press.

Gottlieb, G. (1970). Conceptions of prenatal behavior. In L. R. Aronson (Ed.), *Development and evolution of behavior* (pp. 111-137). San Francisco: Freeman.

Gottlieb, G. (1992). *Individual development and evolution: The genesis of novel behavior.* New York: Oxford University Press.

Gottlieb, J., Alter, M., Gottlieb, B. W., & Wishner, J. (1994). Special education in urban America: It's not justifiable for many. *Journal of Special Education, 27,* 453-465.

Gottlieb, J., & Wishner, J. (1993). *Special education referrals in inner-city schools.* Unpublished manuscript.

Gould, S. J. (1981). *The mismeasure of man.* New York: Norton.

Gould, S. J. (1995). Curveball. In S. Fraser (Ed.), *The bell curve wars: Race, intelligence, and the future of America* (pp. 11-22). New York: Basic Books.

Gould, S. J. (1996). *The mismeasure of man* (Rev. ed.). New York: Norton.

Grant, M. (1916). *The passing of the great race.* New York: Scribner.

Gratz, E., & Pulley, J. L. (1984). A gifted and talented program for migrant students. *Roeper Review, 6,* 147-149.

Grebler, L., Moore, J. W., & Guzman, R. C. (1970). *The Mexican-American people: The nation's second largest minority.* New York: Free Press.

Green, R. B., & Rohwer, W. D. (1971). SES differences on learning and ability tests in Black children. *American Educational Research Journal, 8,* 601-609.

Greene, A. C., Sapp, G. L., & Chissom, B. (1990). Validation of the Stanford-Binet Intelligence Scale: Fourth Edition with Black male students. *Psychology in the Schools, 27,* 35-41.

Gresham, F. M., MacMillan, D. L., & Bocian, K. M. (1996). Learning disabilities, low achievement, and mild mental retardation: More alike than different? *Journal of Learning Disabilities, 29,* 570-581.

Gronlund, N. E., & Linn, R. L. (1990). *Measurement and evaluation in teaching* (6th ed.). New York: Macmillan.

Guadalupe v. Tempe Elementary School District, No. 3, Civ. No. 71-435 (D. Ariz.) (1972).

Guthrie, R. V. (1976). *Even the rat was White: A historical view of psychology.* New York: Harper & Row.

Haggerty, M. E., & Nash, H. B. (1924). Mental capacity of children and paternal occupation. *Journal of Educational Psychology, 15,* 559-572.

Hale-Benson, J. (1986). *Black children: Their roots, culture, and learning styles.* Baltimore, MD: Johns Hopkins University Press.

Hall, C. C. I., Evans, B. J., & Selice, S. (1989). *Black females in the United States: Abstracts of the psychological and behavioral literature, 1967-1987.* (Bibliographies in Psychology, No. 3). Washington, DC: American Psychological Association.

Haller, M. H. (1963). *Eugenics: Hereditarian attitudes in American thought.* New Brunswick, NJ: Rutgers University Press.

Halpern, D. F. (1997). Sex differences in intelligence: Implications for education. *American Psychologist, 58,* 1091-1101.

Hamilton, S. E. (1993). Identifying African American gifted children using a behavioral assessment technique: The Gifted Children Locater. *Journal of Black Psychology, 19,* 63-76.

Haney, W. (1981). Validity, vaudeville, and values: A short history of social concerns over standardized testing. *American Psychologist, 36,* 1021-1034.

Harman, H. H. (1976). *Modern factor analysis* (3rd ed.). Chicago: University of Chicago Press.

Harris, D. F. (1963). *Children's drawings as measures of intellectual maturity: A revision and extension of the Goodenough Draw-a-Man Test.* New York: Harcourt Brace Jovanovich.

Harris, J. J., III, & Ford, D. Y. (1991). Identifying and nurturing the promise of gifted Black American children. *Journal of Negro Education, 60,* 3-18.

Harris, J. R. (1995). Where is the child's environment? A group socialization theory of development. *Psychological Review, 102,* 458-489.

Harrison, P. L. (1999). How does ability testing fit into the blueprint for the future? *Communique, 27,* 31.

Harrison, P. L., Flanagan, D. P., & Genshaft, J. L. (1997). An integration and synthesis of contemporary theories, tests, and issues in the field of intellectual assessment. In D. P. Flanagan, J. L. Genshaft, & P. L. Harrison (Eds.), *Contemporary intellectual assessment: Theories, tests, and issues* (pp. 533-561). New York: Guilford.

Harrison, P. L., Kaufman, A. S., Hickman, J. A., & Kaufman, N. L. (1988). A survey of tests used for adult assessment. *Journal of Psychoeducational Assessment, 6,* 188-198.

Hartley, E. A. (1991). Through Navajo eyes: Examining differences in giftedness. *Journal of American Indian Education, 31,* 53-64.

Hartman, J. S., & Askounis, A. C. (1989). Asian-American students: Are they really a "model minority?" *The School Counselor, 37,* 109-112.

Hartz, H., Adkins, D. M., & Sherwood, R. D. (1984). Predictability of giftedness identification indices for two recognized approaches to elementary school gifted education. *Journal of Educational Research, 77,* 337-342.

Hasegawa, C. (1989). The unmentioned minority. In C. J. Maker & S. W. Schiever (Eds.), *Critical issues in gifted education: Defensible programs for cultural and ethnic minorities* (pp. 192-196). Austin, TX: PRO-ED.

Hauser, R. M. (1995). [Review of the book *The Bell Curve: Intelligence and Class Structure in American Life*]. *Contemporary Sociology: A Journal of Reviews, 24,* 149-153.

Haviland-Jones, J., Gebelt, J. L., & Stapley, J. C. (1997). The question of development of emotion. In P. Salovey & D. J. Sluyter (Eds.), *Emotional development and emotional intelligence: Educational implications* (pp. 233-253). New York: Basic Books.

Hawk, B., Schroeder, S., Robinson, G., Otto, D., Mushak, P., Kleinbaum, D., & Dawson, G. (1986). Relation of lead and social factors to IQ of low-SES children: A partial replication. *American Journal of Mental Deficiency, 91,* 178-183.

Hearnshaw, L. S. (1979). *Cyril Burt, psychologist.* Ithaca, NY: Cornell University Press.

Hebb, D. O. (1949). *The organization of behavior: A neuropsychological theory.* New York: John Wiley.

Heck, A. O. (1930). *Special schools and classes in cities of 10,000 population and more in the United States.* Washington, DC: Government Printing Office.

Heck, A. O. (1953). *Education of exceptional children.* New York: McGraw-Hill.

Hedges, L. V., & Nowell, A. (1998). Black-White test score convergence since 1965. In C. J. Phillips & M. Phillips (Eds.), *The Black-White test score gap* (pp. 149-181). Washington, DC: Brookings Institution.

Helms, J. E. (1992). Why is there no study of cultural equivalence in standardized cognitive ability testing? *American Psychologist, 47,* 1083-1101.

Helms, J. E. (1997). The triple quandary of race, culture, and social class in standardized cognitive ability testing. In D. P. Flanagan, J. L. Genshaft, & P. L. Harrison (Eds.), *Contemporary intellectual assessment: Theories, tests, and issues* (pp. 517-532). New York: Guilford.

Henderson, R. W. (1966). *Environmental stimulation and intellectual development of Mexican-American children.* Unpublished doctoral dissertation, University of Arizona, Tucson.

Henderson, R. W. (1972). Environmental predictors of academic performance of disadvantaged Mexican-American children. *Journal of Consulting and Clinical Psychology, 38,* 297.

Henderson, R. W. (1981). Home environment and intellectual performance. In R. W. Henderson (Ed.), *Parent-child interaction: Theory, research, and prospects* (pp. 3-32). New York: Academic Press.

Henderson, R. W., Bergan, J. R., & Hurt, M., Jr. (1972). Development and validation of the Henderson Environmental Learning Process Scale. *Journal of Social Psychology, 88,* 185-196.

Henderson, R. W., & Garcia, A. B. (1973). The effects of a parent training program on the question-asking behavior of Mexican-American children. *American Educational Research Journal, 10,* 193-201.

Henderson, R. W., & Merritt, C. B. (1968). Environmental background of Mexican-American children with different potentials for school success. *Journal of Social Psychology, 75,* 101-106.

Henderson, R. W., & Swanson, R. A. (1974). Application of social learning principles in a field study. *Exceptional Children, 40,* 53-55.

Henderson, R. W., & Valencia, R. R. (1985). Nondiscriminatory school psychological services: Beyond nonbiased assessment. In J. R. Bergan (Ed.), *School psychology in contemporary society* (pp. 340-377). Columbus, OH: Merrill.

Hendricksen, R. M. (1996). *The Bell Curve*, affirmative action, and the quest for equality. In J. L. Kinchloe, S. R. Steinberg, & A. D. Gresson (Eds.), *Measured lies: The bell curve examined* (pp. 351-365). New York: St. Martin's.

Henley, M., Ramsey, R. S., & Algozzine, R. F. (1999). *Characteristics of and strategies for teaching students with mild disabilities.* Boston: Allyn & Bacon.

Henry, T. S. (1924). Annotated bibliography on gifted children and their education. In G. M. Whipple (Ed.), *The twenty-third yearbook of the Society for the Study of Education: Part I—Report of the Society's Committee on the Education of Gifted Children* (pp. 389-443). Bloomington, IL: Public School Publishing.

Hernandez, R. (1999, June 12). Under U.S. threat, Albany seeks to overhaul special education. *The New York Times*, pp. A1, B4.

Herrnstein, R. J. (1973). *IQ in the meritocracy.* Boston: Little, Brown.

Herrnstein, R. J., & Murray, C. (1994). *The bell curve: Intelligence and class structure in American life.* New York: Free Press.

Hewitt, A. (1930). A comparative study of the intelligence of White and colored children. *Elementary School Journal, 31,* 111-119.

Hicks, R. B., & Pellegrini, R. J. (1966). The meaningfulness of Negro-White differences in intelligence test performance. *Psychological Record, 16,* 43-46.

Higgins, C., & Sivers, C. H. (1958). A comparison of Stanford-Binet and Colored Raven Progressive Matrices IQs for children with low socioeconomic status. *Journal of Consulting Psychology, 22,* 465-468.

High, M. H., & Udall, A. J. (1983). Teacher ratings of students in relation to ethnicity of students and school ethnic balance. *Journal for Education of the Gifted, 6,* 154-166.

Hildreth, G. A. (1933). *A bibliography of mental tests and rating scales.* New York: Psychological Corporation.

Hildreth, G. (1935). Occupational status and intelligence. *Personnel Journal, 13,* 153-157.

Hilliard, A. G., III. (1979). Standardization and cultural bias as impediments to the scientific study and validation of "intelligence." *Journal of Research and Development in Education, 12,* 47-58.

Hilliard, A. G., III. (1984). IQ testing as the emperor's new clothes: A critique of Jensen's *Bias in Mental Testing.* In C. R. Reynolds & R. T. Brown (Eds.), *Perspectives on bias in testing* (pp. 139-169). New York: Plenum.

Hilliard, A. G., III. (1992). The pitfalls and promises of special education practice. *Exceptional Children, 59,* 168-172.

Hirsch, J. (1976). Behavior-genetic analysis and its biosocial consequences. In N. J. Block & G. Dworkin (Eds.), *The IQ controversy: Critical readings* (pp. 156-178). New York: Pantheon Books.

Hobson v. Hansen, 269 F. Supp. 401 (D.C. 1967) aff'd sub. nom., Smuck v. Hobson, 408 F. 2d 175 (D.C. Cir.) (1969).

Hodapp, R. M. (1994). Mental retardation, cultural familial. In R. J. Sternberg (Ed.), *The encyclopedia of human intelligence* (pp. 711-717). New York: Macmillan.

Hollingshead, A. B. (1975). *Four-factor index of social status.* Unpublished manuscript, Yale University, Department of Sociology.

Hollingsworth, L. S. (1929). *Gifted children: Their nature and nurture.* New York: Macmillan.

Hollingsworth, L. S. (1935). The comparative beauty of the faces of highly intelligent children. *Journal of Genetic Psychology, 47,* 268-281.

Hollingsworth, L. S. (1942). *Children above 180 IQ: Stanford-Binet origin and development.* Yonkers-on-the-Hudson, NY: World Book.

Honzik, M. (1967). Environmental correlates for mental growth: Prediction from the family setting at 21 months. *Child Development, 38,* 337-364.

Horn, J. L. (1985). Remodeling old theories of intelligence: Gf-Gc theory. In B. B. Wolman (Ed.), *Handbook of intelligence* (pp. 267-300). New York: John Wiley.

Horn, J. L., & Noll, J. (1998). Human cognitive capabilities: Gf-Gc theory. In D. P. Flanagan, J. L. Genshaft, & P. L. Harrison (Eds.), *Contemporary intellectual assessment* (pp. 53-91). New York: Guilford.

Horowitz, I. L. (1995). The Rushton file. In R. Jacoby & N. Glauberman (Eds.), *The bell curve debate: History, documents, opinions* (pp. 179-200). New York: Times Books.

Hu, A. (1989). Asian Americans: Model minority or double minority? *Amerasia, 15,* 243-257.

Hunt, E. (1997). Nature vs. nurture: The feeling of *vujà dé.* In R. J. Sternberg & E. Grigorenko (Eds.), *Intelligence, heredity, and environment* (pp. 531-551). New York: Cambridge University Press.

Hunt, J. McV. (1961). *Intelligence and experience.* New York: Ronald Press.

Hunter, W. S., & Sommermier, E. (1922). The relation of degree of Indian blood to score on the Otis intelligence test. *Journal of Comparative Psychology, 2,* 257-277.

Indian entities recognized and eligible to receive services from the United States Bureau of Indian Affairs. (1998, December). *Federal Register,* pp. 71941-71946.

Individuals With Disabilities Education Act. (1997). *1997 amendments* [On-line]. Available: http://www.ed.gov/offices/osers/idea/the_law.html

Ironson, G. H. (1982). Use of chi-square and latent trait approaches for detecting item bias. In R. A. Berk (Ed.), *Handbook of methods for detecting test bias* (pp. 117-160). Baltimore, MD: Johns Hopkins University Press.

Irvine, J. T. (1978). Wolof "magical thinking": Culture and conservation revisited. *Journal of Cross-Cultural Psychology, 9,* 300-310.

Izard, C. E. (1980). Cross-cultural perspectives on emotion and emotion communication. In H. C. Triandis & W. Lonner (Eds.), *Handbook of cross-cultural psychology: Basic processes* (Vol. 3, pp. 185-221). Boston: Allyn & Bacon.

Jackson, D. M. (1979). The emerging national and state concern. In H. Passow (Ed.), *The gifted and talented: Their education and development—The seventy-fifth yearbook of the National Society for the Study of Education, Part I* (pp. 45-62). Chicago: University of Chicago Press.

Jackson, D.N. (1975). Intelligence and ideology. *Science, 189,* 1078-1080.

Jackson, G. D. (1975). Another psychological view from the Association of Black Psychologists. *American Psychologist, 30,* 88-93.

Jackson, J. F. (1993). Human behavioral genetics, Scarr's theory, and her views on interventions: A critical review and commentary on their implications for African American children. *Child Development, 64,* 1318-1332.

Jacob K. Javits Gifted and Talented Students Education Act, Public Law 100-297, Title IV, Part B, Section 1101 (1988).

Jacoby, R., & Glauberman, N. (Eds.). (1995). *The bell curve debate: History, documents, opinions.* New York: Times Books.

Jenkins, M. D. (1935). *A socio-psychological study of Negro children of superior intelligence.* Unpublished doctoral dissertation, Northwestern University.

Jenkins, M. D. (1936). A socio-psychological study of Negro children of superior intelligence. *Journal of Negro Education, 5,* 175-190.

Jenkins, M. D. (1943). Case studies of Negro children of Binet IQ 160 and above. *Journal of Negro Education, 12,* 159-166.

Jenkins, M. D. (1948). The upper limit of ability among American Negroes. *Scientific Monthly, 66,* 359-401.

Jenkins, M. D. (1950). Intellectually superior Negro youth: Problems and needs. *Journal of Negro Education, 19,* 322-332.

Jenkins, M. D., & Randall, C. M. (1948). Differential characteristics of superior and unselected Negro college students. *Journal of Social Psychology, 27,* 187-202.

Jensen, A. R. (1969). How much can we boost IQ and scholastic achievement? *Harvard Educational Review, 39,* 1-123.

Jensen, A. R. (1970). Do schools cheat minority children? *Educational Research, 14,* 3-28.

Jensen, A. R. (1972). *Genetics and education.* London: Methuen.

Jensen, A. R. (1973a). *Educability and group differences.* New York: Harper & Row.

Jensen, A. R. (1973b). Level I and Level II abilities in three ethnic groups. *American Educational Research Journal, 10,* 263-276.

Jensen, A. R. (1974). Interaction of Level I and Level II abilities with race and socioeconomic status. *Journal of Educational Psychology, 66,* 99-111.

Jensen, A. R. (1978). Genetic and behavioral effects of nonrandom mating. In R. T. Osborne, C. E. Noble, & N. Weyl (Eds.), *Human variation: The biopsychology of age, race, and sex* (pp. 51-105). New York: Academic Press.

Jensen, A. R. (1980). *Bias in mental testing.* New York: Free Press.

Jensen, A. R. (1981). *Straight talk about mental tests.* New York: Free Press.

Jensen, A. R. (1998). *The g factor: The science of mental ability.* Westport, CT: Praeger.

Jensen, A. R., & Reynolds, C. R. (1982). Race, social class, and ability patterns of the WISC-R. *Personality and Individual Differences, 3,* 423-438.

Jensen, M., & Rosenfeld, L. B. (1974). Influence of mode of presentation, ethnicity, and social class on teachers' evaluations of students. *Journal of Educational Psychology, 66,* 540-574.

Johnsen, S. K., & Corn, A. L. (1987). *SAGES: Screening Assessment for Gifted Elementary Students.* Austin, TX: PRO-ED.

Johnson, D. L., Breckenridge, J., & McGowan, R. (1984). Home environment and early cognitive development in Mexican-American children. In A. W. Gottfried (Ed.), *Home environment and early cognitive development: Longitudinal research* (pp. 151-195). Orlando, FL: Academic Press.

Johnson, D. L., Swank, P., Howie, V. M., Baldwin, C. D., Owen, M., & Luttman, D. (1993). Does HOME add to the prediction of child intelligence over and above SES? *Journal of Genetic Psychology, 154,* 33-40.

Johnston, T. (1987). The persistence of dichotomies in the study of behavioral development. *Developmental Review, 7,* 149-182.

Jones, J. (1995). Back to the future with *The Bell Curve:* Jim Crow, slavery, and *G.* In S. Fraser (Ed.), *The bell curve wars: Race, intelligence, and the future of America* (pp. 80-93). New York: Basic Books.

Jones v. School Board of the City of Alexandria, 179 F. Supp. 280 (originating in Va.) (1961).

Joynson, R. B. (1989). *The Burt affair.* New York: Academic Press.

Judis, J. B. (1995). Hearts of darkness. In S. Fraser (Ed.), *The bell curve wars: Race, intelligence, and the future of America* (pp. 124-129). New York: Basic Books.

Juel-Nielsen, N. (1965). Individual and environment: A psychiatric-psychological investigation of monozygous twins reared apart. *Acta Psychiatrica et Neurologia Scandinavica* (Monograph Supplement No. 183).

Kamin, L. J. (1974). *The science and politics of I.Q.* Hillsdale, NJ: Lawrence Erlbaum.

Kamin, L. J. (1995). Lies, damned lies, and statistics. In R. J. Jacoby & N. Glauberman (Eds.), *The bell curve debate: History, documents, opinions* (pp. 81-105). New York: Times Books.

Kamphaus, R. W., & Reynolds, C. R. (1987). *Clinical and research applications of the K-ABC.* Circle Pines, MN: American Guidance Service.

Kaplan, A. R. (Ed.). (1976). *Human behavioral genetics.* Springfield, IL: Charles C Thomas.

Kaplan, S. N. (1989). The gifted Asian-American child: A general response to a specific issue. In C. J. Maker & S. W. Schiever (Eds.), *Critical issues in gifted education: Defensible programs for cultural and ethnic minorities* (pp. 189-191). Austin, TX: PRO-ED.

Kaufman, A. S. (1990). *Assessing adolescent and adult intelligence.* Boston: Allyn & Bacon.

Kaufman, A. S., & Kaufman, N. L. (1983a). *Kaufman Assessment Battery for Children.* Circle Pines, MN: American Guidance Service.

Kaufman, A. S., & Kaufman, N. L. (1983b). *K-ABC interpretive manual.* Circle Pines, MN: American Guidance Service.

Kaufman, A. S., & Kaufman, N. L. (1990). *Kaufman Brief Intelligence Test.* Circle Pines, MN: American Guidance Service.

Kellaghan, T. (1977). Relationship between home environment and scholastic behavior in a disadvantaged population. *Journal of Educational Psychology, 69,* 754-760.

Kelves, D. J. (1985). *In the name of eugenics: Genetics and the uses of human heredity.* New York: Knopf.

Kempthorne, O. (1978). Logical, epistemological, and statistical aspects of the nature-nurture data interpretation. *Biometrics, 34,* 1-23.

Kennedy, W. A., Van De Riet, V., & White, J. C., Jr. (1963). A normative sample of intelligence and achievement of Negro elementary school children in the southeastern United States. *Monographs of the Society for Research on Child Development, 28,* 1-112.

Keston, M. J., & Jiménez, C. (1954). A study of the performance of English and Spanish editions of the Stanford-Binet intelligence test by Spanish-American children. *Journal of Genetic Psychology, 85,* 263-269.

Khatena, J. (1982). *Educational psychology of the gifted.* New York: John Wiley.

Kim, K. C., & Hurh, W. M. (1983). Korean Americans and the "success" image: A critique. *Amerasia, 10,* 3-21.

Kinchloe, J. L., & Steinberg, S. R. (1996). Who said it can't happen here? In J. L. Kinchloe, S. R. Steinberg, & A. D. Gresson (Eds.), *Measured lies: The bell curve examined* (pp. 3-47). New York: St. Martin's.

Kinchloe, J. L., Steinberg, S. R., & Gresson, A. D. (Eds.). (1996). *Measured lies: The bell curve examined.* New York: St. Martin's.

Kirk, S., McCarthy, J., & Kirk, L. W. (1968). *The Illinois Test of Psycholinguistic Abilities* (Rev. ed.). Urbana: University of Illinois Press.

Kirschenbaum, R. J. (1988). Methods for identifying the gifted and talented American Indian student. *Journal for the Education of the Gifted, 11,* 53-63.

Kirschenbaum, R. J. (1989). Identification of the gifted and talented American Indian student. In C. J. Maker & S. W. Schiever (Eds.), *Critical issues in gifted education: Defensible programs for cultural and ethnic minorities* (pp. 91-101). Austin, TX: PRO-ED.

Kitano, M. K. (1986). Gifted and talented Asian children. In M. K. Kitano & P. C. Chinn (Eds.), *Exceptional Asian children and youth* (pp. 54-60). Reston, VA: Council for Exceptional Children.

Kitano, M. K. (1989). Critique of "Identification of Gifted Asian-American Students." In C. J. Maker & S. W. Schiever (Eds.), *Critical issues in gifted education: Defensible programs for cultural and ethnic minorities* (pp. 163-168). Austin, TX: PRO-ED.

Kitano. M. K., & Chinn, P. C. (Eds.). (1986). *Exceptional Asian children and youth.* Reston, VA: Council for Exceptional Children.

Kleine, P. A. (1990). *For gifted and talented students, school is the place to be—not a place to learn.* Draft of a paper commissioned by the U.S. Department of Education, Office of Gifted and Talented.

Klineberg, O. (1932). A study of psychological differences between "racial" and national groups in Europe. *Archives of Psychology,* No. 132.

Klineberg, O. (1935a). *Negro intelligence and selective migration.* New York: Columbia University Press.

Klineberg, O. (1935b). *Race differences.* New York: Harper.

Klineberg, O. (1963). Negro-White differences in intelligence test performance: A new look at an old problem. *American Psychologist, 18,* 198-203.

Knapp, T. R. (1977). The unit of analysis problems in applications of simple correlational analysis to educational research. *Journal of Educational Statistics, 2,* 171-186.

Knight, B. C., Baker, E. H., & Minder, C. C. (1990). Concurrent validity of the Stanford-Binet: Fourth Edition and Kaufman Assessment Battery for Children with learning-disabled students. *Psychology in the Schools, 27,* 116-120.

Koch, H. L., & Simmons, R. (1926). A study of the test performance of American, Mexican, and Negro children. *Psychological Monographs, 35,* 1-116.

Kodama, H., Shinagawa, F., & Motegi, M. (1978). *WISC-R manual: Standardized in Japan.* New York: Psychological Corporation.

Korkman, M., Kirk, U., & Kemp, S. (1998). *NEPSY: A developmental neuropsychological assessment manual.* New York: Psychological Corporation.

Kozol, J. (1991). *Savage inequalities: Children in America's schools.* New York: Crown.

Kuhlmann, F. (1912). Binet-Simon's system for measuring the intelligence of children [Monograph supplement]. *Journal of Psycho-Asthenics, 1,* 76-92.

Kulik, J. A., & Kulik, C. -L. C. (1997). Ability grouping. In N. Colangelo & G. A. Davis (Eds.), *Handbook of gifted education* (2nd ed., pp. 230-242). Boston: Allyn & Bacon.

Laboratory on Comparative Human Cognition. (1982). Culture and intelligence. In R. J. Sternberg (Ed.), *Handbook of human intelligence* (pp. 642-719). New York: Cambridge University Press.

Lambert, N. M. (1981). Psychological evidence in *Larry P. v. Wilson Riles:* An evaluation for the defense. *American Psychologist, 36,* 937-952.

Lamke, T. A., Nelson, M. J., & Kelso, P. C. (1961). *Henmon-Nelson Test of Mental Ability–Revised.* Boston: Houghton Mifflin.

Lander, E. S., & Botstein, D. (1989). Mapping Mendellian factor underlying quantitative traits using RFLP linkage maps. *Genetics, 121,* 185-199.

Lane, C. (1995). Tainted sources. In R. Jacoby & N. Glauberman (Eds.), *The bell curve debate: History, documents, opinions* (pp. 125-139). New York: Times Books.

Laosa, L. M. (1981). Maternal behavior: Sociocultural diversity in modes of family interaction. In R. W. Henderson (Ed.), *Parent-child interaction: Theory, research, and prospects* (pp. 125-167). New York: Academic Press.

Laosa, L. M. (1987, August). *Population generalizability and ethical dilemmas in research, policy, and practice: Preliminary considerations.* Paper presented at the meeting of the American Psychological Association, New York.

Laosa, L. M. (1990). Population generalizability, cultural sensitivity, and ethical dilemmas. In C. B. Fisher & W. W. Tryon (Eds.), *Ethics in applied developmental psychology* (pp. 227-251). Norwood, NJ: Ablex.

Laosa, L. M., & Henderson, R. W. (1991). Cognitive socialization and competence: The academic development of Chicanos. In R. R. Valencia (Ed.), *Chicano school failure and success: Research agendas for the 1990s* (pp. 164-199, Stanford Series on Education and Public Policy). London: Falmer.

Laosa, L. M., & Sigel, I. E. (Eds.). (1982). *Families as learning environments for children.* New York: Plenum.

Larry P. v. Riles, 343 F. Supp. 1306 (N.D. Cal. 1972, order granting preliminary injunction), aff'd 502 F. 2d 63 (9th Cir. 1974), 495 F. Supp. 926 (N.D. Cal. 1979, decision on merits), aff'd No. 80-427 (9th Cir. Jan. 23, 1984), No. C-71-2270 R.F.P. (Sept. 25, 1986, order modifying judgment).

Larson, P. D. (1989). Administrative implications in developing programs for gifted Asian-American students. In M. K. Kitano & P. C. Chinn (Eds.), *Exceptional Asian children and youth* (pp. 197-200). Reston, VA: Council for Exceptional Children.

Leacock, E. (1969). *Teaching and learning in city schools.* New York: Basic Books.

Lee, T. F. (1993). *Gene future: The promise and perils of the new biology.* New York: Plenum.

Leong, F. T. L. (1986). Counseling and psychotherapy with Asian-Americans: Review of the literature. *Journal of Counseling Psvchology, 33,*196-206.

Leong, F. T. L., & Whitfield, J. R. (1992). *Asians in the United States: Abstracts of the Psychological and Behavioral Literature, 1967-1991.* (Bibliographies in Psychology, No. 11). Washington, DC: American Psychological Association.

Lerner, J. (1993). *Learning disabilities: Theories, diagnosis, and teaching strategies* (6th ed.). Boston: Houghton Mifflin.

Lerner, R. M. (1980). Concepts of epigenesis: Descriptive and explanatory issues. *Human Development, 23,* 63-72.

Lerner, R. M. (1991). Changing organism-context relations as the basic process of development: A developmental contextual perspective. *Developmental Psychology, 27,* 27-32.

Lesser, G. S., Fifer, G., & Clark, D. H. (1965). Mental abilities of children from different social class and cultural groups. *Monographs of the Society for Research in Child Development, 30*(4, No. 102).

Lesser, J. H. (1985-1986). Always "outsiders": Asians, naturalization, and the Supreme Court. *Amerasia, 12,* 83-100.

Lester, G., & Kelman, M. (1997). State disparities in the diagnosis and placement of pupils with learning disabilities. *Journal of Learning Disabilities, 30,* 599-607.

Levine, M. S. (1977). *Canonical analysis and factor comparison* (Quantitative Applications in the Social Sciences, Series No. 7-001). Beverly Hills, CA: Sage.

Levins, R., & Lewontin, R. (1985). *The dialectical biologist.* Cambridge, MA: Harvard University Press.

Lewontin, R. C. (1970). Race and intelligence. *Bulletin of the Atomic Scientists, 26,* 2-8.

Lewontin, R. C. (1973). Race and intelligence. In C. Senna (Ed.), *The fallacy of I.Q.* (pp. 1-17). New York: Joseph Okpatu.

Lewontin, R. C. (1975). Genetic aspects of intelligence. *Annual Review of Genetics, 9,* 387-405.

Lewontin, R. C. (1976). Race and intelligence. In H. J. Block & G. Dworkin (Eds.), *The IQ controversy: Critical readings* (pp. 78-92). New York: Pantheon Books.

Lewontin, R. C., Rose, S., & Kamin, L. J. (1984). *Not in our genes: Biology, ideology, and human nature.* New York: Pantheon.

Lindstrom, F. B., Hardert, R. A., & Johnson, L. L. (Eds.). (1995). *Kimball Young on sociology in transition, 1912-1968: An oral account by the 35th president of the ASA.* Lanham, MD: University Press of America.

Linklater, M. (1995). The curious laird of Nigg. In R. Jacoby & N. Glauberman (Eds.), *The bell curve debate: History, documents, opinions* (pp. 140-143). New York: Times Books.

Linn, R. L., & Harnisch, D. L. (1981). Interactions between item content and group membership on achievement test items. *Journal of Educational Measurement, 18,* 109-118.

Locurto, C. (1991). *Sense and nonsense about IQ: The case for uniqueness.* New York: Praeger.

Loehlin, J. C. (1981). [Review of the book *Identical Twins Reared Apart: A Reanalysis*]. *Acta Geneticae Medicae et Gemellogiae, 30,* 297-298.

Loehlin, J. C. (1989). Partitioning environmental and genetic contributions to behavioral development. *American Psychologist, 44,* 1285-1292.

Loehlin, J. C., Horn, J. M., & Willerman, L. (1989). Modeling IQ change: Evidence from the Texas Adoption Project. *Child Development, 60,* 993-1004.

Loehlin, J. C., Lindzey, G., & Spuhler, J. N. (1975). *Race differences in intelligence.* San Francisco: Freeman.

Lomotey, K. (Ed.). (1990). *Going to school: The African-American experience.* Albany: State University of New York Press.

Long, H. H. (1934). The intelligence of colored elementary pupils in Washington, D.C. *Journal of Negro Education, 6,* 205-222.

Long, H. H. (1935). Test results of third-grade Negro children selected on the basis of socioeconomic status. *Journal of Negro Education, 4,* 192-212.

Long, H. H. (1957). The relative learning capacity of Negroes and Whites. *Journal of Negro Education, 26,* 121-134.

Longstreth, L. E. (1978). A comment on "Race, IQ, and the Middle Class" by Trotman: Rampant false conclusions. *Journal of Educational Psychology, 70,* 469-472.

López, E. C. (1997). The cognitive assessment of limited English proficient and bilingual children. In D. P. Flanagan, J. L. Genshaft, & P. L. Harrison (Eds.), *Contemporary intellectual assessment: Theories, tests, and issues* (pp. 503-516). New York: Guilford.

Lugg, C. A. (1996). Social Darwinism as public policy. In J. L. Kinchloe, S. R. Steinberg, & A. D. Gresson (Eds.), *Measured lies: The bell curve examined* (pp. 367-378). New York: St. Martin's.

Luria, A. (1973). *The working brain: An introduction to neuropsychology.* New York: Basic Books.

Lynn, M. (1989). Criticisms of an evolutionary hypothesis about race differences: A rebuttal to Rushton's reply. *Journal of Research in Personality, 23,* 21-34.

Lynn, R., & Hampson, S. (1985-1986). The structure of Japanese abilities: An analysis in terms of the hierarchical model of intelligence. *Current Psychological Research and Reviews, 4,* 309-322.

Macionis, J. (1994). *Sociology* (4th ed.). Englewood Cliffs, NJ: Prentice Hall.

Mackenzie, B. (1984). Explaining race differences in IQ: The logic, the methodology, and the evidence. *American Psychologist, 39,* 1214-1233.

Mackintosh, N. J. (Ed.). (1995). *Cyril Burt: Fraud or framed?* Oxford, UK: Oxford University Press.

MacLear, M. (1922). Sectional differences as shown by academic ratings and Army tests. *School and Society, 15,* 676-678.

MacMillan, D. (1977). *Mental retardation in school and society.* Boston: Little, Brown.

MacMillan, D. L., & Meyers, C. E. (1980). *Larry P.:* An educational interpretation. *School Psychology Review, 9,* 136-148.

MacMillan, D. L., Siperstein, G. N., & Gresham, F. M. (1996). Commentary: A challenge to the viability of mild mental retardation as a diagnostic category. *Exceptional Children, 62,* 356-371.

MacPhee, D., Ramey, C., & Yeates, K. O. (1984). Home environments and early cognitive development: Implications for intervention. In A. W. Gottfried (Ed.), *Home environment and early cognitive development: Longitudinal research* (pp. 343-369). Orlando, FL: Academic Press.

Madden, R., & Gardner, E. F. (1969). *Stanford Early Achievement Test.* New York: Harcourt, Brace, & World.

Madden, R., Gardner, E. F., Rudman, H. C., Karlsen, B., & Merwin, J. C. (1974). *Stanford Achievement Test.* New York: Harcourt Brace Jovanovich.

Maddux, C. D., & Johnson, L. (1998). Computer-assisted assessment. In H. Booney Vance (Ed.), *Psychological assessment of children: Best practices for school and clinical settings* (2nd ed., pp. 87-105). New York: John Wiley.

Maheady, L., Towne, R., Algozzine, B., Mercer, J., & Ysseldyke, J. (1983). Minority overrepresentation: A case for alternative practices prior to referral. *Learning Disabilities Quarterly, 6,* 448-456.

Mainstream Science on Intelligence. (1994, December 13). *The Wall Street Journal,* p. A18.

Maker, C. J., & Schiever, S. W. (Eds.). (1989a). *Critical issues in gifted education: Defensible programs for cultural and ethnic minorities.* Austin, TX: PRO-ED.

Maker, C. J., & Schiever, S. W. (1989b). Summary of American Indian section. In C. J. Maker & S. W. Schiever (Eds.), *Critical issues in gifted education: Defensible programs for cultural and ethnic minorities* (pp. 142-148). Austin, TX: PRO-ED.

Margolin, L. (1994). *Goodness personified: The emergence of gifted children.* New York: Aldine de Gruyter.

Marjoribanks, K. (1972a). Environment, social class, and mental abilities. *Journal of Educational Psychology, 63,* 103-109.

Marjoribanks, K. (1972b). Ethnic environmental influences on mental abilities. *American Journal of Sociology, 78,* 323-327.

Marjoribanks, K. (1974). Another view of the relation of environments to mental abilities. *Journal of Educational Psychology, 66,* 460-463.

Marjoribanks, K. (1977). *Environmental correlates of Australian children's performances.* Unpublished manuscript, University of Adelaide.

Marjoribanks, K. (1978). Family and school environmental correlates of school-related affective characteristics: An Australian study. *Journal of Social Psychology, 106,* 181-189.

Marjoribanks, K. (1979). *Families and their learning environments: An empirical analysis.* London: Routledge & Kegan Paul.

Marks, R. (1981). *The idea of IQ.* Lanham, MD: University Press of America.

Massey, D. G., & Denton, N. A. (1993). *American apartheid: Segregation and the making of the underclass.* Cambridge, MA: Harvard University Press.

Matthew, J. L., Golin, A. K., Moore, M. W., & Baker, C. (1992). Use of SOMPA in identification of gifted African-American children. *Journal for the Education of the Gifted, 15,* 344-356.

McAnulty, E. A. (1929). Distribution of intelligence in the Los Angeles elementary schools. *Los Angeles Educational Research Bulletin, 8,* 6-8.

McArdle, J. J., & Goldsmith, H. H. (1990). Some alternative structural equation models for multivariate biometric analyses. *Behavior Genetics, 20,* 569-608.

McArdle, J. J., & Prescott, C. A. (1997). Contemporary models for the biometric genetic analysis of intellectual abilities. In D. P. Flanagan, J. L. Genshaft, & P. L. Harrison (Eds.), *Contemporary intellectual assessment: Theories, tests, and issues* (pp. 403-436). New York: Guilford.

McCarney, S. B., & Henage, D. (1987-1990). *Gifted Evaluation Scale.* Columbia, MO: Hawthorne Evaluation Services.

McCarthy, D. (1972). *McCarthy Scales of Children's Abilities.* New York: Psychological Corporation.

McCartney, K., Bernieri, F., & Harris, M. J. (1990). Growing up and growing apart: A developmental meta-analysis of twin studies. *Psychological Bulletin, 107,* 226-237.

McDougall, W. (1921). *Is America safe for democracy?* New York: Scribner.

McGrew, K. S., & Flanagan, D. P. (1998). *The Intelligence Test Desk Reference (ITDR): Gf-Gc Cross-Battery Assessment.* Boston: Allyn & Bacon.

McGue, M., Bouchard, T. J., Jr., Lykken, D. T., & Finkel, D. (1991). On genes, environment, and experience. *Behavioral and Brain Sciences, 14,* 400-401.

McKenzie, J. (1986). The influence of identification practices, race, and SES on the identification of gifted students. *Gifted Child Quarterly, 30,* 93-95.

McKusick, V. (1994). *Mendelian inheritance in man* (10th ed.). Baltimore, MD: Johns Hopkins University Press.

McShane, D. (1980). A review of scores of American Indian children on the Wechsler Intelligence Scale. *White Cloud Journal, 2,* 18-22.

McShane, D., & Cook, V. J. (1985). Transcultural intellectual assessment: Performance by Hispanics on the Wechsler scales. In B. B. Wolman (Ed.), *Handbook of intelligence: Theories, measurement, and applications* (pp. 737-785). New York: John Wiley.

Melesky, T. J. (1985). Identifying and providing for the Hispanic gifted child. *National Association of Bilingual Education Journal, 9,* 43-56.

Mercer, J. R. (1970). Sociological perspectives on mild mental retardation. In H. C. Haywood (Ed.), *Social-cultural aspects of mental retardation* (pp. 378-391). New York: Appleton-Century-Crofts.

Mercer, J. R. (1977). Identifying the gifted Chicano child. In J. L. Martinez, Jr. (Ed.), *Chicano Psychology* (pp. 155-173). New York: Academic Press.

Mercer, J. R. (1979). In defense of racially and culturally non-discriminatory assessment. *School Psychology Digest, 8,* 89-115.

Mercer, J. R. (1984). What is a racially and culturally nondiscriminatory test? A sociological and pluralistic perspective. In C. R. Reynolds & R. T. Brown (Eds.), *Perspectives on bias in mental testing* (pp. 293-356). New York: Plenum.

Mercer, J. R. (1988). Ethnic differences in IQ scores: What do they mean? *Hispanic Journal of Behavioral Sciences, 10,* 199-218.

Mercer, J. R. (1989). Alternative paradigms for assessment in a pluralistic society. In J. Banks & C. A. McKee Banks (Eds.), *Multicultural education: Issues and perspectives* (pp. 289-304). Boston: Allyn & Bacon.

Mercer, J. R., & Brown, W. C. (1973). Racial differences in IQ: Fact or fiction? In C. Sienna (Ed.), *The fallacy of IQ* (pp. 56-113). New York: Joseph Okpatu.

Mercer, J., & Lewis, J. (1979). *System of Multicultural and Pluralistic Assessment: Technical manual.* New York: Psychological Corporation.

Messick, S. (1981). Constructs and their vicissitudes in educational and psychological measurement. *Psychological Bulletin, 89,* 575-588.

Messick, S. (1989). Validity. In R. L. Linn (Ed.), *Educational measurement* (3rd ed., pp. 13-103). New York: Macmillan.

Miller, A. (1995). Professors of hate. In R. Jacoby & N. Glauberman (Eds.), *The bell curve debate: History, documents, opinions* (pp. 162-178). New York: Times Books.

Miller, J. G. (1997). A cultural-psychology perspective on intelligence. In R. J. Sternberg & E. Grigorenko (Eds.), *Intelligence, heredity, and environment* (pp. 269-302). New York: Cambridge University Press.

Modgil, S., & Modgil, C. (Eds.). (1987). *Arthur Jensen: Consensus and controversy.* New York: Falmer.

Montgomery, D. (1989). Identification of giftedness among American Indian people. In C. J. Maker & S. W. Schiever (Eds.), *Critical issues in gifted education: Defensible programs for cultural and ethnic minorities* (pp. 79-90). Austin, TX: PRO-ED.

Moore, E. G. J. (1986). Family socialization and the IQ test performance of traditional and transracially adopted Black children. *Developmental Psychology, 22,* 317-326.

Moore, J. (1989). Is there a Hispanic underclass? *Social Science Quarterly, 70,* 265-284.

Moreno, J. F. (Ed.). (1999). *The elusive quest for equality: 150 years of Chicano/Chicana education.* Cambridge, MA: Harvard Educational Review.

Moreno, S. (1973). [White, Chicano, and Black rates of representation in gifted programs in California public schools.] Unpublished data, San Diego State University.

Morse, J. (1914). A comparison of White and colored children measured by the Binet scale of intelligence. *Popular Science Monthly, 84,* 75-79.

Mosychuk, H. (1969). *Differential home environments and mental ability patterns.* Unpublished doctoral dissertation, University of Alberta.

Muñoz-Sandoval, A. F., Cummins, J., Alvarado, C. G., & Ruef, M. L. (1998). *Bilingual Verbal Ability Tests.* Itasca, IL: Riverside.

Murphey, D. D. (1995). Rethinking the American dream: Reactions of the media to *The Bell Curve. Journal of Social, Political, and Economic Studies, 20,* 93-128.

Murphy, J. (1990). *The reform of American public education in the 1980s: Perspectives and cases.* Berkeley, CA: McCutchan.

Murray, C. (1988). The coming of custodial democracy. *Commentary, 86*(3), 9-14.

Murray, C. (1995). *The Bell Curve* and its critics. *Commentary, 99*(5), 23-30.

Murray, H. (1938). *Explorations in personality.* New York: Oxford University Press.

Myers, G. C. (1922). *A Pantomime Group Intelligence Test.* New York: Newson.

Naglieri, J. A., & Das, J. P. (1997). *Cognitive Assessment System: Administration and scoring manual.* Itasca, IL: Riverside.

National/State Leadership Training Institute, Office of Ventura County Superintendent of Schools. (1981). *Balancing the scale for the disadvantaged gifted* (proceedings from the Fourth National Conference on Disadvantaged Gifted/Talented). Ventura, CA: Author.

Neff, W. S. (1938). Socioeconomic status and intelligence: A critical survey. *Psychological Bulletin, 35,* 727-757.

Neisser, U. (1998a). Introduction: Rising test scores and what they mean. In U. Neisser (Ed.), *The rising curve: Long-term gains in IQ and related measures* (pp. 3-22). Washington, DC: American Psychological Association.

Neisser, U. (Ed.). (1996). *The rising curve: Long-term gains in IQ and related measures.* Washington, DC: American Psychological Association.

Neisser, U., Boodoo, G., Bouchard, T. J., Jr., Boykin, W. A., Brody, N., Ceci, S. J., Halpern, D. F., Loehlin, J. C., Perloff, R., Sternberg, R. J., & Urbina, S. (1996). Intelligence: Knowns and unknowns. *American Psychologist, 51,* 77-101.

Newland, T. E. (1976). *The gifted in socioeducational perspective.* Englewood Cliffs, NJ: Prentice Hall.

Newman, H. H., Freeman, F. N., & Holzinger, K. J. (1937). *Twins: A study of heredity and environment.* Chicago: University of Chicago Press.

Nichols, P. L. (1970). *The effects of heredity and environment on intelligence test performance in 4 and 7 year old White and Negro sibling pairs.* Doctoral dissertation, University of Michigan. (University Microfilms, No. 71-18, 874)

Nichols, P. L., & Anderson, V. E. (1973). Intellectual performance, race, and socioeconomic status. *Social Biology, 20,* 367-374.

Nisbett, R. E. (1998). Race, genetics, and IQ. In C. Jencks & M. Phillips (Eds.), *The Black-White test score gap* (pp. 86-102). Washington, DC: Brookings Institution.

Nunley, M. (1995). *The Bell Curve:* Too smooth to be true [Book review]. *American Behavioral Scientist, 39,* 74-83.

Oakes, J. (1985). *Keeping track: How schools structure inequality.* New Haven, CT: Yale University Press.

Oakland, T. (1978). Predictive validity of readiness tests for middle and lower socioeconomic status Anglo, Black, and Mexican American children. *Journal of Educational Psychology, 70,* 574-582.

Oakland, T. (1979a). Research on the Adaptive Behavior Inventory for Children and the Estimated Learning Potential. *School Psychology Digest, 8,* 63-70.

Oakland, T. (1979b). Research on the AIBC and the ELP: A revisit to an old topic. *School Psychology Digest, 8,* 212-213.

Oakland, T., & Glutting, J. J. (1990). Examiner observations of children's WISC-R test-related behaviors: Possible socioeconomic status, race, and gender effects. *Psychological Assessment, 2,* 86-90.

Ochoa, S. H., Powell, M. P., & Robles-Pina, R. (1996). School psychologists' assessment practices with bilingual and limited-English-proficient students. *Journal of Psychoeducational Assessment, 14,* 250-275.

Ogbu, J. U. (1978). *Minority education and caste: The American system in cross-cultural perspective.* New York: Academic Press.

Ogbu, J. U. (1983). Minority status and schooling in plural societies. *Comparative Education Review, 27,* 168-190.

Ogbu, J. U. (1986). The consequences of the American caste system. In U. Neisser (Ed.), *The school achievement of minority children* (pp. 19-56). Hillsdale, NJ: Lawrence Erlbaum.

Ogbu, J. U. (1987). Variability in minority responses to schooling: Nonimmigrants vs. immigrants. In G. Spindler & L. Spindler (Eds.), *Interpretive ethnography of education: At home and abroad* (pp. 255-278). Hillsdale, NJ: Lawrence Erlbaum.

Ogbu, J. U. (1991). Immigrant and involuntary minorities in comparative perspective. In M. A. Gibson & J. U. Ogbu (Eds.), *Minority status and schooling: A comparative study of immigrant and involuntary minorities* (pp. 3-33). New York: Garland.

Ogbu, J. U. (1994). Culture and intelligence. In R. J. Sternberg (Ed.), *Encyclopedia of human intelligence* (Vol. 2, pp. 328-338). New York: Macmillan.

Ohio State University, Center for Human Resource Research. (n.d.). *National Longitudinal Survey of Youth.* Columbus: Author.

Olmedo, E. L., & Walker, V. R. (1990). *Hispanics in the United States: Abstracts of the psychological and behavioral literature, 1980-1989.* (Bibliographies in Psychology, No. 8). Washington, DC: American Psychological Association.

Omi, M., & Winant, H. (1994). *Racial formation in the United States from the 1960s to the 1990s* (2nd ed.). New York: Routledge & Kegan Paul.

Ong, P. M. (1984). Chinatown unemployment and the ethnic labor market. *Amerasia, 11,* 35-54.

Ortiz, V. Z., & González, A. (1989). Validation of a short form of the WISC-R with accelerated and gifted Hispanic students. *Gifted Child Quarterly, 33,* 152-155.

Ortiz, V. Z., & Volloff, W. (1987). Identification of gifted and accelerated Hispanic students. *Journal for the Education of the Gifted, 11,* 45-55.

Osborne, R. T., & Gregor, A. J. (1968). Racial differences in heritability estimates for tests of spatial ability. *Perceptual and Motor Skills, 27,* 735-739.

Osborne, R. T., & McGurk, F. C. J. (1982a). Summary and conclusions. In R. T. Osborne & F. C. J. McGurk (Eds.), *The testing of Negro intelligence* (Vol. 2, pp. 290-297). Athens, GA: Foundation for Human Understanding.

Osborne, R. T., & McGurk, F. C. J. (Eds.). (1982b). *The testing of Negro intelligence* (Vol. 2). Athens, GA: Foundation for Human Understanding.

Osborne, R. T., & Miele, F. (1969). Racial differences in environmental influences on numerical ability as determined by heritability estimates. *Perceptual and Motor Skills, 28,* 533-538.

Otis, A. S., & Lennon, R. T. (1967). *Otis-Lennon Mental Ability Test.* New York: Harcourt, Brace, & World.

Otis, A. S., & Lennon, R. T. (1982). *Manual for the School Ability Test.* New York: Psychological Corporation.

Otto, H. J. (1931). Administrative practices followed in the organization of elementary schools. *American School Board Journal, 83*, 35-36.

Overton, T. (1997). *Assessment in special education: An applied approach.* Upper Saddle River, NJ: Merrill/Prentice Hall.

Oyama, S. (1985). *The ontogeny of information: Developmental systems and evolution.* New York: Cambridge University Press.

Packard, A. (1986-present). Review of the MIDAS: Multiple Intelligence Developmental Assessment Scales. In *Buro's mental measurements yearbook* (WinSPIRS database).

Padilla, A. M., & Aranda, P. (1974). *Latino mental health: Bibliography and abstracts.* Rockville, MD: Alcohol, Drug Abuse, and Mental Health Administration.

Parents in Action on Special Education v. Joseph P. Hannon, No. 74C 3586 (N.D. Ill.) (1980).

Paschal, F. C., & Sullivan, L. R. (1925). Racial differences in the mental and physical development of Mexican children. *Comparative Psychology Monographs, 3*, 1-76.

Passow, A. H. (1989). Needed research and development in educating high ability children. *Roeper Review, 11*, 223-229.

Patton, J. M. (1992). Assessment and identification of African-American learners with gifts and talents. *Exceptional Children, 59*, 150-159.

Pearl, A. (1991). The big picture and Chicano school failure. In R. R. Valencia (Ed.), *Chicano school failure and success: Research and policy agendas for the 1990s* (pp. 271-300, Stanford Series on Education and Public Policy). London: Falmer.

Pearl, A. (1997). Cultural and accumulated environmental deficits model. In R. R. Valencia (Ed.), *The evolution of deficit thinking: Educational thought and practice* (pp. 113-131, Stanford Series on Education and Public Policy). London: Falmer.

Pearson, R. (1991). *Race, intelligence, and bias in academe.* Washington, DC: Scott-Townsend.

Pendarvis, E. D., Howley, A. A., & Howley, C. B. (1990). *The abilities of gifted children.* Englewood Cliffs, NJ: Prentice Hall.

Pérez, S. M., & De La Rosa Salazar, D. (1993). Economic, labor force, and social implications of Latino educational population trends. *Hispanic Journal of Behavioral Sciences, 15*, 188-229.

Perrine, J. (1989). Situational identification of gifted Hispanic students. In C. J. Maker & S. W. Schiever (Eds.), *Critical issues in gifted education: Defensible programs for cultural and ethnic minorities* (pp. 5-18). Austin, TX: PRO-ED.

Pettigrew, T. F. (1964a). Negro American intelligence: A new look at an old controversy. *Journal of Negro Education, 33*, 6-25.

Pettigrew, T. F. (1964b). *A profile of the Negro Americans.* Princeton, NJ: Van Nostrand.

Pfeiffer, A. B. (1989). Purpose of programs for gifted and talented and highly motivated American Indian students. In C. J. Maker & S. W. Schiever (Eds.), *Critical issues in gifted education: Defensible programs for cultural and ethnic minorities* (pp. 102-106). Austin, TX: PRO-ED.

Phillips, B. A. (1914). The Binet test applied to colored children. *Psychological Clinic, 8*, 190-196.

Phillips, D. C., & Kelley, M. C. (1975). Hierarchical theories of development in education and psychology. *Harvard Educational Review, 45*, 351-375.

Piaget, J. (1929). *The child's conception of the world* (J. Tomlinson & A. Tomlinson, Trans.). New York: Harcourt, Brace, & World. (Original work published 1926)

Pintner, R. (1919). A non-language group intelligence test. *Journal of Applied Psychology, 3*, 199-214.

Pintner, R. (1927). *Non-Language Mental Tests.* Columbus: Ohio State University Press.

Pintner, R. (1931). *Intelligence testing: Methods and results* (2nd ed.). New York: Henry Holt.

Plessy v. Ferguson, 16 S.C+. 1138 (1896).

Plomin, R. (1983). Developmental behavioral genetics. *Child Development, 54*, 253-259.

Plomin, R. (1997). Identifying genes for cognitive abilities and disabilities. In R. J. Sternberg & E.–Grigorenko (Eds.), *Intelligence, heredity, and environment* (pp. 89-104). New York: Cambridge University Press.

Plomin, R., & Bergeman, C. S. (1991). The nature of nurture: Genetic influences on "environmental" measures. *Behavioral and Brain Sciences, 14,* 373-386.

Plomin, R., & Daniels, D. (1987). Why are children in the same family so different from one another? *Behavioral and Brain Sciences, 10,* 1-16.

Plomin, R., & DeFries, J. C. (1980). Genetics and intelligence: Recent data. *Intelligence, 4,* 15-24.

Plomin, R., DeFries, J. C., McClearn, G. E., & Rutter, M. (1997). *Behavioral genetics* (3rd ed.). New York: Freeman.

Plomin, R., & Loehlin, J. C. (1989). Direct and indirect heritability estimates: A puzzle. *Behavior Genetics, 19,* 331-342.

Plomin, R., & Neiderhiser, J. M. (1991). Quantitative genetics, molecular genetics, and intelligence. *Intelligence, 15,* 369-387.

Plomin, R., & Petrill, S. A. (1997). Genetics and intelligence: What's new? *Intelligence, 24,* 53-77.

Plomin, R., & Rende, R. (1990). Human behavioral genetics. *Annual Review of Psychology, 42,* 161-190.

Plucker, J. A. (1996). Gifted Asian-American students: Identification, curricula, and counseling concerns. *Journal for the Education of the Gifted, 19,* 315-343.

Poortinga, Y. H. (1983). Psychometric approaches to intergroup comparison: The problem of equivalence. In S. H. Irvine & J. W. Berry (Eds.), *Human assessment and cultural factors* (pp. 237-258). New York: Plenum.

Prasse, D. P., & Reschly, D. J. (1986). *Larry P.:* A case of segregation, testing, or program efficacy? *Exceptional Children, 52,* 333-346.

Prendes-Lintel, M. (1989-present). Review of the Batería Woodcock Psico-Educativa en Español. In *Buro's mental measurements yearbook* (WinSPIRS database).

Pressey, L. W. (1920). The influence of (a) inadequate schooling and (b) poor environment upon results with tests of intelligence. *Journal of Applied Psychology, 4,* 91-96.

Pressey, S. L., & Ralston, R. (1919). The relation of the general intelligence of school children to the occupation of their fathers. *Journal of Applied Psychology, 4,* 366-373.

Prewitt Diaz, J. O. (1988). Assessment of Puerto Rican children in bilingual education programs in the United States: A critique of Lloyd M. Dunn's monograph. *Hispanic Journal of Behavioral Sciences, 10,* 237-252.

Price, J. St. Clair. (1929). The intelligence of Negro college freshmen. *School and Society, 30,* 749-754.

Price, J. St. Clair. (1934). Negro-White differences in general intelligence. *Journal of Negro Education, 3,* 424-452.

Prifitera, A., Weiss, L. G., & Saklofske, D. H. (1998). The WISC-III in context. In A. Prifitera & D. Saklofske (Eds.), *WISC-III clinical use and interpretation: Scientist-practitioner perspectives* (pp. 1-38). New York: Academic Press.

Proctor, L. S. (1929). *A case study of thirty superior colored children of Washington, D.C.* Unpublished master's thesis, The University of Chicago.

Provine, W. B. (1986). Geneticists and race. *American Zoologist, 26,* 857-887.

Race, H. V. (1918). A study of a class of children of superior intelligence. *Journal of Educational Psychology, 9,* 91-98.

Ramey, C., Farran, D. C., & Campbell, F. A. (1979). Predicting IQ from mother-child interactions. *Child Development, 50,* 804-814.

Ramey, C., MacPhee, D., & Yeates, K. (1982). Preventing developmental retardation: A general systems model. In L. Bond & J. Joffe (Eds.), *Facilitating infant and early childhood development* (pp. 343-401). Hanover, NH: University Press of New England.

Ramey, C., Mills, P., Campbell, F., & O'Brien, C. (1975). Infants' home environments: A comparison of high risk families from the general population. *American Journal of Mental Deficiency, 80,* 40-42.

Randalls, E. H. (1929). _A comparative study of the intelligence test results of Mexican American and Negro children in two elementary schools._ Unpublished master's thesis, University of Southern California.

Ratliff, Y. P. (1960). _Spanish-speaking and English-speaking children in southwest Texas: A comparative study of intelligence, socio-economic status, and achievement._ Unpublished master's thesis, The University of Texas at Austin.

Raven, J. C., Court, J. H., & Raven, J. (1990). _Raven manual: Coloured Progressive Matrices_ (1990 ed.). Oxford, UK: Oxford Psychologists Press.

Raven, J. C., Court, J. H., & Raven, J. (1992). _Raven manual: Standard Progressive Matrices_ (1992 ed.). Oxford, UK: Oxford Psychologists Press.

Raven, J. C., Court, J. H., & Raven, J. (1994). _Raven manual: Advanced Progressive Matrices_ (1994 ed.). Oxford, UK: Oxford Psychologists Press.

Reed, A., Jr. (1994). Looking backward [Review of the book _The Bell Curve: Intelligence and Class Structure in American Life_]. _Nation, 259,_ 654-662.

Reed, T. E. (1969). Caucasian genes in American Negroes. _Science, 165,_ 762-768.

Reis, S. M., Westberg, K., Kulikowich, J., Caillard, F., Hebert, T., Purcell, J., Rogers, J., & Smist, J. (1992, April). Modifying regular classroom instruction with curriculum compacting. In J. S. Renzulli (Chair), _Regular classroom practices with gifted students: Findings from the National Research Center on the Gifted and Talented_ (symposium conducted at the meeting of the American Educational Research Association, San Francisco).

Reiss, A. J., Duncan, O. D., Hatt, P. K., & North, C. C. (1961). _Occupations and social status._ New York: Free Press.

Renzulli, J. S. (1978). What makes giftedness? Reexamining a definition. _Phi Delta Kappan, 60,_ 180-184, 261.

Renzulli, J. S. (1986). The three-ring conception of giftedness: A developmental model for creative productivity. In R. J. Sternberg & J. E. Davidson (Eds.), _Conceptions of giftedness_ (pp. 53-92). New York: Cambridge University Press.

Renzulli, J. S. (1991). The National Research Center on the Gifted and Talented: The dream, the design, and the destination. _Gifted Child Quarterly, 35,_ 73-80.

Renzulli, J. S., Reid, B. D., & Gubbins, E. J. (1991). _Setting an agenda: Research priorities for the gifted and talented through the year 2000._ Storrs, CT: National Research Center on the Gifted and Talented.

Renzulli, J. S., & Reis, S. M. (1991). The reform movement and the quiet crisis in education. _Gifted Child Quarterly, 35,_ 26-35.

Renzulli, J. S., & Reis, S. M. (1994). Research related to the Schoolwide Enrichment Model. _Gifted Child Quarterly, 38,_ 2-14.

Renzulli, J. S., & Reis, S. M. (1997). The Schoolwide Enrichment Model: New directions for developing high-end learning. In N. Colangelo & G. A. Davis (Eds.), _Handbook of gifted education_ (2nd ed., pp. 136-154). Boston: Allyn & Bacon.

Renzulli, J. S., Smith, L. H., White, A. J., Callahan, C. M., & Hartman, R. K. (1976). _Scales for Rating the Behavioral Characteristics of Superior Students._ Wethersfield, CT: Creative Learning Press.

Reschly, D. (1979). Nonbiased assessment. In G. Phye & D. Reschly (Eds.), _School psychology: Perspectives and issues_ (pp. 215-253). New York: Academic Press.

Reschly, D. J. (1981). Psychological testing in educational classification and placement. _American Psychologist, 36,_ 1094-1102.

Reschly, D. J. (1988). Minority MMR overrepresentation and special education reform. _Exceptional Children, 54,_ 316-323.

Reschly, D. J. (1990). Aptitude tests in educational classification and placement. In G. Goldstein & M. Hersen (Eds.), _Handbook of psychological assessment_ (2nd ed., pp. 148-172). New York: Pergamon.

Reschly, D. J., & Ward, S. M. (1991). Uses of adaptive behavior measures and overrepresentation of Black students in programs for students with mild mental retardation. *American Journal on Mental Retardation, 96,* 257-268.

Resnick, D. P., & Goodman, M. (1993). American culture and the gifted. In P. O. Ross (Ed.), *National excellence: A case for developing America's talent* (pp. 109-121). Washington, DC: U.S. Department of Education, Office of Educational Research and Improvement.

Resnick, L. B. (1979). The future of IQ testing in education. *Intelligence, 3,* 241-253.

Reyes, P., & Valencia, R. R. (1993). Educational policy and the growing Latino student population: Problems and prospects. *Hispanic Journal of Behavioral Sciences, 16,* 258-283.

Reynolds, A. (1933). *The education of Spanish-speaking children in five southwestern states* (Bulletin No. 11). Washington, DC: Government Printing Office.

Reynolds, C. R. (1982a). Methods for detecting construct and predictive bias. In R. A. Berk (Ed.), *Handbook of methods for detecting test bias* (pp. 199-227). Baltimore, MD: Johns Hopkins University Press.

Reynolds, C. R. (1982b). The problem of bias in psychological assessment. In C. R. Reynolds & T. B. Gutkin (Eds.), *The handbook of school psychology* (pp. 178-208). New York: John Wiley.

Reynolds, C. R., & Brown, R. T. (Eds.). (1984). *Perspectives on bias in mental testing.* New York: Plenum.

Reynolds, C. R., & Kaiser, S. M. (1990). Bias in assessment of aptitude. In C. R. Reynolds & R. W. Kamphaus (Eds.), *Handbook of psychological and educational assessment of children: Intelligence and achievement* (pp. 611-653). New York: Guilford.

Reynolds, C. R., & Kaufman, A. S. (1990). Assessment of children's intelligence with the Wechsler Intelligence Scale for Children–Revised (WISC-R). In C. R. Reynolds & R. W. Kamphaus (Eds.), *Handbook of psychological and educational assessment of children: Intelligence and achievement* (pp. 127-165). New York: Guilford.

Rhodes, L. (1992). Focusing attention on the individual in identification of gifted Black students. *Roeper Review, 14,* 108-110.

Richert, E. S. (1985a). Identification of gifted children in the United States: The need for pluralistic assessment. *Roeper Review, 8,* 68-72.

Richert, E. S. (1985b). The state of the identification of gifted students in the United States. *Gifted Education International, 3,* 47-51.

Richert, E. S. (1987). Rampant problems and promising practices in the identification of disadvantaged gifted students. *Gifted Child Quarterly, 31,* 149-154.

Richert, E. S. (1997). Excellence with equity in identification and programming. In N. Colangelo & G. A. Davis (Eds.), *Handbook of gifted education* (2nd ed., pp. 75-88). Boston: Allyn & Bacon.

Riley, R. (1993). *National excellence: A case for developing America's talent.* Washington, DC: U.S. Department of Education.

Riojas Clark, E. (1981). A double minority: The gifted Mexican American child. In T. H. Escobedo (Ed.), *Education and Chicanos: Issues and research* (Monograph No. 8, pp. 21-23). Los Angeles: University of California, Los Angeles, Spanish Speaking Mental Health Center.

Robbins, R. (1991). American Indian gifted and talented students: Their problems and proposed solutions. *Journal of American Indian Education, 31,* 15-24.

Robinson, M. L., & Meenes, M. (1947). The relationship between test intelligence of third grade Negro children and the occupations of their parents. *Journal of Negro Education, 16,* 136-141.

Robinson, W. S. (1950). Ecological correlations and the behavior of individuals. *American Sociological Review, 15,* 351-357.

Rogers, M. R. (1998). Psychoeducational assessment of culturally and linguistically diverse children and youth. In H. B. Vance (Ed.), *Psychological assessment of children: Best practices for school and clinical settings* (2nd ed., pp. 355-385). New York: John Wiley.

Roid, G. H., & Miller, L. J. (1997). *Leiter International Performance Scale–Revised.* Wood Dale, IL: Stoelting.

Romero, M. E. (1994). Identifying giftedness among Keresan Pueblo Indians: The Keres study. *Journal of American Indian Education, 34,* 35-58.

Ross, A. C. (1982). Brain hemispheric functions and the Native American. *Journal of American Indian Education, 22,* 2-5.

Ross, A. C. (1989). Brain hemispheric functions and the Native American. *Journal of American Indian Education, 29,* 72-76.

Ross, P. O. (1997). Federal policy on gifted and talented education. In N. Colangelo & G. A. Davis (Eds.), *Handbook of gifted education* (2nd ed., pp. 553-559). Boston: Allyn & Bacon.

Rowe, D. C. (1994). *The limits of family influence: Genes, experience, and behavior.* New York: Guilford.

Rushton, J. P. (1995). *Race, evolution, and behavior: A life history perspective.* New Brunswick, NJ: Transaction Books.

Russo, C. J., Ford, D. Y., & Harris, J. J., III. (1993). The educational rights of gifted students: Lost in the legal shuffle? *Roeper Review, 16,* 67-71.

Saarni, C. (1997). Emotional competence and self-regulation in childhood. In P. Salovey & D. J. Sluyter (Eds.), *Emotional development and emotional intelligence: Educational implications* (pp. 35-66). New York: Basic Books.

Salovey, P., & Sluyter, D. J. (1997). *Emotional development and emotional intelligence: Educational implications.* New York: Basic Books.

Salvia, J., & Ysseldyke, J. E. (1988). *Assessment in special and remedial education* (4th ed.). Boston: Houghton Mifflin.

Samuda, R. J. (1998). *Psychological testing of American minorities: Issues and consequences* (2nd ed.). Thousand Oaks, CA: Sage.

Sánchez, G. I. (1932). Group differences in Spanish-speaking children: A critical review. *Journal of Applied Psychology, 16,* 549-558.

Sánchez, G. I. (1934). Bilingualism and mental measures. *Journal of Applied Psychology, 18,* 765-772.

Sandoval, J. (1998a). Critical thinking in test interpretation. In J. Sandoval, C. L. Frisby, K. F. Geisinger, J. D. Scheuneman, & J. R. Grenier (Eds.), *Test interpretation and diversity: Achieving equity in assessment* (pp. 31-49). Washington, DC: American Psychological Association.

Sandoval, J. (1998b). Test interpretation in a diverse future. In J. Sandoval, C. L. Frisby, K. F. Geisinger, J. D. Scheuneman, & J. R. Grenier (Eds.), *Test interpretation and diversity: Achieving equity in assessment* (pp. 387-402). Washington, DC: American Psychological Association.

Sandoval, J., & Duran, R. P. (1998). Language. In J. Sandoval, C. L. Frisby, K. F. Geisinger, J. D. Scheuneman, & J. R. Grenier (Eds.), *Test interpretation and diversity: Achieving equity in assessment* (pp. 181-211). Washington, DC: American Psychological Association.

Sandoval, J., & Irvin, M. G. (1990). Legal and ethical issues in the assessment of children. In C. R. Reynolds & R. W. Kamphaus (Eds.), *Handbook of psychological and educational assessment of children: Intelligence and achievement* (pp. 86-104). New York: Guilford.

San Miguel, G., Jr. (1987). *"Let them all take heed": Mexican Americans and their campaign for educational equality in Texas, 1910-1981.* Austin: The University of Texas Press.

San Miguel, G., Jr., & Valencia, R. R. (1998). From the Treaty of Guadalupe Hidalgo to *Hopwood:* The educational plight and struggle of Mexican Americans in the Southwest. *Harvard Educational Review, 68,* 353-412.

Sarason, S. B., & Doris, J. (1979). *Educational handicap, public policy, and social history.* New York: Free Press.

Sapon-Shevin, M. (1994). *Playing favorites: Gifted education and the disruption of community.* Ithaca: State University of New York Press.

Sattler, J. M. (1988). *Assessment of children* (3rd ed.). San Diego: Author.

Sattler, J. M. (1990). *Assessment of children* (3rd ed., rev.). San Diego: Author.

Sattler, J. M. (1992). *Assessment of children* (3rd ed., rev. and updated). San Diego: Author.

Sattler, J. M., & Kuncik, T. M. (1976). Ethnicity, socioeconomic status, and pattern of WISC scores as variables that affect psychologists' estimates of "effective intelligence." *Journal of Clinical Psychology, 32,* 362-366.

Scarr, S. (1981). Reply: Some myths about heritability and IQ. In S. Scarr (Ed.), *Race, social class, and individual differences in I.Q.* (pp. 256-260). Hillsdale, NJ: Lawrence Erlbaum.

Scarr, S. (1992). Developmental theories for the 1990s: Development and individual differences. *Child Development, 63,* 1-19.

Scarr, S. (1993). Biological and cultural diversity: The legacy of Darwin for development. *Child Development, 64,* 1333-1353.

Scarr, S. (1997). Behavior-genetic and socialization theories of intelligence: Truce and reconciliation. In R. J. Sternberg & E. Grigorenko (Eds.), *Intelligence, heredity, and environment* (pp. 3-41). New York: Cambridge University Press.

Scarr, S., & Carter-Saltzman, L. (1982). Genetics and intelligence. In R. J. Sternberg (Ed.), *Handbook of human intelligence* (pp. 792-896). New York: Cambridge University Press.

Scarr, S., Pakstis, A. J., Katz, S. H., & Barker, W. B. (1977). Absence of a relationship between degree of White ancestry and intellectual skills within a Black population. *Human Genetics, 39,* 69-86.

Scarr, S., & Weinberg, R. A. (1976). IQ test performance of Black children adopted by White parents. *American Psychologist, 31,* 726-739.

Scarr, S., & Weinberg, R. A. (1977). Intellectual similarities within families of both adopted and biological children. *Intelligence, 1,* 170-191.

Scarr, S., Weinberg, R. A., & Waldman, I. D. (1993). IQ correlations in transracial adoptive families. *Intelligence, 17,* 541-555.

Scarr-Salapatek, S. (1971). Race, social class, and IQ. *Science, 174,* 1285-1295.

Schaefer, E., Bell, R., & Bayley, N. (1959). Development of a maternal behavior research instrument. *Journal of Genetic Psychology, 95,* 83-104.

Schönemann, P. H. (1994). Heritability. In R. J. Sternberg (Ed.), *Encyclopedia of human intelligence* (pp. 528-536). New York: Macmillan.

Scott, D. M. (1994). Cognitive conceit [Review of the book *The Bell Curve: Intelligence and Class Structure in American Life*]. *Social Policy, 25,* 50-59.

Scott, M. S., Perou, R., Urbano, R., Hogan, A., & Gold, S. (1992). The identification of giftedness: A comparison of White, Hispanic, and Black families. *Gifted Child Quarterly, 36,* 131-139.

Scruggs, T. E., & Cohn, S. J. (1983). A university-based summer program for a highly able but poorly achieving Indian child. *Gifted Child Quarterly, 27,* 90-93.

Sedgwick, J. (1995). Inside the Pioneer Fund. In R. Jacoby & N. Glauberman (Eds.), *The bell curve debate: History, documents, opinions* (pp. 144-161). New York: Times Books.

Shearer, B. C. (1994-1996). *The MIDAS: Multiple Intelligence Developmental Assessment Scales.* Kent, OH: Multiple Intelligences Research and Consulting.

Sheldon, W. H. (1924). The intelligence of Mexican children. *School and Society, 19,* 139-142.

Shepard, L. A. (1981). Identifying bias in test items. In B. F. Green (Ed.), *New directions for testing and measurement: Issues in testing—Coaching, disclosure, and ethnic bias* (No. 11, pp. 79-104). San Francisco: Jossey-Bass.

Shepard, L. A. (1982). Definitions of bias. In R. A. Berk (Ed.), *Handbook of methods for detecting test bias* (pp. 9-30). Baltimore, MD: Johns Hopkins University Press.

Shields, J. (1962). *Monozygotic twins brought up apart and brought up together.* London: Oxford University Press.

Shuey, A. M. (1958). *The testing of Negro intelligence.* Lynchburg, VA: J. P. Bell.

Shuey, A. M. (1966). *The testing of Negro intelligence* (2nd ed.). New York: Social Science Press.

Siegel, L. S. (1984). Home environmental influences on cognitive development in preterm and full-term children during the first 5 years. In A. W. Gottfried (Ed.), *Home environment and early cognitive development: Longitudinal research* (pp. 197-234). Orlando, FL: Academic Press.

Siegel, L. S. (1989). IQ is irrelevant to the definition of learning disabilities. *Journal of Learning Disabilities, 22,* 469-486.

Sigman, M., & Whaley, S. E. (1998). The role of nutrition in the development of intelligence. In U. Neisser (Ed.), *The rising curve: Long-term gains in IQ and related measures* (pp. 155-182). Washington, DC: American Psychological Association.

Sims, V. M. (1927). *Sims Score Card for Socio-Economic Status*. Bloomington, IL: Public School Publishing.

Sisk, D. A. (1989). Identifying and nurturing talent among the American Indians. In C. J. Maker & S. W. Schiever (Eds.), *Critical issues in gifted education: Defensible programs for cultural and ethnic minorities* (pp. 128-132). Austin, TX: PRO-ED.

Skrabanek, P. (1989). The mismeasure of Burt [Review of the book *The Burt Affair*]. *Lancet, 8667*, 856-857.

Slavin, R.E. (1986). *Educational psychology: Theory into practice*. Englewood Cliffs, NJ: Prentice Hall.

Slavin, R. E. (1991). Are cooperative learning and "untracking" harmful to the gifted? Response to Allan. *Educational Leadership, 48*, 68-71.

Slavin, R. E. (1997). *Educational psychology: Theory and practice* (5th ed.). Boston: Allyn & Bacon.

Slavin, R. E., Madden, N. A., & Stevens, R. J. (1990). Cooperative learning models for the 3 R's. *Educational Leadership, 47*, 22-29.

Sleeter, C. E. (1986). Learning disabilities: The social construction of a special education category. *Exceptional Children, 53*, 46-54.

Slosson, R. L. (1963). *Slosson Intelligence Test and Slosson Oral Reading Test*. New York: Slosson Educational Publications.

Smith, M. S., & Bissell, J. S. (1970). Report analysis: The impact of Head Start. *Harvard Educational Review, 14*, 51-104.

Snyderman, M., & Rothman, S. (1988). *The IQ controversy: The media and public policy*. New Brunswick, NJ: Transaction Books.

Sorokin, P. A. (1956). *Fads and foibles in modern sociology and related sciences*. Chicago: Henry Regnery.

Sowell, T. (1995). Ethnicity and IQ. In S. Fraser (Ed.), *The bell curve wars: Race, intelligence, and the future of America* (pp. 70-79). New York: Basic Books.

Spearman, C. (1927). *The abilities of man*. New York: Macmillan.

Specht, L. F. (1919). A Terman class in Public School No. 64, Manhattan. *School and Society, 9*, 393-398.

Sperrazzo, G., & Williams, W. L. (1959). Racial differences on Progressive Matrices. *Journal of Consulting Psychology, 23*, 273-274.

State Education Department. (1999). *1999 agenda for reforming education for students with learning disabilities*. New York: Office of Vocational and Educational Services for Individuals with Disabilities.

Stedman, L. M. (1924). *Education of gifted children*. Yonkers-on-the-Hudson, NY: World Book.

Steele, C. M., & Aronson, J. (1998). Stereotype threat and the test performance of academically successful African Americans. In C. Jencks & M. Phillips (Eds.), *The Black-White test score gap* (pp. 401-427). Washington, DC: Brookings Institution.

Steen, R. G. (1996). *DNA and destiny: Nature and nurture in human behavior*. New York: Plenum.

Stern, W. (1914). *The psychological methods of testing intelligence*. Baltimore, MD: Warwick & York.

Sternberg, R. J. (1986). A triarchic theory of intellectual giftedness. In R. J. Sternberg & J. E. Davidson (Eds.), *Conceptions of giftedness* (pp. 223-243). New York: Cambridge University Press.

Sternberg, R. J. (1988). *The triarchic mind: A new theory of intelligence*. New York: Viking.

Sternberg, R. J. (1996). *Successful intelligence: How practical and creative intelligence determines success in life*. New York: Simon & Schuster.

Sternberg, R. J. (1997a). The triarchic theory of intelligence. In D. P. Flanagan, J. L. Genshaft, & P. L. Harrison (Eds.), *Contemporary intellectual assessment: Themes, tests, and issues* (pp. 92-104). New York: Guilford.

Sternberg, R. J. (1997b). A triarchic view of giftedness: Theory and practice. In N. Colangelo & G. A. Davis (Eds.), *Handbook of gifted education* (2nd ed., pp. 43-53). Boston: Allyn & Bacon.

Sternberg, R. J., & Berg, C. A. (1986). Quantitative integration: Definitions of intelligence: A comparison of the 1921 and 1986 symposia. In R. J. Sternberg & D. K. Detterman (Eds.), *What is intelligence? Contemporary viewpoints on its nature and definition* (pp. 155-162). Norwood, NJ: Ablex.

Sternberg, R. J., & Davidson, J. E. (Eds.). (1986a). *Conceptions of giftedness.* New York: Cambridge University Press.

Sternberg, R. J., & Davidson, J. E. (1986b). Conceptions of giftedness: A map of the terrain. In R. J. Sternberg & J. E. Davidson (Eds.), *Conceptions of giftedness* (pp. 3-18). New York: Cambridge University Press.

Sternberg, R. J., & Grigorenko, E. (Eds.). (1997). *Intelligence, heredity, and environment.* New York: Cambridge University Press.

Sternberg, R. J., & Kaufman, J. C. (1998). Human abilities. *Annual Review of Psychology, 49,* 479-502.

Stoolmiller, M. (1999). Implications of the restricted range of family environments for estimates of heritability and nonshared environment in behavioral genetic adoption studies. *Psychological Bulletin, 125,* 392-409.

Storfer, M. D. (1990). *Intelligence and giftedness: The contributions of heredity and early environment.* San Francisco: Jossey-Bass.

Strong, A. C. (1913). Three hundred fifty White and colored children measured by the Binet-Simon measuring scale of intelligence. *Pedagogical Seminary, 20,* 485-515.

Stroud, J. B. (1928). A study of the relation of intelligence test scores of public school children to the economic status of their parents. *Pedagogical Seminary and Journal of Genetic Psychology, 35,* 105-111.

Sue, S., & Morishima, J. K. (1982). *The mental health of Asian Americans.* San Francisco: Jossey-Bass.

Sumner, F. C. (1925). Environic factors which prohibit creative scholarship among Negroes. *School and Society, 22,* 294-296.

Sunne, D. (1917). A comparative study of White and Negro children. *Journal of Applied Psychology, 1,* 71-83.

Suzuki, L. A., & Gutkin, T. B. (1993, August). *Racial/ethnic ability patterns on the WISC-R and theories of intelligence.* Poster presented at the meeting of the American Psychological Association, Toronto.

Suzuki, L. A., & Valencia, R. R. (1997). Race-ethnicity and measured intelligence: Educational implications. *American Psychologist, 52,* 1103-1114.

Swanson, R. A., & Henderson, R. W. (1976). Achieving home-school continuities in the socialization of an academic motive. *Journal of Experimental Education, 44,* 38-44.

Takaki, R. (1990). *Iron cages: Race and culture in nineteenth century America.* New York: Knopf.

Tallent-Runnels, M. K., & Martin, M. R. (1992). Identifying Hispanic gifted children using the Screening Assessment for Gifted Elementary Students. *Psychological Reports, 70,* 939-942.

Tanaka, K. (1989). A response to "Are We Meeting the Needs of Gifted Asian-Americans?" In C. J. Maker & S. W. Schiever (Eds.), *Critical issues in gifted education: Defensible programs for cultural and ethnic minorities* (pp. 174-178). Austin, TX: PRO-ED.

Tannenbaum, A. J. (1983). *Gifted children: Psychological and educational perspectives.* New York: Macmillan.

Tannenbaum, A. J. (1990). Defensible? Venerable? Vulnerable? *Gifted Child Quarterly, 34,* 84-86.

Tate, D., & Gibson, G. (1980). Socioeconomic status and Black and White intelligence revisited. *Social Behavior and Personality, 8,* 233-237.

Taussig, F. (1909). *Principles of economics* (Vol. 2). New York: Macmillan.

Taussig, F. (1920). *Principles of economics* (2nd ed.). New York: Macmillan.

Taylor, H. F. (1980). *The IQ game: A methodological inquiry into the heredity-environment controversy.* New Brunswick, NJ: Rutgers University Press.

Taylor, H. F. (1992). Intelligence. In E. F. Borgatta & M. L. Borgatta (Eds.), *Encyclopedia of sociology* (pp. 941-949). New York: Macmillan.

Taylor, L. J., & Skanes, G. R. (1976). Level I and Level II intelligence in Inuit and White children from similar environments. *Journal of Cross-Cultural Psychology, 7,* 157-168.

Taylor, P. S. (1929). *Mexican labor in the United States: Racial school statistics* (University of California Publications in Economics, No. 5). Los Angeles: University of California Press.

Terman, L. M. (1916). *The measurement of intelligence: An explanation of and a complete guide for the use of the Stanford revision and extension of the Binet-Simon scales.* Boston: Houghton Mifflin.

Terman, L. M. (1917). The Stanford revisions and extensions of the Binet-Simon scale for measuring intelligence. In *Educational Psychology Monographs* (No. 18). Baltimore, MD: Warwick & York.

Terman, L. M. (1920). The use of intelligence tests in the grading of school children. *Journal of Educational Research, 1,* 20-32.

Terman, L. M. (1922). *Intelligence tests and school reorganization.* Yonkers-on-the-Hudson, NY: World Book.

Terman, L. M. (1925). *Genetic studies of genius: Vol. 1. Mental and physical traits of a thousand gifted children.* Stanford, CA: Stanford University Press.

Terman, L. M. (1926). *Genetic studies of genius: Vol. 1. Mental and physical traits of a thousand gifted children* (2nd ed.). Stanford, CA: Stanford University Press.

Terman, L. M., Burks, B. S., & Jensen, D. W. (1930). *Genetic studies of genius: Vol. 3. Follow-up studies of a thousand gifted children.* Stanford, CA: Stanford University Press.

Terman, L. M., & Merrill, M. A. (1937). *Measuring intelligence.* Boston: Houghton Mifflin.

Terman, L. M., & Merrill, M. A. (1960). *Stanford-Binet Intelligence Scale: 1960 norms edition.* Boston: Houghton Mifflin.

Terman, L. M., & Merrill, M. A. (1973). *Stanford-Binet Intelligence Scale: 1972 norms edition.* Boston: Houghton Mifflin.

Terman, L. M., & Oden, M. H. (1947). *Genetic studies of genius: Vol. 4. The gifted child grows up.* Stanford, CA: Stanford University Press.

Terman, L. M., & Oden, M. H. (1959). *Genetic studies of genius: Vol. 5. The gifted group at mid-life— Thirty-five years' follow-up of a superior group.* Stanford, CA: Stanford University Press.

Terman, L. M., Thorndike, E. L., Freeman, F. N., Colvin, S. S., Pintner, R., & Ruml, B. (1921). Intelligence and its measurement: A symposium. *Journal of Educational Psychology, 12,* 123-147, 195-216, 271-275.

Texas Education Agency. (1998). *Enrollment trends in Texas public schools.* Austin, TX: Author.

Thomas, W. B. (1982). Black intellectuals' critique of early mental testing: A little known saga of the 1920s. *American Journal of Education, 90,* 258-292.

Thorndike, R. L., Hagen, E. P., & Sattler, J. M. (1986a). *Stanford-Binet Intelligence Scale: Fourth Edition.* Itasca, IL: Riverside.

Thorndike, R. L., Hagen, E. P., & Sattler, J. M. (1986b). *Technical manual for the Stanford-Binet Intelligence Scale: Fourth Edition.* Itasca, IL: Riverside.

Throssel, S. G. (1989). It's about time. In C. J. Maker & S. W. Schiever (Eds.), *Critical issues in gifted education: Defensible programs for cultural and ethnic minorities* (pp. 113-115). Austin, TX: PRO-ED.

Thurstone, L. L. (1941). *Factorial studies of intelligence.* Chicago: University of Chicago Press.

Tindal, S. (1986-present). Review of the Slosson Full-Range Intelligence Test. In *Buro's mental measurements yearbook* (WinSPIRS database).

Tittle, C. K. (1982). Use of judgmental methods in item bias studies. In R. A. Berk (Ed.), *Handbook of methods for detecting test bias* (pp. 31-63). Baltimore, MD: Johns Hopkins University Press.

Tonemah, S. (1987). Assessing American Indian gifted and talented students' abilities. *Journal for the Education of the Gifted, 10,* 181-194.

Traub, R. (1989-present). Review of the Gifted Evaluation Scale. In *Buro's mental measurements yearbook* (WinSPIRS database).

Trevisan, M. S. (1986-present). Review of the MIDAS: Multiple Intelligence Developmental Assessment Scales. In *Buro's mental measurements yearbook* (WinSPIRS database).

Trimble, J. E. & Bagwell, W. M. (1995). *North American Indians and Alaskan Natives: Abstracts of the psychological and behavioral literature, 1967-1994.* (Bibliographies in Psychology, No. 15). Washington, DC: American Psychological Association.

Trotman, F. K. (1977). Race, IQ, and the middle class. *Journal of Educational Psychology, 69,* 266-273.

Trotman, F. K. (1978). Race, IQ, and rampant misrepresentations: A reply. *Journal of Educational Psychology, 70,* 478-481.

Trueba, H. T. (1988). Comments on L. M. Dunn's *Bilingual Hispanic Children on the U.S. Mainland: A Review of Research on Their Cognitive, Linguistic, and Scholastic Development. Hispanic Journal of Behavioral Sciences, 10,* 253-262.

Trueba, H. T. (1991). From failure to success: The roles of culture and cultural conflict in the academic achievement of Chicano students. In R. R. Valencia (Ed.), *Chicano school failure and success: Research and policy agendas for the 1990s* (pp. 151-163, Stanford Series on Education and Public Policy). London: Falmer.

Tsai, D. M. (1992). *Family impact on high achieving Chinese-American students: A qualitative analysis.* Unpublished doctoral dissertation, University of Connecticut, Storrs.

Tucker, W. H. (1994). *The science and politics of racial research.* Urbana: University of Illinois Press.

Tyack, D. (1974). *The one best system: A history of American urban education.* Cambridge, MA: Harvard University Press.

U.S. Department of Commerce, Bureau of the Census. (1989). *The March 1988 Current Population Survey* [Machine-readable file]. Washington, DC: Author.

U.S. Department of Education. (1991a). *America 2000.* Washington, DC: Author.

U.S. Department of Education. (1991b). *Final report: Gifted and talented education programs for eighth-grade public school students* (National Educational Longitudinal Study No. 88). Washington, DC: Author.

U.S. Department of Education. (1993). *National excellence: A case for developing America's talent.* Washington, DC: Author.

U.S. Department of Education, Office for Civil Rights. (1991). *1988 elementary and secondary school civil rights survey: State and national summaries* (Report No. ASI 4804-33). Washington, DC: Author.

U.S. Department of Education, Office for Civil Rights. (1997). *Fall 1994 elementary and secondary school civil rights compliance report.* Washington, DC: Author.

U.S. Department of Education, Office of Educational Research and Improvement. (1998). *Talent and diversity: The emerging world of limited English proficient students in gifted education.* Washington, DC: Author.

U.S. Department of Health, Education, and Welfare. (1972). *Education of the gifted and talented.* Washington, DC: Author.

U.S. Department of the Interior, Bureau of Education. (1926). *Cities reporting the use of homogeneous grouping and of the Winnetka technique and the Dalton plan* (City School Leaflet No. 22). Washington, DC: Government Printing Office.

Valencia, R. R. (1979). Comparison of intellectual performance of Chicano and Anglo third-grade boys on the Raven's Coloured Progressive Matrices. *Psychology in the Schools, 16,* 448-453.

Valencia, R. R. (1984). Concurrent validity of the Kaufman Assessment Battery for Children in a sample of Mexican American children. *Educational and Psychological Measurement, 44,* 365-372.

Valencia, R. R. (1985a). *Chicanos and intelligence testing: A descriptive state of the art.* Unpublished manuscript, Stanford University.

Valencia, R. R. (1985b). Stability of the Kaufman Assessment Battery for Children for a sample of Mexican American children. *Journal of School Psychology, 23,* 189-193.

Valencia, R. R. (Ed.). (1991). *Chicano school failure and success: Research and policy agenda for the 1990s* (Stanford Series on Education and Public Policy). London: Falmer.

Valencia, R. R. (Ed.). (1997a). *The evolution of deficit thinking: Educational thought and practice* (Stanford Series on Education and Public Policy). London: Falmer.

Valencia, R. R. (1997b). Genetic pathology model of deficit thinking. In R. R. Valencia (Ed.), *The evolution of deficit thinking: Educational thought and practice* (pp. 41-112, Stanford Series on Education and Public Policy). London: Falmer.

Valencia, R. R. (1997c). Latinos and education: An overview of sociodemographic characteristics and schooling conditions. In M. Yepes-Baraya (Ed.), *ETS Invitational Conference on Latino Education Issues: Conference proceedings* (pp. 13-37). Princeton, NJ: Educational Testing Service.

Valencia, R. R. (1999). Educational testing and Mexican American students: Problems and prospects. In J. F. Moreno (Ed.), *The elusive quest for equality: 150 years of Chicano/Chicana education* (pp. 123-139). Cambridge, MA: Harvard Educational Review.

Valencia, R. R., & Aburto, A. (1991). The uses and abuses of educational testing: Chicanos as a case in point. In R. R. Valencia (Ed.), *Chicano school failure and success: Research and policy agendas for the 1990s* (pp. 203-251, Stanford Series on Education and Public Policy). London: Falmer.

Valencia, R. R., & Chapa, J. (Eds.). (1993). Latino population growth and demographic trends: Implications for education [Special issue]. *Hispanic Journal of Behavioral Sciences, 15*(2).

Valencia, R. R., & Cruz, J. (1981). *Mexican American mothers' estimations of their preschool children's cognitive performance* (Report No. 90-C-1777). Washington, DC: U.S. Department of Health, Education, and Welfare; Office of Human Development Services; Administration for Children, Youth, and Families.

Valencia, R. R., Henderson, R. W., & Rankin, R. J. (1981). Relationship of family constellation and schooling to intellectual performance of Mexican American children. *Journal of Educational Psychology, 73*, 524-532.

Valencia, R. R., Henderson, R. W., & Rankin, R. J. (1985). Family status, family constellation, and home environmental variables as predictors of cognitive performance of Mexican American children. *Journal of Educational Psychology, 77*, 323-331.

Valencia, R. R., & Rankin, R. J. (1983). Concurrent validity and reliability of the Kaufman version of the McCarthy scales for a sample of Mexican American children. *Educational and Psychological Measurement, 43*, 915-925.

Valencia, R. R., Rankin, R. J., & Henderson, R. W. (1986, May). *Path analysis of sociocultural and family constellation variables on intellectual performance of Anglo, Black, and Mexican American children.* Paper presented at the meeting of the Western Psychological Association, Seattle, WA.

Valencia, R. R., Rankin, R. J., & Livingston, R. (1995). K-ABC and content bias: Comparisons between Mexican American and White children. *Psychology in the Schools, 32*, 153-169.

Valencia, R. R., & Solórzano, D. G. (1997). Contemporary deficit thinking. In R. R. Valencia (Ed.), *The evolution of deficit thinking: Educational thought and practice.* (pp. 160-210, Stanford Series on Education and Public Policy). London: Falmer.

Valien, P., & Horton, C. (1954). Some demographic characteristics of outstanding Negro women. *Journal of Negro Education, 23*, 406-420.

Van Alstyne, D. (1929). *The environment of 3 year old children.* Unpublished doctoral dissertation, Columbia University.

Van Alstyne, D. (1961). *Van Alstyne Picture Vocabulary Test.* New York: Harcourt, Brace, & World.

Vance, H. B., & Awadh, A. M. (1998). Best practices in assessment of children: Issues and trends. In H. B. Vance (Ed.), *Psychological assessment of children: Best practices for school and clinical settings* (2nd ed., pp. 1-10). New York: John Wiley.

Vance, H. B., Huelsman, C. B., & Wherry, R. J. (1976). The hierarchical factor structure of the Wechsler Intelligence Scale for Children as it relates to disadvantaged White and Black children. *Journal of Genetic Psychology, 95*, 287-293.

Vandenberg, S. G. (1969). A twin study of spatial ability. *Multivariate Behavioral Research, 4*, 273-294.

Vandenberg, S. G. (1970). A comparison of heritability estimates of U.S. Negro and White high school students. *Acta Geneticae Medicae et Gemellologiae, 19*, 280-284.

Van Sickle, J. H. (1910). Provision for gifted children in public schools. *Elementary School Teacher, 10*, 357-366.

VanTassel-Baska, J. L. (1991). Research on special populations of gifted learners. In M. C. Wang, M. C. Reynolds, & H. J. Walberg (Eds.), *Handbook of special education: Research and practice: Vol. 4. Emerging programs* (pp. 77-101). New York: Pergamon.

VanTassel-Baska, J. L., Patton, J., & Prillaman, D. (1989). Disadvantaged gifted learners: At-risk for educational attention. *Focus on Exceptional Children, 3,* 1-15.

Vernon, P. E. (1950). *The structure of human abilities.* London: Methuen.

Vernon, P. E. (1969). *Intelligence and the cultural environment.* London: Methuen.

Vernon, P. E. (1979). *Intelligence: Heredity and environment.* San Francisco: Freeman.

Vernon, P. E., Jackson, D. N., & Messick, S. (1988). Cultural influence on patterns of abilities in North America. In S. H. Irvine & J. W. Berry (Eds.), *Human abilities in cultural context* (pp. 208-231). New York: Cambridge University Press.

Vraniak, D. (1993). *Technical report to the Bureau of Indian Affairs (BIA).* Madison: University of Wisconsin Press.

Vraniak, D. (1999, August). [Discussant.] In L. Suzuki (Chair), *Intelligence 2000: A multicultural mosaic* (symposium conducted at the meeting of the American Psychological Association, Boston).

Vraniak, D. A., Suzuki, L. A., Lee, C., Kubo, T., Pieterse, A., Short, E. L., Lin, F. C., Kim, E. S., Yahav, E., & Choi, S. (1998, August). *Intelligence exposed: A multicultural redress.* Poster presented at the meeting of the American Psychological Association, San Francisco.

Vroon, P. A. (1980). Intelligence on myths and measurement. In G. E. Stelmach (Ed.), *Advances in psychology 3* (pp. 27-44). New York: North-Holland.

Wachs, T. D. (1979). Proximal experience and early cognitive-intellectual environment. *Merrill-Palmer Quarterly, 25,* 3-41.

Wachs, T. D. (1984). Proximal experience and early cognitive-intellectual environment: The social environment. In A. W. Gottfried (Ed.), *Home environment and early cognitive development: Longitudinal research* (pp. 273-328). Orlando, FL: Academic Press.

Wachs, T. D., & Gruen, G. E. (1982). *Early experience and human development.* New York: Plenum.

Wachs, T. D., & Plomin, R. (Eds.). (1991). *Conceptualization and measurement of organism-environment interaction.* Washington, DC: American Psychological Association.

Wachs, T. D., Uzgiris, I., & Hunt, J. McV. (1971). Cognitive development in infants of different age levels and from different environmental backgrounds: An explanatory investigation. *Merrill-Palmer Quarterly, 17,* 283-317.

Wade, N. (1976). IQ and heredity: Suspicion of fraud beclouds classic experiment. *Science, 194,* 916-919.

Wahlsten, D. (1990). Insensitivity of the analysis of variance to heredity-environment interaction. *Behavioral and Brain Sciences, 13,* 109-161.

Wahlsten, D., & Gottlieb, G. (1997). The invalid separation of effects of nature and nurture: Lessons from animal experimentation. In R. J. Sternberg & E. Grigorenko (Eds.), *Intelligence, heredity, and environment* (pp. 163-192). New York: Cambridge University Press.

Walberg, H. J., & Marjoribanks, K. (1976). Family environment and cognitive models. *Review of Educational Research, 76,* 527-551.

Wallace, B. (1975). Genetics and the great IQ controversy. *American Biology Teacher, 37,* 12-18, 50.

Wallin, J. E. W. (1912). *Experimental studies of mental defectives.* Baltimore, MD: Warrick & York.

Ward, J. A., & Hetzel, H. R. (1980). *Biology today and tomorrow.* St. Paul, MN: West Publishing.

Warner, W. L. (1941). Introduction. In A. Davis, B. B. Gardner, & M. R. Gardner (Eds.), *Deep South: A social anthropology study of caste and class* (pp. 4-6). Chicago: University of Chicago Press.

Warner, W. L. (1949). *Social class in America.* New York: Harper.

Warner, W. L., Meeker, M., & Eells, K. (1949). *Social class in America.* Chicago: Science Research Associates.

Warner, W. L., Meeker, M., & Eells, K. (1960). *Social class in America: A manual of procedures for the assessment of social class.* New York: Harper & Row.

Warner, W. L., & Srole, L. (1945). *The social systems of American ethnic groups.* New Haven, CT: Yale University Press.

Wechsler, D. (1949). *Manual for the Wechsler Intelligence Scale for Children.* New York: Psychological Corporation.

Wechsler, D. (1967). *Wechsler Preschool and Primary Scale of Intelligence.* New York: Psychological Corporation.

Wechsler, D. (1974). *Manual for the Wechsler Intelligence Scale for Children–Revised.* New York: Psychological Corporation.

Wechsler, D. (1991). *Manual for the Wechsler Intelligence Scale for Children–Third Edition.* New York: Psychological Corporation.

Weinberg, R. A. (1989). Intelligence and IQ: Landmark issues and great debates. *American Psychologist, 44,* 98-104.

Weinberg, R. A., Scarr, S., & Waldman, I. D. (1992). The Minnesota Transracial Adoption Study: A follow-up of IQ performance at adolescence. *Intelligence, 16,* 117-135.

Weintrob, J., & Weintrob, R. (1912). The influence of environment on mental ability as shown by the Binet-Simon tests. *Journal of Educational Psychology, 3,* 577-583.

Weiss, J. (1969). *The identification and measurement of home environmental factors related to achievement motivation and self-esteem.* Unpublished doctoral dissertation, University of Chicago.

Weiss, L. G., & Prifitera, A. (1995). An evaluation of differential prediction of WIAT Achievement scores from WISC-III FSIQ across ethnic and gender groups. *Journal of School Psychology, 33,* 297-304.

Weiss, L. G., Prifitera, A., & Roid, G. (1993). The WISC-III and the fairness of predicting achievement across ethnic and gender groups. *Journal of Psychoeducational Assessment* (monograph series, Advances in Psychological Assessment, Wechsler Intelligence Scale for Children–Third Edition), pp. 35-42.

Whipple, G. M. (Ed.). (1924). *The twenty-third yearbook of the National Society for the Study of Education: Part I–Report of the Society's Committee on the Education of Gifted Children.* Bloomington, IL: Public School Publishing.

White, B. L., & Carew, J. V. (1973). *Experience and environment.* Englewood Cliffs, NJ: Prentice Hall.

White, K. R. (1982). The relation between socioeconomic status and academic achievement. *Psychological Bulletin, 91,* 461-481.

Whiteman, M., Brown, B., & Deutsch, M. (1967). Some effects of social class and race on children's language and intellectual abilities. In M. Deutsch (Ed.), *The disadvantaged child: Studies of the social environment and the learning process.* New York: Basic Books.

Wickwire, P. J. N. (1971). *The academic achievement and language development of American children of Latin heritage: Factors of intellect, home educational environment, and personality.* Unpublished doctoral dissertation, The University of Texas at Austin.

Wilgosh, L., Mulcahy, R., & Watters, B. (1986). Identifying gifted Canadian Inuit children using conventional IQ measures and nonverbal (performance) indicators. *Canadian Journal of Special Education, 2,* 67-189.

Wilkerson, D. A. (1936). Review of M. D. Jenkins: A socio-psychological study of Negro children of superior intelligence. *Journal of Negro Education, 5,* 126-131.

Wilkinson, C., & Ortiz, A. (1986). *Characteristics of limited English proficient and English proficient learning disabled Hispanic students at initial assessment and at reevaluation.* Austin: The University of Texas, Department of Special Education, Handicapped Minority Research Institute. (ERIC Document Reproduction Service No. 283 314)

Williams, R. L. (1971). Abuses and misuses in testing Black children. *Counseling Psychologist, 2,* 62-77.

Williams, T. (1976). Abilities and environments. In W. H. Sewell, R. M. Hauser, & D. L. Featherman (Eds.), *Schooling and achievement in American society* (pp. 61-102). New York: Academic Press.

Willig, A. C. (1988). A case of blaming the victim: The Dunn monograph on *Bilingual Hispanic Children on the U.S. Mainland. Hispanic Journal of Behavioral Sciences, 10,* 219-236.

Wilson, A. B. (1967). Educational consequences of segregation in a California community. In U.S. Commission on Civil Rights, *Racial isolation in the public schools* (Vol. 2, appendixes). Washington, DC: Government Printing Office.

Wilson, M. S., & Reschly, D. J. (1996). Assessment in school psychology training and practice. *School Psychology Review, 25,* 9-23.

Winner, E. (1996). *Gifted children: Myths and realities.* New York: Basic Books.

Winner, E. (1997). Exceptionally high intelligence and schooling. *American Psychologist, 52,* 1070-1081.

WinSPIRS. (1989-present). *Buro's mental measurements yearbook* [Computer software].

Witty, P. A., & Jenkins, M. D. (1934). The educational achievement of a group of gifted Negro children. *Journal of Educational Psychology, 25,* 585-597.

Witty, P. A., & Jenkins, M. D. (1935). The case of "B": A gifted Negro girl. *Journal of Social Psychology, 6,* 117-124.

Wober, M. (1974). Towards an understanding of Kiganda concept of intelligence. In J. W. Berry & P. R. Dasen (Eds.), *Culture and cognition.* London: Methuen.

Wolf, R. M. (1964). *The identification and measurement of environmental variables related to intelligence.* Unpublished doctoral dissertation, University of Chicago.

Wolf, R. M. (1966). The measurement of environments. In A. Anastasi (Ed.), *Testing problems in perspective* (Rev. ed., pp. 491-503). Washington, DC: American Council on Education.

Wolfe, A. (1995). Has there been a cognitive revolution in America? The flawed sociology of *The Bell Curve.* In S. Fraser (Ed.), *The bell curve wars: Race, intelligence, and the future of America* (pp. 109-123). New York: Basic Books.

Wolff, J. L. (1978). Utility of socioeconomic status as a control in racial comparisons of IQ. *Journal of Educational Psychology, 70,* 473-477.

Woliver, R., & Woliver, G. M. (1991). Gifted adolescents in the emerging minorities: Asian and Pacific Islanders. In M. Bireley & J. Genshaft (Eds.), *Understanding the gifted adolescent: Educational, developmental, and multicultural issues* (pp. 248-257). New York: Teachers College Press.

Woltereck, R. (1909). Weitere experimentelle Untersuchungen über das Wesen quantitativer Artunterschieder bei Daphniden [Further experimental studies on the essence of genus differences among daphnids]. *Verhandlungen der Deutschen Zoologischen Gesellschaft, 19,* 110-173.

Wong, S. V., & Wong, P. R. (1989). Teaching strategies and practices for the education of gifted Cantonese students. In C. J. Maker & S. W. Schiever (Eds.), *Critical issues in gifted education: Defensible programs for cultural and ethnic minorities* (pp. 182-188). Austin, TX: PRO-ED.

Woo, E. (1989). Personal reflections on the purpose of special education for gifted Asian-Americans. In C. J. Maker & S. W. Schiever (Eds.), *Critical issues in gifted education: Defensible programs for cultural and ethnic minorities* (pp. 179-181). Austin, TX: PRO-ED.

Woodcock, R. W. (1982). *Batería Woodcock Psico-Educativa en Español.* Itasca, IL: Riverside.

Woodcock, R. W. (1994). Norms. In R. J. Sternberg (Ed.), *The encyclopedia of human intelligence* (pp. 770-775). New York: Macmillan.

Woodcock, R. W., & Johnson, M. B. C. (1989). *Woodcock-Johnson Psychoeducational Battery–Revised.* Itasca, IL: Riverside.

Woodcock, R. W., & Muñoz-Sandoval, A. F. (1996). *Batería Woodcock-Muñoz–Revisada.* Itasca, IL: Riverside.

Woods, F. A. (1914). Racial origins of successful Americans. *Popular Science Monthly, 84,* 397-402.

Woods, S. B., & Achey, V. H. (1990). Successful identification of gifted racial/ethnic group students without changing classification requirements. *Roeper Review, 13,* 21-36.

Woolfolk, A. E. (1995). *Educational psychology* (6th ed.). Boston: Allyn & Bacon.

Wright, R. (1995, January 2). Has Charles Murray read his own book? *New Republic,* p. 6.

Wulbert, M., Inglis, S., Kriegsman, E., & Mills, B. (1975). Language delay and associated mother-child interactions. *Developmental Psychology, 11,* 61-70.

Yarrow, L., Rubenstein, J., & Pederson, F. (1975). *Infant and environment.* Washington, DC: Hemisphere.

Yee, A. (1968). Interpersonal attitudes of teachers and advantaged and disadvantaged pupils. *Journal of Human Resources, 3,* 327-332.

Young, K. (1922). *Mental differences in certain immigration groups: Psychological tests of South Europeans in typical California schools with bearing on the educational policy and on the problems of racial contacts in this country* (Vol. 1, No. 11). Eugene: University of Oregon Press.

Ysseldyke, J. E., Algozzine, B., Shinn, M. R., & McGue, M. (1982). Similarities and differences between low achievers and students classified as learning disabled. *Journal of Special Education, 16,* 73-85.

Zappia, I. A. (1989). Identification of gifted Hispanic students. In C. J. Maker & S. W. Schiever (Eds.), *Critical issues in gifted education: Defensible programs for cultural and ethnic minorities* (pp. 19-26). Austin, TX: PRO-ED.

Zarske, J. A., & Moore, C. L. (1982). Recategorized WISC-R scores for non-handicapped, learning disabled, educationally disadvantaged, and regular classroom Navajo children. *School Psychology Review, 11,* 319-323.

Zenderland, L. (1990). Burt again [Review of the book *The Burt Affair*]. *Science, 248,* 884-886.

Zettel, J. (1979). State provisions for educating the gifted and talented. In H. Passow (Ed.), *The gifted and talented: Their education and development—The seventy-fifth yearbook of the National Society for the Study of Education, Part I* (pp. 63-74). Chicago: University of Chicago Press.

Zigler, E., & Hodapp, R. M. (1986). *Understanding mental retardation.* New York: Cambridge University Press.

Zuckerman, M., & Brody, N. (1988). Oysters, rabbits, and people: A critique of "Race Differences in Behaviour" by J. P. Rushton. *Personality and Individual Differences, 9,* 1025-1033.

Name Index

Subject Index

About the Authors

Richard R. Valencia, Ph.D., is Professor in the Department of Educational Psychology, College of Education, The University of Texas at Austin. He earned his B.A. (psychology), M.A. (educational psychology), and Ph.D. (early childhood education) from the University of California, Santa Barbara. Dr. Valencia's research and scholarly interests include the historical, social, psychological, testing, legal, demographic, and policy aspects of minority schooling; and cultural bias in intelligence tests. He has published extensively in the area of minority education, particularly regarding Mexican American students (e.g., historical aspects, cognitive development, school failure and success, testing issues). He is the editor of *Chicano School Failure and Success: Research and Policy Agendas for the 1990s* (Falmer, 1991), which received an Outstanding Academic Book award (*CHOICE*, 1993). He also is editor of *The Evolution of Deficit Thinking: Educational Thought and Practice* (Falmer, 1997). For more than a decade, he was associated editorially with the *Journal of Educational Psychology* (associate editor, 1987-1990; editorial board member, 1984-1986, 1990-1995). He also has served on the editorial boards of the *Review of Educational Research*. He currently serves on the editorial boards of the *American Educational Research Journal* and the *Hispanic Journal of Behavioral Sciences,* and *Educational Psychologist.*

Lisa A. Suzuki, Ph.D., is Assistant Professor in the Department of Applied Psychology, School of Education, New York University. She obtained her B.A. (psychology) from Whitman College, her M.Ed. (counselor education) from the University of Hawaii–Manoa, and her Ph.D. (counseling psychology) from the University of Nebraska–

Lincoln. She is senior editor (with P. J. Meller and J. G. Ponterotto) of the *Handbook of Multicultural Assessment* (1996), coeditor (with J. G. Ponterotto, M. Casas, and C. Alexander) of the *Handbook of Multicultural Counseling* (Sage, 1995), and coeditor (with M. Kopala) of *Using Qualitative Methods in Psychology* (Sage, 1999). She has published and presented on topics related to multicultural issues in assessment and in particular racial/ethnic group differences in measured intelligence. She previously worked as a school counselor and psychological examiner for the Department of Education in the state of Hawaii. Over the years, she has administered more than 300 intelligence tests to diverse populations. It was her observations during this time that sparked her interest in pursuing a greater understanding of the cognitive abilities and intelligences of diverse racial/ethnic populations. She currently serves on the editorial board of the *Journal of Multicultural Counseling and Development* and is an editorial consultant for the *Asian Journal of Counseling.*